EXPLORING ANIMAL BEHAVIOR IN LABORATORY AND FIELD

An Hypothesis–Testing Approach to the Development, Causation, Function, and Evolution of Animal Behavior

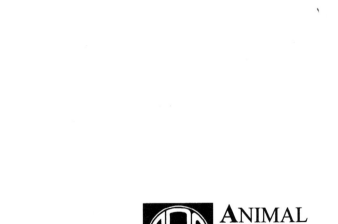

EXPLORING ANIMAL BEHAVIOR IN LABORATORY AND FIELD

An Hypothesis-Testing Approach to the Development, Causation, Function, and Evolution of Animal Behavior

Edited by

Bonnie J. Ploger
Hamline University, USA

Ken Yasukawa
Beloit College, USA

ACADEMIC PRESS

An imprint of Elsevier Science

Amsterdam   Boston   London   New York   Oxford   Paris   San Diego
San Francisco   Singapore   Sydney   Tokyo

Senior Publishing Editor	Charles R. Crumly, Ph.D.
Senior Project Manager	Julio Esperas
Product Manager	Anne O'Mara
Cover Design	Monty Lewis
Copyeditor	Connie Day
Production Services	Matrix Productions
Composition	International Typesetting and Composition
Printer	Edwards Brothers

This book is printed on acid-free paper. ∞

ON THE COVER: An adult great egret (Ardea alba) in breeding condition preens its feathers. Preening not only removes feather lice but also repairs small tears in feathers that interfere with their aerodynamic properties. Preening and other types of grooming behavior are useful for learning to describe behavior (chapter 1) because such behavior is frequent, easy to see, and consists of repeatable elements that are often fairly easy to sketch and describe. In a variety of birds, including egrets, some courtship displays probably evolved from preening movements. Possible evolutionary sequences of such displays can be explored by mapping the displays of related species onto a phylogenetic tree (chapter 35). Photograph by Bonnie Ploger.

Given a choice of large and small nuts of various nutritive values and differing shell thickness, this female Eastern gray squirrel (Sciurus carolinensis) seems to be thwarting the experiment by taking two different types of nut simultaneously! Her attempt ultimately failed because, unlike some ground squirrels such as chipmunks, gray squirrels do not have cheek pouches sufficiently large to carry multiple nuts of the sizes shown. Squirrels are excellent subjects for studying the economics of foraging (chapter 19). While foraging, squirrels are vulnerable to predators, but may reduce such risks by responding to alarm calls given by conspecifics or even other species that have detected predators (chapter 26). Photograph by Bonnie Ploger.

This male strawberry poison frog (Dendrobates pumilio) was photographed while climbing onto its display perch. Males display by giving loud clicking calls while perching conspicuously on slightly elevated sites near the ground that they may defend for a week or more. After attracting a female, a male cares for the fertilized eggs until they hatch and the female returns to carry each tadpole to a separate, tiny pool formed by leaves of bromeliad plants. The vocal displays of frogs and toads provide good opportunities for investigating variation in male courtship in the field (chapter 30). Tadpoles make interesting subjects for investigations of behavioral thermoregulation (chapter 6), aggregation and kin recognition (chapter 17). Photograph by Bonnie Ploger.

In many fish, it is the male that builds the nest, as this male dwarf gourami (Colisa lalia) is doing by blowing bubbles to form a foamy mass in the duckweeds at the surface of the water. With its thread-like pelvic fins extended forward, this male is also displaying to attract females and defend his nest from rival males. Male characteristics such as size, color and behavior can influence the outcome of territorial disputes between males, and male ability to attract females. In many common aquarium fish, these characteristics are easily manipulated to study mate choice (chapter 31). Photograph by Emory Matts.

ACADEMIC PRESS
An imprint of Elsevier Science
525 B Street, Suite 1900, San Diego, California 92101-4495, USA
http://www.academicpress.com

Academic Press
84 Theobald's Road, London WC1X 8RR, UK
http://www.academicpress.com

Library of Congress Control Number: 2002104257
International Standard Book Number: 0-12-558330-3

PRINTED IN THE UNITED STATES OF AMERICA
02 03 04 05 06 07 EB 9 8 7 6 5 4 3 2 1

CONTENTS

Appendices

CONTRIBUTORS

Daniel J. Albrecht, Department of Biology, Rocky Mountain College, 1511 Poly Drive, Billings, Montana, 59101

Kathryn L. Anderson, Princeton High School, 807 South 8th Avenue, Princeton, MN 55371, USA

Jonathan R. Banks, School of Biological Sciences, University of Manchester, 3.614 Stopford Building, Oxford Road, Manchester M13 9PT, UK

Robert C. Beason, Department of Biology, University of Louisiana at Monroe, 700 University Avenue, Monroe, LA 71209, USA

Penny L. Bernstein, Biological Sciences, Kent State University Stark Campus, 6000 Frank Avenue, Canton, OH 44720, USA

Nancy L. Buschhaus, Department of Biological Sciences, University of Tennessee at Martin, Martin, TN 38238, USA

T. Dic Charge, Department of Biological Sciences, University of Lethbridge, Lethbridge, AB T1K 3M4, Canada

Nicola S. Clayton, Experimental Psychology, University of Cambridge, Downing Street, Cambridge CB2 3EB, UK

Lisa E. Cody, Animal Behavior Graduate Group, University of California, Davis, CA 95616, USA

Mary Crowe, Department of Biology, Coastal Carolina University, P. O. Box 261954, Conway, SC 29528, USA

Ester Desfilis, Instituto Cavanilles de Biodiversidad y Biología Evolutiva, Universidad de Valencia, Apdo. 2085, 46071—Valencia, Spain

Gerhard Engler, School of Biological Sciences, University of Manchester, 3.614 Stopford Building, Oxford Road, Manchester M13 9PT, UK

Enrique Font, Instituto Cavanilles de Biodiversidad y Biología Evolutiva, Universidad de Valencia, Apdo. 2085, 46071—Valencia, Spain

Don C. Forester, Department of Biological Sciences, Towson University, Towson, MD 21252, USA

Lynn L. Gillie, Division of Mathematics and Natural Science, Elmira College, One Park Place, Elmira, NY 14901, USA

Terry Glover, Division of Social and Behavioral Sciences, Bloomfield College, Bloomfield, NJ 07003, USA

Sylvia L. Halkin, Department of Biological Sciences, Central Connecticut State University, New Britain, CT 06050-4010, USA

Judith A. Halstead, Department of Chemistry, Skidmore College, 815 N. Broadway, Saratoga Springs, NY 12866, USA

Peggy S. M. Hill, Faculty of Biological Sciences, University of Tulsa, Tulsa, OK 74104, USA

Wendy L. Hill, Department of Psychology, Lafayette College, Easton, PA 18042, USA

Chad D. Hoefler, Department of Entomology and Program in Organismic and Evolutionary Biology, University of Massachusetts, Amherst, MA 01003, USA

Anne Houtman, Department of Biology, Soka University, Aliso Viejo, CA 92656, USA

Alastair Inman, Department of Biology, Knox College, Galesburg, IL 61401, USA

Elizabeth M. Jakob, Department of Psychology, University of Massachusetts, Amherst, MA 01003, USA

Jasmin M. Keramaty, Department of Biology, Skidmore College, 815 N. Broadway, Saratoga Springs, NY 12866, USA

C. O'Neil Krekorian, San Diego State University, Department of Biology, 5500 Campanile Drive, San Diego, CA 92182-4614, USA

Cheryl A. Logan, Departments of Psychology and Biology, University of North Carolina at Greensboro, P. O. Box 26170, Greensboro, NC 27401-6170, USA

Janice R. Matthews, Entomology Department, The University of Georgia, Athens, GA 30602, USA

Robert W. Matthews, Entomology Department, The University of Georgia, Athens, GA 30602, USA

Betty McGuire, Department of Biological Sciences, Smith College, Northampton, MA 01063, USA

Gail R. Michener, Department of Biological Sciences, University of Lethbridge, Lethbridge, AB T1K 3M4, Canada

Robert D. Montgomerie, Department of Biology, Queen's University, Kingston, ON K7L 3N6, Canada

Kathleen N. Morgan, Department of Psychology, Wheaton College, Norton, MA 02766, USA

Alison M. Mostrom, University of the Sciences in Philadelphia, Department of Biological Sciences, 600 South 43rd Street, Philadelphia, PA 19104-4495, USA

D. James Mountjoy, Department of Biology, Knox College, Galesburg, IL 61401, USA

Ronald L. Mumme, Department of Biology, Allegheny College, 520 North Main Street, Meadville, PA 16335, USA

James O. Palmer, Department of Biology, Allegheny College, 520 North Main Street, Meadville, PA 16335, USA

Bonnie J. Ploger, Department of Biology, Hamline University, 1536 Hewitt Avenue, St. Paul, MN 55104, USA

Susan M. Rankin, Department of Biology, Allegheny College, 520 North Main Street, Meadville, PA 16335, USA

Monica Raveret Richter, Department of Biology, Skidmore College, 815 N. Broadway, Saratoga Springs, NY 12866, USA

Kierstin Savastano, Department of Chemistry, Skidmore College, 815 N. Broadway, Saratoga Springs, NY 12866, USA

Joanna E. Scheib, Department of Psychology, University of California, Davis, CA 95616, USA

Michelle Pellissier Scott, Department of Zoology, University of New Hampshire, Durham, NH 03824, USA

Philip K. Stoddard, Department of Biological Sciences, Florida International University, Miami, FL 33199 USA

Jennifer J. Templeton, Department of Biology, Knox College, Galesburg, IL 61401, USA

George H. Waring, Department of Zoology, Southern Illinois University at Carbondale, Carbondale, IL 62901, USA

Harrington Wells, Faculty of Biological Sciences, University of Tulsa, Tulsa, OK 74104, USA

Howard H. Whiteman, Department of Biological Sciences, Murray State University, Murray, KY 42071, USA

Ken Yasukawa, Department of Biology, Beloit College, 700 College Street, Beloit, WI 53511, USA

Günther K. H. Zupanc, School of Biological Sciences, University of Manchester, 3.614 Stopford Building, Oxford Road, Manchester M13 9PT, UK

PREFACE

In the past 15 years or so there has been a national movement in science education toward designing science curricula, including laboratory courses, that engage students actively in all phases of scientific investigation, from formulating research questions through interpreting final results of student-designed experiments (Costenson & Lawson 1986; National Science Foundation 1990; Sundberg 1997; Wright & Govindarajan 1992; McNeal & D'Avanzo 1997). Laboratory sections of animal behavior courses are particularly useful for encouraging student inquiry, because students can often conduct sophisticated experiments, including original research, without specialized technical equipment. The editors and many other members of the Animal Behavior Society use an inquiry approach when we teach animal behavior laboratories. The contributors to this book have designed their exercises to teach you the principles and methods of animal behavior and to encourage you to design, conduct, and analyze your own experiments— that is, to engage you in inquiry-based learning.

The idea for this book developed at the 1994 meeting of the Animal Behavior Society (ABS), an international society of biologists, psychologists, and anthropologists involved in the study of animal behavior. For this meeting, Bonnie Ploger organized a workshop on teaching animal behavior, as part of the society's efforts to increase educational outreach at the college and precollege level. During the workshop, presenters demonstrated their favorite, most successful animal behavior laboratory exercises, while participants played the roles of students during most of these demonstrations, thereby testing the exercises presented. Workshop participants included ABS members teaching at the undergraduate and graduate levels at small colleges and research universities, as well as a number of local high school teachers invited to participate as part of our outreach efforts.

In addition to the demonstrations, the workshop also included handouts of further exercises that Ken Yasukawa had been collecting over the years from ABS members. The workshop was so successful, and demand for lab handouts was so high, that we decided to develop a laboratory manual of favorite animal behavior labs by ABS members. The result is this book, *Exploring Animal Behavior in Laboratory and Field*, and its companion manual for instructors, *Teaching Animal Behavior in Laboratory and Field*, both of which have both been endorsed by the Animal Behavior Society.

Exploring Animal Behavior in Laboratory and Field includes exercises that were demonstrated at the 1994 ABS workshop and at a similar education workshop organized by Ted Burk for the 1995 ABS meeting, as well as additional exercises developed and tested by ABS members in their classrooms. All labs have been peer-reviewed by ABS members. Some of the exercises in this manual may seem familiar to instructors because they are based on "classics" that have been used successfully in animal behavior courses for decades, while being modified and updated with new theoretical and methodological advances. Other exercises in this manual are completely new. The topics covered include descriptive ethology, causation and development of behavior, and behavioral ecology. Both field and laboratory exercises are included on a broad variety of taxonomic groups: arthropods, fish, amphibians, reptiles, birds, and mammals, including humans. Exercises illustrate issues of current theoretical importance and up-to-date methods used by biologists, psychologists, and anthropologists who study animal behavior.

ORGANIZATION

Classification of Exercises in the Table of Contents

Each chapter presents an animal behavior exercise that has been used by its author(s). The exercises are organized into the following five parts: Describing Behavior, Causation, Development, Adaptation and Evolution, and Appendices. Some exercises fit more than one category, however. For instructors, we have indicated these by listing the secondary category to which the exercise belongs as a bullet item in the table of contents of the instructor's manual. Field and laboratory exercises are similarly distinguished by bullets in that table of contents.

Exercises in Part 1, Describing Behavior, and those in other sections that engage you in learning to observe and describe behavior are designed to help you learn to apply methods that you have read about in more detail in supplemental readings. Before doing such exercises, you should read about the relevant topics in a good introductory handbook of methods, such as Martin & Bateson (1993) or Lehner (1996).

In this manual, you will find exercises that use a variety of teaching styles. For instructors, these are classified according to the following pedagogical approaches in the table of contents of the instructor's manual. In *Traditional* labs, you will follow a predetermined protocol to test particular hypotheses explicitly stated in the exercise. In *Inquiry* labs, you first brainstorm to generate your own hypotheses and then design your own experiments to test your hypotheses. In *Combined Pedagogy* labs, exercises involve both traditional and inquiry approaches. These labs typically begin with a traditional activity in which you test a particular hypothesis by following a detailed protocol. After you have learned the basic techniques and have become familiar with the study animals, you develop your own hypotheses, which you then test in the inquiry portion of the lab. Exercises that your instructor can easily adapt to inquiry-based approaches are indicated as *Traditional*[*] in the table of contents of the instructor's manual.

Inquiry-Based Pedagogical Style

We have observed that students are sometimes initially uncomfortable with an inquiry approach because it is unfamiliar, requires considerable effort to develop ideas and design experiments, and may arouse anxiety among those who feel science classes are about finding the "right" answer (Sundberg 1997). If you are one of these students, we encourage you to hang in there. Science isn't about the "right" answer but about the process by which answers are generated; it isn't a collection of facts but rather a research-based process of inquiry. That is why we have attempted to engage you actively in all phases of scientific investigation, from design to reporting.

The following analogy, suggested by Alison Gopnik (1999), demonstrates the value of this approach:

Imagine If We Taught Baseball the Way We Teach Science

- Until they were 12, children would read about baseball technique and occasionally hear inspirational stories of the great baseball players.
- They would answer quizzes about baseball rules.
- Conservative coaches would argue that we ought to make children practice fundamental baseball skills, throwing the ball to second base 20 times in a row, followed by tagging first base 70 times.
- Others would reply that the economic history of the reserve clause proved that there was, in fact, no such thing as "objectively accurate" pitching.
- Undergraduates might be allowed, under strict supervision, to reproduce famous historic baseball plays.
- But only in graduate school would they, at last, actually get to play a game.

If we taught baseball this way, we might expect about the same degree of success in the Little League World Series that we currently see in science performance.

Organization of Each Exercise

Each exercise is divided into the following sections: "Introduction," "Materials," "Procedure," "Hypotheses and Predictions," "Data Recording and Analyses," and "Questions for Discussion." Exercises with multiple activities, such as those designated as Combined Pedagogy, may deviate from this organization somewhat. The "Introduction" sections for most labs will provide you with good examples of how to write an Introduction section in a scientific paper, although some introductions were written with a different purpose and so do not provide such examples.

Please note that most exercises include statistical analysis. We routinely introduce our students to statistical analysis in our animal behavior courses. To support your efforts in incorporating statistical analysis into your analyses, we have provided Appendix C to introduce the basics of statistical description and inference and to provide you with guidelines for selecting statistical tests. Your instructor can provide examples using results of statistical tests of sample data, such as are provided in the instructor's manual, *Teaching Animal Behavior in Laboratory and Field*.

Whenever you design a study of animal behavior, it is very important for you to consider the proper care and use of animals. Members of the Animal Behavior Society are strongly committed to animal welfare and are bound by the guidelines of the society, which we have included for your information in Appendix A. For example, anyone who submits a paper to the journal *Animal Behaviour* must stipulate that the study was conducted in compliance with the "Guidelines for the Treatment of Animals in Behavioural Research and Teaching." In addition, most colleges and universities have their own methods of ensuring that animals are used properly and humanely. For example, your school may have an Institutional Animal Care and Use Committee (IACUC) or some other oversight group. Similarly, research or teaching activities that use human subjects must meet specific ethical standards. We have included in Appendix B a portion of the "Ethical Use of Human Subjects" policy of the American Psychological Association. As with animal research, your college or university has its own oversight group for research on human subjects, such as the Institutional Review Board (IRB), which must approve protocols to be used in research projects. Please be sure to consult the relevant appendix for guidelines and to ask your instructor whether you need approval from a local oversight group when you plan your research projects.

Finally, within most exercises you will find important vocabulary terms in boldface. These terms are defined in the Glossary at the end of the book. You should try to use these key terms as you discuss the topics in each exercise and as you design a study, gather and analyze the data, and report your results to your peers. As in foreign-language instruction, it is important for you speak the language—that is, to "speak animal behavior." Developing your language skills will help you to understand animal behavior, and it will facilitate your attempts to design your own studies.

GOALS

Our primary goal in *Exploring Animal Behavior in Laboratory and Field* is to provide exercises that engage you actively in all phases of scientific investigation, from formulating research questions through interpreting final results. We hope that your interest in animal behavior, combined with the ability to conduct sophisticated experiments and even original research, will help to facilitate your development as a research scientist. Our secondary goal is to share with you the collective teaching expertise and experience of the members of the Animal Behavior Society. In a sense we have attempted to formalize the process by which we arrived at the inquiry approach in our own teaching, as well as the informal network of information exchange that has existed among teachers of animal behavior for decades. We hope that you will find our efforts worthwhile.

ACKNOWLEDGMENTS

Bonnie gives special thanks to Alison Mostrom and Sylvia Halkin for contributing ideas for the format of the book, critiquing the book proposal, and providing many helpful suggestions during the planning of the book. Your enthusiasm about the project, repeated offers of help, and periodic queries about progress helped keep the book developing during the nomadic phase of Bonnie's career. Bonnie thanks Alison for her close friendship, which has supported her throughout this process. Ken thanks his former student, Rebecca Brooks, with whom he edited a collection of exercises in animal behavior designed for use in advanced high school biology courses. Those exercises were tested by Bec at Brimfield High School in Illinois and are available at http://www.animalbehavior.org/ABS/Education/Labs/index. phtml.

We thank all of the participants in the Education Workshops at the 1994 and 1995 annual meetings of the Animal Behavior Society, including all of the presenters (Charlie Baube, Ted Burk, Jed Burtt, Pat DeCoursey, Joseph Dulka, Linda Fink, Sylvia Halkin, Jasmin Keramaty, Mary Ketchum, Barbara Kirkevold, Scott Knight, Bob Matthews, Laura McMahon, Kathy Morgan, Terry O'Connor, Tom Rambo, Monica Richter, Ron Rutowski, Beth Stevens, and Lauren Wentz) and all coauthors with these presenters.

Without your presentations, ideas, and enthusiasm, this book project would not have begun. We must thank Ted Burk for organizing the 1995 workshop—a big job, as we all know! We also thank him for sending us copies of the lab exercises from that workshop and for providing information about how to contact those authors.

Bonnie thanks the following people for helping her organize the 1994 workshop by providing local logistic support and arranging financial support from the Animal Behavior Society: Lee Drickamer, Jim Ha, Christy Kimpo, Bob Matthews, Terry O'Connor, and Zuleyma Tang-Martinez. Thanks, too, to all of the ABS members who contributed lab handouts for the teaching materials display at the 1994 workshop, and to Lynette Hart for providing demonstration materials for this display. Sylvia Halkin helped contact potential contributors to the teaching materials display, and we thank her for this as well as for helping Ken to organize the display.

Many people contributed valuable comments to earlier drafts of each exercise during the peer review process, including Allison Alberts, George Barlow, Grant Blank, Jack Bradbury, Mike Breed, Jane Brockmann, Matthew Campbell, MarthaLeah Chaiken, Anne Clark, Ethan Clotfelter, Pat DeCoursey, Terry Derting, Darrin Good, Rick Howard, Bob Jeanne, Cathy Marler, Trish McConnell, Liza Moscovice, Scott Sakaluk, Nathan Sanders, Trish Schwagmeyer, Peter Sherman, Paul Switzer, Harry Tiebout, Steve Trumbo, Bruce Waldman, Fred Wasserman, Lauren Wentz, Linda Whittingham, and Kathy Wynne-Edwards. We also thank seven such reviewers who wish to remain anonymous.

We thank Dianna Kile, Noel McKee, and Joe Stoup for secretarial assistance and Arno Damerow for his computer expertise. Thanks to Chuck Crumly for his enthusiastic support of this project from its inception and for his help in the final production of the book.

Bonnie thanks the following people for their major influence on the development of her teaching: Jane Brockmann, Peter Feinsinger, Doug Mock, and Susan M. Smith. Bonnie also thanks her colleagues in the biology department at Hamline University, and former colleagues at Grinnell College, for the many stimulating conversations about pedagogy that have been so helpful. She especially thanks Pres Martin for his support of her inquiry-based approach to teaching and for leading the Hamline biology department into adopting this approach for the introductory sequence of required courses. Bonnie also thanks her husband, Don Allan, for his support, encouragement, and, most important, deepest friendship. She also thanks him for sharing his infectious enthusiasm for observing squirrel behavior. Ken thanks his colleagues in the biology department at Beloit College and the members of the BioQUEST Curriculum Consortium for their steadfast commitment to the inquiry approach to learning and for their uncommonly collaborative spirit. He also thanks the many participants in the Behavior Interdisciplinary Seminar (BIDS) at the University of Wisconsin—Madison for providing a home away from home, lots of stimulating conversation, and a network of dedicated animal behaviorists. And of course he thanks Sondra Fox, who is always his best critic and advocate, for countless conversations about pedagogy and student learning.

LITERATURE CITED

Costenson, K. & Lawson, A. E. 1986. Why isn't inquiry used in more classrooms? *American Biology Teacher,* **48**, 150–158.

Gopnik, A. 1999. Small Wonders. *New York Review of Books,* **May 6, 1999**.

Lehner, P. N. 1996. Handbook of Ethological Methods. 2nd ed. Cambridge: Cambridge University Press.

Martin, P. & Bateson, P. 1993. *Measuring Behaviour: An Introductory Guide.* 2nd ed. Cambridge: Cambridge University Press.

McNeal, A. P. & D'Avanzo, C. 1997. *Student-Active Science: Models of Innovation in College Science Teaching.* New York: Saunders.

National Science Foundation. 1990. *Report of the National Science Foundation Workshop on Undergraduate Laboratory Development.* Washington, DC: National Science Foundation.

Sundberg, M. D. 1997. Assessing the effectiveness of an investigative laboratory to confront common misconceptions in life sciences. In: *Student-Active Science: Models of Innovation in College Science Teaching* (ed. A. P. McNeal & C. D'Avanzo), pp. 141–161. New York: Saunders.

Wright, E. L. & Govindarajan, G. 1992. A vision of biology education for the 21st Century. *American Biology Teacher,* **54**, 269–274.

INTRODUCTION

Humans have always been interested in animals and their behavior. During our early history, much of this interest was grounded in practical need. Animals were an important source of food, so a thorough, working knowledge of how potential food items behaved was extremely important to successful hunting. Animals also assumed important roles in our early rituals and religious beliefs. Today, people are still drawn to animals—we surround ourselves with them. We keep them in our houses, we watch them for entertainment and recreation, we use them to do work, we raise them for food and clothing, we hunt them, we use them to test products, and we use them to answer questions in an attempt to improve the human condition.

Many people assume that animals are just like us and so endow animals with human feelings and emotions. We say that our dogs act guilty when we find them on our beds or that our cats are jealous of our children. We think of our pets as members of our families and of wild animals as crafty or cruel or courageous. Although such interpretations of an animal's behavior are acceptable and even useful for most people, they also create some problems, especially the assumption that animals are "just like us." Why is that a problem? Because animals, like humans, are unique—they *aren't* just like us. In some ways we are like other animals, and in other ways we are very different from them. A deeper understanding of animals requires us to think of them as organisms with their own attributes. Instead of assuming that animals think like us, we must instead think like them if we are going to understand them. And we must be willing to assume, at least for the moment, that animal thinking is very different from human thinking.

Some people dedicate their lives to understanding how and why animals do the things they do. Many are trained as veterinarians, but their interest is usually more medical than behavioral. Veterinarians try to ensure the health and safety of the animals in their charge. Usually, this involves yearly check-ups and tending to sick or injured animals. Only occasionally does a vet become concerned with an animal's behavior—and then it is usually when the animal is doing something that its owner doesn't like. Animal breeders and trainers are also very much interested in how and why animals do what they do. In some cases, a knowledge of animal behavior helps breeders or trainers to do their jobs, and in others the job is really to make the animal do something in a certain way or at a certain time. Professional hunters, naturalists, nature photographers, and wildlife artists must also understand animals, because animals are the subjects of their work.

A few people dedicate their lives to understanding animal behavior not because they need to know about animals to do their jobs but because it is their job to understand animals. These people use scientific methods to study the behavior of animals, and they are highly trained in such specific fields as biology, psychology, and anthropology. Although they go by various names (for example, ethologists, behavioral ecologists, comparative psychologists, and behavioral primatologists), we will refer to them as animal behaviorists, scientists who study the behavior of animals. Animal behaviorists are an extremely diverse lot—men and women of many different nationalities, races, and ethnic origins; scientists trained in many different fields of scientific inquiry—but they have one thing in common, an interest in and a curiosity about how and why animals do what they do.

We have all been exposed to their work, either directly or indirectly. For example, animals are increasingly becoming the subjects of movies and TV shows—some cable channels specialize in them. Sometimes these shows are inaccurate or excessively **anthropomorphic** (presenting animals as though they were human), but they are usually based on work done by animal behaviorists, and sometimes they are produced by animal behaviorists. Some of the current practices of wildlife managers, conservation officers, zoo operators, and other animal keepers are grounded in sound principles taken from animal behavior. And, of course, some animal behaviorists, such as Jane Goodall and Diane Fosse, have become famous in our popular culture.

Our purpose in this book is to expose college students to the questions, methods, and approaches used by scientists who study animal behavior. We have three reasons for writing this book. First, animal behavior has the potential to be very important in our everyday lives. We depend on animals in so many ways, but so few of us understand animals on their own terms. For this reason, educated people should be aware of the information provided by animal behaviorists and their ways of doing things. Second, animal behavior is an excellent example of the workings of science.

The scientific study of animal behavior uses the same principles as the other sciences and applies them to a group of organisms that are simultaneously familiar and mysterious. Our society is increasingly driven by scientific information and methods, so it is crucial for educated people to understand the scientific method. And finally, in our experience, animal behavior is interesting and fun for students to study. Exercises based on the study of animal behavior are therefore an excellent way to motivate students, to get them to think critically, and to show them the power and intrigue of science, as well as its limits and pitfalls.

THE STUDY OF ANIMAL BEHAVIOR

According to Niko Tinbergen, who won the 1972 Nobel Prize in medicine along with fellow animal behaviorists Konrad Lorenz and Karl von Frisch, the scientific study of animal behavior has four components: causation, development, evolution, and function. We can describe these components in the form of four questions. (1) What causes an animal to perform a certain behavior? (2) How does the behavior change as an animal develops from conception through death, but especially during its early life? (3) What is the evolutionary history of the behavior? (4) How does the behavior help the animal to survive and reproduce successfully?

Causation of Behavior

Answers to the question of what causes an animal to perform a certain behavior can take many forms. An experimental psychologist might try to identify the stimuli (the events in the animal's external environment) that immediately precede the behavior and might then experiment with these stimuli and their components to understand the stimulus–response relationship of the behavior. A cognitive psychologist might try to understand the mental processes that are triggered by an appropriate stimulus and that lead to the performance of the behavior. A neurobiologist might study the anatomy and physiology of single neurons (nerve cells) or of the entire neural pathway associated with the perception of the appropriate stimulus, its processing in the nervous system, and the production of the appropriate motor patterns (the behavior). An endocrinologist might study the role that hormones play in preparing an animal to respond, for example by organizing the nervous system during development and by priming it to respond rapidly. An ethologist might view causation in terms of systems of motivation and might construct conceptual models of different systems that identify the appropriate stimuli (**releasers**) and the mechanism that produces the behavior (**action pattern**). In a sense, although each of these approaches uses different tools and concepts, all of them are concerned with the way in which a behavior is triggered or elicited. Animal behaviorists refer to such investigations as research on **proximate mechanisms** because these triggers are the most closely related to the behavior itself. At the simplest level, we could study reflexes such as the knee jerk, but usually animal behaviorists are interested in more complex behavior.

Development of Behavior

A different sort of proximate mechanism is one that enables the behavior to come into being as an animal develops from its earliest stages of growth. In many cases, behavior must be learned and practiced before it is performed appropriately and correctly, so a portion of the study of development is concerned with the kinds of experience that are necessary and whether the timing of that experience makes a difference in **learned behavior**. Animal behaviorists interested in learned behavior might study normal development to identify potentially important kinds of experience and then might experimentally withhold, alter, or shift in time that experience to see whether such manipulations alter normal development. In other cases, behavior seems to appear rather suddenly and seems to be correct and complete the first time it is performed. Prior experience seems to have little to do with the development of **innate behavior**, although nervous and muscle systems must be properly developed if the behavior is to be performed at all. Such behavior will develop normally even if animal behaviorists withhold the usual early experiences of young animals. Although animal behaviorists have sometimes made a big deal about the differences between innate and learned behavior, most now think that the two don't really represent completely different developmental processes. Instead, the development of behavior can be viewed as a range of different degrees of innateness or learnedness. Although extremes do exist, most behaviors fall somewhere between the two extremes.

Evolution of Behavior

At the opposite end of the spectrum of studies of animal behavior are attempts to identify the evolutionary history of a particular behavior. Because the process of **evolution** (descent with modification) occurs over long periods (sometimes millions of years), experimental manipulations cannot shed any light on evolutionary questions (at least not quickly enough for an animal behaviorist to learn the result). Therefore, evolutionary questions about behavior are addressed by comparing the behavior of different species of animals. Of course, this **comparative method** requires (1) that the animals being compared are members of an evolutionary lineage and (2) that we know what that lineage is. Fortunately for animal behaviorists, there are ways to get this information. Today, evolutionary biologists can take information on some aspect of a group of animals—say, the structure of some organ or the sequence of a particular bit of DNA—and, with the help of computer programs, come up with an evolutionary "family tree" of these species. Such a **phylogeny** is a hypothesized evolutionary relationship, and if several different analyses that are based on different information (for example, different bits of DNA and different anatomical structures) all produce the same tree, then we can have some confidence that the proposed phylogeny is correct. If an animal behaviorist then looks at a particular behavior in these same species of animals, it may be possible to determine where the behavior arose (in which species of animal it first appeared) and how in evolved within the phylogeny (whether it arose separately more than once and whether it was passed on to descendant species). It is even possible for an evolutionary biologist to construct a phylogeny using behavioral characteristics alone, using the same methods and computer programs. Because the evolutionary history of a behavior is far removed from its performance by an individual animal but is nevertheless an important aspect of that behavior, such analyses are called studies of **ultimate mechanisms**.

Function of Behavior

Many students of animal behavior focus on trying to determine whether and how a behavior affects an animal's ability to survive and reproduce. This approach is called the **functional analysis** of animal behavior, and it also is a study of ultimate mechanisms because the functions help to shape the evolutionary responses in subsequent generations. In this approach, a particular behavior is examined to determine whether it enhances an animal's ability to survive and/or to reproduce and, if it does, how it enhances survival and/or reproduction. Although this approach can become quite complex, for our purposes here we will limit such considerations to advantages (functions) to individuals—that is, how the behavior helps those individuals who perform it to live longer, or get mates faster, or get more mates, or have more young, or enhance the survival of their young. If the behavior improves an individual's performance in any of these aspects of **reproductive success**, when compared to individuals who don't perform the behavior, then the behavior is said to serve that particular function (for example, one might say the behavior functions to improve survival).

WORDS OF CAUTION

Several words of caution are appropriate here. One involves the ways in which animal behaviorists talk about behavior vs what they really mean to say. Often animals are said to follow behavioral **strategies** and to make **decisions** or **choices** among stated alternatives. Such statements make it seem as though animals are making conscious, rational decisions the way we do when faced with a set of alternatives. These statements, however, do not imply conscious, purposeful behavior; they refer to functional or evolutionary "choices" or, rather, comparisons among behavioral alternatives. Thus, when animal behaviorists say that an animal "chooses" a foraging strategy that provides the maximum rate of food intake, what they really mean is that of the different foraging strategies available to the animal, the rate-maximizing one provides the highest survival and reproductive success, and that is why we observe it in the animal. In other words, animals aren't aware of the different possibilities, and they don't consciously choose one strategy over another. Instead, the animals that adopt the best strategy (in terms of survival and reproductive success) produce the most offspring, which inherit genes "for" and—accordingly follow—that strategy Animals that happen to adopt a poorer strategy leave fewer offspring, so eventually, that strategy disappears.

Such functional "choices" mean that the behavior in question must have a genetic basis, but what does *that* mean exactly? What it surely does *not* mean is that behavior is "controlled" by genes. All behavior occurs as a result of an interaction between an animal's genetic constitution and the environment in which that genetic constitution operates. All it really means is that different individuals have different genetic constitutions and that these differences are responsible for some of the differences in their behavior. An explanation, called the "cake analogy," which was proposed by animal behaviorist Richard Dawkins, is instructive. Suppose a friend gives you a recipe for a great-tasting cake. You copy the recipe and use your copy to make the cake, but your cake tastes terrible. You double-check your recipe and discover that you made a small mistake in copying the original recipe—you inadvertently wrote "one T salt" instead of "one t salt." As a result of your small error, your cake contained a tablespoon of salt instead of a teaspoon of salt. No wonder it tasted terrible. But does this mean that the single ingredient (one T salt) "controls" the production of a terrible-tasting cake? Of course not. The entire recipe, along with how you mixed the ingredients, certain properties of your oven, and many other factors "controlled" the cake. What is clear, however, is that a small difference in the recipe (or in an animal's genetic constitution) can produce an important difference in the expression of the recipe (or in an animal's behavior).

A very different problem is the "just so story," named by Stephen Jay Gould in honor of Kipling's stories for children. It is very easy for creative animal behaviorists and college biology students to think up ways in which a behavior might be caused, or develop, or evolve, or function. Remember, however, that just because the idea "makes sense" doesn't mean it is correct. In other words, a "good story" is not necessarily a good explanation. Science involves the proposing and testing of **hypotheses**, and each idea must be treated as an hypothesis to be tested. You must think of ways to test your idea (your proposed hypothesis), and the tests should be attempts to disprove your idea, rather than to prove it.

In many cases, several hypotheses might exist to explain a particular behavior. We might be tempted to test each separately, for the sake of simplicity, but there is a danger in such simplicity. The danger is that results that support one hypothesis are likely to support others as well, and when this happens, we cannot eliminate any of these hypotheses (remember, we should try to disprove each hypothesis). An alternative, "stronger" method is to test many or all of the hypotheses at the same time. This method of multiple working hypotheses has the advantage of eliminating many or all of the hypotheses at once, so that our understanding can be increased rapidly. It also requires, however, that each hypothesis makes at least one unique prediction (that is, a prediction that none of the other working hypotheses makes), and sometimes such critical predictions simply do not exist.

We should also warn you of another potential problem. You are attempting to answer questions about an animal's behavior, but you are using your own senses to gather information about that behavior. Unfortunately, your senses do not produce prefect reproductions of the world around you, and your senses may be quite different from those of your animal subjects. Some animals can see light that we cannot see, and some can hear sounds that we cannot hear. Some animals have senses that we don't seem to have at all—some can detect electric or magnetic fields, for example. In other words, the sensory world of your subject animal may be very different from you own, so the animals may be using cues that you cannot perceive at all.

A related problem has to do with expectations, rather than sensory capabilities. In many cases, we know that people see "what they want to see" rather than what really happens. A clever animal behaviorist once described this phenomenon as "I wouldn't have seen it if I didn't believe it."

A final problem can be called the "eyewitness" phenomenon. If five people are eyewitnesses to a crime, it is quite common for them to give five different descriptions of the perpetrator. Sometimes the accounts can differ radically— one witness might say the assailant was a white older man, another that she was black teenager. In addition, a particularly persuasive and confident witness can often sway the others, even when the confident one is wrong.

Some of the challenges facing animal behaviorists involve finding ways to overcome these and other problems inherent in scientific research. Remember that although the scientific method attempts to be objective, it is practiced by people, who cannot be objective. Each of us has opinions and beliefs that affect the ways we see and interpret things.

WORDS OF ENCOURAGEMENT

On the other hand, despite the challenges facing students of animal behavior, progress has been steady and rapid. Many species of animals have been studied, from invertebrates to humans, and both proximate and ultimate mechanisms have received lots of attention. We know quite a bit about neural and physiological mechanisms, about the ways in which animals learn, about how behavior develops, and about what functions behaviors serve, and we are beginning to see the evolutionary history of behavior unfolding. It's an exciting time for animal behavior because our field seems to be coming into its own. One of the important next steps that animal behaviorists must take is spreading the word to others. We hope to show you, through these exercises, that animal behavior is an interesting and important field of science. Although most of you won't become animal behaviorists, all of you can learn to appreciate what animal behavior has to offer, and all of you can learn something about real science as you learn about animal behavior.

PART 1

Describing Behavior

CHAPTER 1

Learning to Describe and Quantify Animal Behavior

BONNIE J. PLOGER
Department of Biology, Hamline University, St. Paul, MN 55104, USA

INTRODUCTION

Research in animal behavior typically focuses on providing insight into the causation, development, adaptive value, or evolution of particular behavior patterns. To investigate such issues, researchers must be able to observe and describe the details of behavior, including detecting differences and similarities among the actions performed under differing social and environmental circumstances by individual animals differing in character-istics such as age, sex, and species. To make such comparisons requires quan-tification of behavior.

To quantify behavior requires breaking up behavior into distinguishable units that can be named. Obvious units include actions such as walking and sleeping. Can you think of others? What about quick actions such as individual steps during walking or individually repeated scratches of an itch? In order for observers to quantify behavioral acts, each type of act that is measured must be clearly defined. Before deciding which behavior patterns to define and measure, you must first spend time observing and describing a variety of behavior patterns seen in your study species so that

ISBN 0-12-558330-3

you can decide what questions are reasonable to research in your species. Thus a period of initial description is the critical first step in any research on animal behavior. This exercise introduces the skills you will need to describe animal behavior sufficiently to develop reasonable research questions and to investigate these questions quantitatively.

Clear definitions are needed in any scientific study, but in some disciplines the definitions are so well established that new studies can simply use the old terms rather than including definitions. For example, when you study anatomy, you can use widely known terms such as *biceps* and *vertebrae*. Because few such conventions exist for behavior, and because the behavior of many animals has never been described at all, studies of behavior need to include descriptions and definitions of terms, just as early anatomical studies did. Without clear definitions, researchers would not be able to repeat the observations and experiments of others or to interpret the validity of published work, because it would not be clear exactly what behavior was observed and measured. Researchers must therefore name, classify, and describe all of the behavior patterns that are measured in a study. Description of behavior was the primary goal of many early studies of **ethology**, the zoological, evolutionary approach to animal behavior that includes examination, generally under natural conditions, of the proximate causes and development of behavior, as well as its ultimate evolution and adaptive value. The purpose of many of these early ethological studies was to create a complete **ethogram**, a dictionary of the names and descriptions of *all* of the behavior patterns that constitute a species's behavioral repertoire (Lehner 1987). Descriptions in an ethogram typically include not only written information but also drawings that illustrate the postures and movements involved in an action pattern.

Complete ethograms are rarely published anymore in the behavior literature, but they are often sought by zoos to help define normal behavior and to distinguish it from pathology due to illness or problems with animal care. Most studies today name and define only those action patterns that are being measured for a specific purpose, such as to study the causation, development, adaptive value, or evolution of some aspects of behavior. Such studies include only partial ethograms or a few **operational definitions** that describe exactly how the researcher(s) measured particular behavior patterns.

In this exercise you will first watch ducks to learn how to observe, describe, and quantify behavior. You will thus be learning the skills you will need for creating initial descriptions of behavior, the critical first phase of any research project. As you develop these skills, you will also improve your ability to observe; you are likely to begin to see new behavior patterns that you did not notice at first, to notice differing contexts in which a behavior pattern is performed, and to detect subtle differences between similar behavior patterns. Thus, as you learn to describe behavior, you will actually start seeing more. In the second part of the exercise, you will have the opportunity to apply what you have learned to conducting an actual, though short, research study. In this study you will describe and quantify behavior of an animal species of your choosing in order to test an hypothesis of your own that compares at least two different types of animals or situations in which the behavior occurs.

MATERIALS

Each pair of students should have a stopwatch or its equivalent in wrist-watch form. One pair of binoculars per student is helpful but not essential. Each person should have a notebook and a pencil or pen for recording data.

PROCEDURE

Getting Started

In this exercise you will alternate between making behavioral observations and discussing your observations with your classmates by answering questions in the "Questions for Discussion" section at the end of this chapter. During each observation period, you will be recording data in your data notebooks for later use in answering the questions in the worksheet at the end of this chapter. Before beginning to record data, you should take a few minutes to organize your notebook and to become familiar with the sorts of information that you should include in it.

A good way to organize your notebook is to leave the first few pages blank for an index that you can develop to help you find information in your later notes. For the index to be useful, you will need to number all the pages in your notebook, if they are not numbered already. On the inside cover of your notebook, be sure to write your name, address, and phone or e-mail address so that someone finding your notebook can return it to you if you lose it. I like leaving the last pages of my notebook available for other types of notes, which I write back to front. Such notes may include ideas for future research projects, data transcribed from tape recordings, and information I may want in the field, such as phone numbers for equipment repair shops.

After the index pages, you should enter your observations for each day. Make each day's observations easy to find by making the date stand out in an obvious way. You can do this by starting each day's observations on a new page with the date at the top or by drawing a horizontal line across the page just above each new date. Write dates in an unambiguous way, such as "10 Sep 02" (this so-called continental dating is the convention in animal behavior literature) rather than in Arabic numerals such as 10/9/02, which could be interpreted to mean 10 September or October 9. Before beginning your observations, you should record the time when you arrived at the study site, a description of general weather conditions, and any other relevant information (such as the presence of people or predators) that might affect the animals you are studying. If the weather or other relevant conditions change while you are observing your study subjects, you should record this information, because such changes may be responsible for changes in their behavior.

During any one day of observations, including today, you will have periods when you are watching the animals (**observation periods**) and other times when you are doing something else, such as resting, discussing ideas with other observers, or planning your next observation period. During each observation period, you should record the time frequently, so that you will know approximately when and how often different behavior patterns occurred. I record times in the left margin of my notes and record my running notes of behavior to the right of these times. To distinguish the starting and ending times of observation periods from other times that you may note while observing behavior, write in your notes explicit phrases such as "1330 Start Observations" and "1335 End Observations." Clear recording of the starting and ending times of each observation period is necessary so that you can determine the rates at which different actions occurred and so that you can be sure you are sampling different individuals or behavior patterns for similar amounts of time. Note that the aforementioned times use the 24-hour clock, which is the convention in the animal behavior literature, because it saves time and avoids the ambiguity that can otherwise arise if you forget to include "a.m." or "p.m." In your notes, always use the 24-hour clock—for example, by recording times as 0530 and 1400, not as 5:30 a.m. and 2:00 p.m., respectively.

When describing behavior, be objective by writing what you actually *saw*, not your interpretation of the action. If you record something in your notes and later think you made a mistake, do not erase or otherwise obliterate the original note that you think is in error. Instead, draw a single line through the note, leaving it readable, and include a comment about why you think the information is in error. Doing this is important because you may later discover that what you thought was in error was actually correct, as is likely to occur if the behavior was surprising to you. By clearly marking suspected errors, you will also preserve a record of your errors and improvements. This will be useful because you might need to exclude from your data analysis the data collected during observation periods when you made a lot of errors, as is typical in the first day(s) of observing a new species or when an observer is particularly fatigued.

Make your notes as descriptive and detailed as possible. Avoid vague general descriptions, such as "The duck swam to the left" because they are of little value. For example, instead of writing "The ducks fought," write "Duck A's bill struck duck B at the back of the neck, appearing to touch the feathers only slightly or missing entirely." Note that the latter description includes important information about which individuals were involved in the interaction, who appeared to be the aggressor, and the observer's uncertainty about whether contact was made. This is the type of detail for which you should strive.

It is critically important that when observing animals, you be as unobtrusive as possible. Your goal is to study the natural behavior of animals, not their responses to you! Some animals, especially those living free in the wild, may be so sensitive to your presence that you will need to hide from their view, but even captive domestic and zoo animals may alter their behavior because of your movements. Therefore, when watching animals, sit or stand quietly, avoid gesturing with your hands, and be as still as possible.

Identifying Anthropomorphism

Keep in mind that when observing and describing animal behavior, it is easy to misinterpret the actions of animals by assuming that they are just like humans, with human thought processes and emotions. Such **anthropomorphism** is very common. But animals are not just like us. Indeed, many species differ dramatically from each other and from us in their sensory abilities, behavioral responses, and ability to learn, which may or may not involve thought. To gain greater insight into the nature of anthropomorphic thinking and the problems that it causes, you should engage in a class discussion of this topic before beginning your observations of animals. In this class discussion, answer the first and second questions in the "Questions for Discussion" section at the end of this chapter.

Classifying Behavior

With your partner, watch a group of ducks for 15 minutes. Remember that this 15 minutes should be your observation period, so it should not include other activities such as finding your notebook in your backpack or discussing plans with your partner. Also remember to record the actual times when you begin and end your observation period. As you watch the ducks, classify the behavior that you see into as many different categories as you can. Record your observations in your data notebook. To make your observations, record all the behavioral acts that you can, by any individuals. This is *ad libitum* **sampling** (Altmann 1974; Martin & Bateson 1993).

Some of the behavior that you observe will occur as **events**, which are clearly distinguishable, discrete acts of fairly short duration that can be recorded instantaneously and can be quantified by counting each occurrence (Altmann 1974; Martin & Bateson 1993). For example, one quack by a duck is an event, quantifiable as the number of quacks per minute. Other behavior patterns will occur as **states**, actions that last for a relatively long time and can best be quantified by timing the duration of the action (Altmann 1974; Martin & Bateson 1993). For example, sleeping is a state quantifiable as the amount of time spent sleeping in 24 hours. Your categories should include at least three states and as many different types of events as possible.

After the 15-minute observation period, meet with the rest of the class to discuss your observations. Your discussion should include answering the third through sixth questions in the "Questions for Discussion" section at the end of this chapter.

Quantifying Tail-Wags

After the class discussion, join your partner again to read this section about tail-wags, and then collect quantitative data on this behavior. The tail-wag (Fig. 1.1) is one of the most frequent behavior patterns in mallard ducks (*Anas platyrhynchos*; Hailman & Dzelzkalns 1974) and is common in other duck species as well. Therefore, you should be able to see the action frequently enough to quantify several aspects of the behavior and observe some of its variations. When you first observe tail-wagging, it may seem quite simple and uninteresting. Do not be deceived! If you watch carefully, you will find a surprising amount of variation.

The high frequency of tail-wagging also suggests that the behavior serves one or more important functions. McKinney (1965) considered tail-wagging to function primarily to remove water from the tail. Noting that dry tails are also wagged in a variety of different contexts, McKinney suggested that tail-wagging might also help reset misplaced tail feathers, facilitate reinversion of cloacal lips after defecation, aid in removing oil-gland secretions during preening, and assist in retracting the penis after copulation (most male birds lack a penis; but ducks and geese are exceptions). Hailman & Baylis (1991), assisted by undergraduate ethology students, found evidence that repositioning misplaced tail feathers following strenuous activity may be a more important function of tail-wagging than removal of water, when they observed that tail-wagging occurred with similar frequency in mallards taking off and landing on land and on water. Analysis of behavioral sequences in mallards led to another hypothesis for the function of tail-wagging. By creating an ethogram of 116 behavior patterns and collecting behavioral sequences of these actions, Hailman & Dzelzkalns (1974) and undergraduate ethology students discovered that tail-wagging frequently occurs both before and after courtship and other displays.

Tail-wag (TW)   Head-flick (HF)   Head-shake (HS)

Wing-shuffle and tail-fan (WTF)   Foot-shake (FS)   Foot-pecking (FP)

Wing-and-leg-stretch (WLS)   A   B

Wing-shake (WS)   Body-shake (BS)   C

Figure 1.1 **Some behavior patterns involved in self-maintenance of mallard ducks that are generally performed on land. From Mckinney 1965. Use the abbreviations provided when recording behavioral sequences involving any of these actions.**

This led them to propose that tail–wagging may serve as "punctuation" for the communication of messages. You may have noticed that in the aforementioned two studies coauthored by Hailman, undergraduate students in ethology courses collected the data that were published. The work you do in a course concerning animal behavior could be publishable!

This is another good reason to learn the basic skills of this exercise well and to be meticulously careful in your data collection.

With your partner, watch the ducks for 10 minutes to quantify the following aspects of tail-wagging: the total number of bouts of tail-wagging in your observation period and, for each bout of tail-wagging, the duration of each bout and the frequency of tail-wagging (number of tail-wags per bout). Remember to be sure that this 10 minutes is for observations only and that you record the starting and ending times in your data notebook. One of you should take the role of observer, quietly telling your partner the relevant information while that person records this information in your data notebook. To make your observations, watch all of the ducks continuously and record data for each occurrence of tail-wagging by any individuals. This is behavior sampling with **continuous recording** (Martin & Bateson 1993).

Tail-wagging involves several side-to-side shakes of the slightly fanned tail (McKinney 1965; see Fig. 1.1). Note that this definition and the drawing in Figure 1.1 do not provide sufficient information to quantify the behavior—for example, to compare the number of tail-wags collected by different observers working independently. The ambiguity arises because a tail-wag could start with the tail aligned with the center line of the body (that is, aligned in the medial plane) and end when the tail returned to the starting position (1) after being shifted laterally in one direction, (2) after being shifted laterally both to the left and to the right by passing through the starting position, or (3) after reaching the point farthest from the starting position.

Before you begin, you will need to define what you mean by a tail-wag and to decide how you will count each wag. To standardize methods among all observers, we will all use the same definition and method of counting tail-wags. The easiest way to count tail-wags is to count each lateral movement, left or right. Thus, if the tail moves as follows: "left, center, right, center, left, center, right, center" you would count 4 lateral movements. Then you would divide by 2 to get the total number of tail-wags, where we define a tail-wag as beginning with the tail aligned in the medial plane and ending when the tail returns to this position after being shifted laterally *both* to the left and to the right. This is our operational definition of a tail-wag. Note that we could just as easily have defined a tail-wag as being lateral movement to one side. In any study of behavior, you must explicitly state your operational definitions so that others can repeat your experiment. In the case of tail-wags, had we not agreed on the same operational definition, some of you would have counted twice as many tail-wags as others while watching the same individual at the same time.

After the 10-minute observation period, meet with the rest of the class to discuss your observations. Your discussion should include answering the seventh and eighth questions in the "Questions for Discussion" section at the end of this chapter. If time is available after the class discussion, record your answers to the questions in Part A of the worksheet. Otherwise, answer these questions later, before turning in your Worksheet.

Defining Behavior Categories

Working *alone* for a 15-minute observation period, collect data needed to describe five different behavior patterns that are all distinct events involved in care of the body. Such **self-maintenance behavior**, also called **comfort movements**, include bathing, preening, stretching, scratching, and shaking movements (McKinney 1965), but you should be able to break some of these large categories into several finer ones. See Figures 1.1 and 1.2 for examples of abbreviations and names of behavior patterns involved in self-maintenance. Your five behavior patterns should differ from those shown in Figures 1.1 and 1.2.

To make your observations, watch all the ducks, sketching and making notes about each occurrence of your target behavior patterns, whether these occur for the same or different individuals. This is **behavior sampling** of five different behavior patterns.

Because the actions occur quickly and may vary in form and timing, for a complete description of each behavior pattern, you will need to sketch and take notes on the same action repeatedly. It often helps to concentrate on only one part of the animal's body at a time. Thus, for a single behavior pattern your notebook may contain many separate, quick sketches and descriptions of different body parts in different stages of the movement. After taking the animal apart like this, you must then put it back together, because a good description of any behavior pattern must include the whole animal. Therefore, be sure that you have sketches and notes about all different parts of the body so that you can create a composite sketch and description of the whole animal during the action.

Use arrows and sequentially labeled sketches to convey the sequence of movements of behavior patterns when appropriate, as shown in Figure 1.1 (for example, head-flick and body-shake). When sketching while watching the action, however, do not try to make slow, careful sketches like those shown in Figures 1.1 and 1.2. Instead, make quick, simple drawings that include only a few critical lines. Variations on stick figures will usually suffice. See Figure 1.3 for examples of the types of quick sketches you might make while watching some of the behavior patterns illustrated in Figures 1.1 and 1.2.

It is of critical importance to sketch and make notes *while* you are watching the behavior. Do not simply watch the action repeatedly and then wait until the end to try to describe it from memory, because memory is unreliable, as is well known from the highly variable accounts of different eyewitnesses relying on their memories of the same crime.

Bill-Cleaning (BC)

A

B

Wing-thrashing (WT)

A

B

C

D

E

Head-dipping (HD)

A

B

C

Swimming-shake (SS)

Figure 1.2 **Some self-maintenance activities performed by mallard ducks while swimming and bathing. From Mckinney 1965. Use the abbreviations provided when recording behavioral sequences involving any of these actions.**

Instead of relying on your memory, sketch and jot down notes while watching the action. You can do this by glancing at the page only once in a while or, better yet, writing without looking at the page at all so that you can watch your subject continuously while it is performing the action.

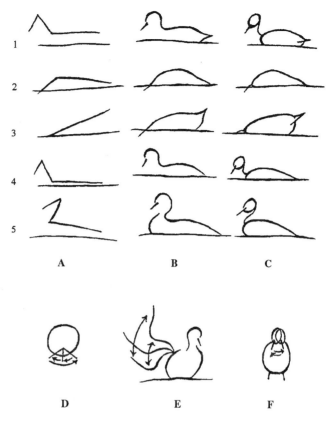

Figure 1.3 **Examples of various methods for making quick sketches of some of the behavior patterns illustrated in Figures 1.1 and 1.2. A. Angled-lines sketch of head-dipping. B. Calligraphic sketch of head-dipping. C. Ball-and-stick sketch of head-dipping. D. Use of straight lines to illustrate a tail-wag. E. Calligraphic sketch of wing-thrashing. F. Ball-and-stick sketch of head-shake.**

Immediately after the animal you are watching stops exhibiting the behavior pattern you were recording, stop watching for a few moments so that you can add any necessary comments about the observation you just made. Such comments should include definitions of any abbreviations you used, explanations about parts of drawings that are hard to understand, and any other relevant comments that you did not have time to write while watching the animal. Also, if you missed seeing something, clearly state that you didn't see the action or the details of movement of a particular body part. For example, while recording a tail–wag, you may have sketched the tail as being held horizontal to the ground but have failed to notice the position of the head and whether the tail feathers were spread apart or held close together.

You will need to use the sketches and notes recorded in your data notebook to create the composite sketches of each of the five behavior patterns that are required for Part B of the worksheet. Your notebook should contain sufficiently descriptive primitive sketches and notes that someone else reading your notebook would be able to find all the information that you used to create the more detailed, formal composite sketches and descriptions that are required for the worksheet. To determine the sample size for your composite sketch of a particular behavior pattern, count the total number of separate sketches and written descriptions of the same action that you recorded in your notebook as occurring at different times. The notes in your data notebook should be sufficiently clear that a naïve reader could check your answers in Part B of the worksheet by making a similar count of sketches and notes for each of your five behavior patterns.

After the 15-minute observation period, join the rest of the class to discuss the last two questions in the "Questions for Discussion" section. After this discussion, you may want to change the names of some of your behavior categories before completing Part B of the worksheet. Be sure to record any such name changes clearly in your data notebook.

Censusing with Scan Sampling

With your partner, you will conduct a 6-minute census of activities involved in self-maintenance. Before doing the census, you will need to find four ducks that you can identify repeatedly as individuals. To distinguish individuals, look for differences in the patterning of dark vs light pigmentation on the bill and legs, differences in plumage patterns, and missing wing or tail feathers. Look also for metal or plastic leg bands, which are usually present on captive birds and may also occur on wild birds that have been captured and released by biologists. Detectable gender, age, and size differences may also be useful in distinguishing individuals. If you are distinguishing some individuals by plumage or bill colors, sketches of the key distinguishing features will help you identify individuals quickly. In your data notebook, record the distinguishing features of each of your four individuals, and assign them names that you can remember. I prefer to use names that remind me of their features, such as Right-Leg Band, or Head Spot. Avoid using human names because these are likely to lead you to anthropomorphic thinking. Also avoid gender-specific language because this may lead to false assumptions. For example, juveniles of either sex may have plumage similar to that of adult females, and in some domestic varieties, such as white, Peking ducks, the sexes will be indistinguishable to you.

To do your census, scan these four individuals every 2 minutes for 6 minutes, starting at time = 0 minutes. For each scan, record the action pattern of each individual at the instant that you looked at that individual. This is **scan sampling** with **instantaneous recording**. Record the behavior patterns of each individual at each time in Table 1.1 of Part C of the worksheet by entering the abbreviations for the behavior patterns that you observed. Use the abbreviations that you developed for the five behavior patterns that you defined in your last observation period, as well as those given in Figures 1.1 and 1.2. If at the instant of your scan, a duck is doing a new action not shown in Figure 1.1 or 1.2 or in your earlier descriptions, then quickly sketch or jot notes of the behavior in your data notebook at the moment of the scan. In the interval following this scan, describe the behavior in more detail, assign it an abbreviation, and enter this in Table 1.1. To complete Part C of the worksheet, you will need to use these notes to create a formal sketch and description of any behavior patterns occurring in your scans that were not previously described or shown in Figure 1.1 or 1.2. As much as possible, use the 2 minutes between scans to complete Part C of the worksheet.

You should notice that when using scan sampling, you will be recording states as events. When you use other sampling methods, you may also find it appropriate to record states as events, depending on your research question (Altmann 1974; Martin & Bateson 1993). For example, you might want to know whether the frequency of preening differs when it is raining from when it is sunny. For such a comparison, instead of measuring the duration of a preening bout, you might measure the number of preening bouts per hour by simply counting each start of a preening bout in the hour.

Sequencing Behavior

Working alone, watch one duck continuously for a 5-minute observation period. This is **focal–animal sampling** (Altmann 1974; Martin & Bateson 1993). During each bout of self-maintenance behavior, sequentially record every event occurring in the bout (that is, make a continuous recording). Also record the approximate starting and ending times of each bout of self-maintenance behavior to estimate sequence durations.

To record the sequence of events in a bout, write the abbreviations rather than the names of the behavior patterns in your data notebook. Record only the abbreviations for behavior categories that you named earlier, when defining behavior categories and censusing with scan sampling. If new behavior patterns occur during a sequence, simply record these as OTH for "Other." For example, as soon as your focal duck began a self-maintenance activity such as preening, you would write down each and every time an event occurred, in order of occurrence, until the duck ceased to perform self-maintenance behavior. If the same event were repeated several times in succession, then you would record its abbreviation the corresponding number of times in your notes, rather than writing its abbreviation only once. Be sure to use arrows to indicate how to read your

FS → FS → FS → FP → FS → FP → WLS → WS → BS → TW → TW →

TW → TW → TW → TW → TW → WTF → BS → HS → HS → HS → BS →

TW → TW → TW.

Figure 1.4 **Example of a sequence of behavior. Abbreviations are those of behavior patterns shown in Figure 1.1. This hypothetical sequence does not represent a sequence that you would expect to observe. Arrows indicate temporal sequence.**

sequence (see the example in Fig. 1.4). To save time while recording the sequence, you may want to add the arrows immediately *after* you finish recording the sequence. Complete Part D of the worksheet by using the data that you collected on the *longest* sequence that you observed.

Keep in mind that your sequence information is based on only a single animal and so is not sufficient for drawing general conclusions about duck behavior. Of course, no study would do focal-animal sampling on just one individual or present data for only a single sequence. You would instead observe many individuals, one at a time, for similar amounts of time, until you had watched as many focal animals performing as many sequences as necessary to achieve a sample size sufficient for your planned statistical analysis. For example, each of the pairs of ethology students who contributed their data to the Hailman & Dzelzkalns (1974) study of mallard tail-wags observed a different bird. They gathered data on 42 sequences of the display called "grunt-whistle" and on so many sequences of other displays that they observed a total of 6307 action patterns, including 1085 occurrences of tail-wagging.

Making Comparisons of Behavior

Form a group of 3 or 4 students to do this last activity. Find a location where all of you can observe at least eight individuals of another species, preferably *not* ducks. If you are at a zoo, find an exhibit that meets this criterion. If you are at a field site, look for common, easily observed species such as bees, squirrels, pigeons, or sparrows. In rural areas, farm animals may be observable in sufficient numbers.

After deciding what species to observe, watch your animals for a while and record information about their common behavior patterns. Then brainstorm with your group to decide on a research question to ask. Your question should involve a comparison of the behavior of different groups of animals. You can define your comparison groups (independent variables) by characteristics that are intrinsic to the animals (e.g., male vs female, juvenile vs adult, dominant vs subordinate) or by characteristics that result from interactions of the animals with their environments (e.g., near vs far from people, near vs far from conspecific neighbors, in center vs at edge of group, in group vs alone). Develop operational definitions for each of your comparison groups that enable you to distinguish between the groups.

Decide what dependent variable (effect) you will measure to determine whether the comparison groups differ. Your dependent variable might be a rate (e.g., bites/min), a duration (e.g., length of a bout of tail-wagging, or time spent within 1 meter of people), a latency (e.g., time from researcher presenting a stimulus until first occurrence of the behavior), or intensity (e.g., partly bared teeth vs fully bared teeth).

If your research question requires recording distances, you may find it easier to use the size of the animals as a unit of measure, rather than recording in meters (or in feet, which should later be converted to meters). Of course, this method will work only if the animals that you are observing are quite uniform in size. For example, you could write, "The animals stood one body length apart." If you are unsure of a distance, you should not write anything like the preceding statement (which would falsely indicate certainty), nor should you write nothing. Instead, you should write something like "The animals were about one body length apart" or, better yet, "The animals were between 1 and 2 body lengths apart" or "1.0–1.5 meters," if you measured in meters.

Decide what sampling methods (focal-animal sampling, behavior sampling, or scan sampling) and what recording methods (continuous or instantaneous recording) you will use. Discuss the methods with all members of your group to be sure that everyone will make observations in the same way. Then record values of the dependent variable until you have observed *at least four individuals from each comparison group* (independent variable) performing the behavior (dependent variable). Record your data in the data notebook of one of your group members. Later, complete Part E of the worksheet.

HYPOTHESES AND PREDICTIONS

What general research question did you investigate in your comparison of the behavior of two kinds of animals? What differences did you predict between the two species of animals?

DATA RECORDING AND ANALYSES

You may be required to analyze the results of your comparison of behavior and to present these results as an oral or written report. Even if you do not have enough data for such analyses, your instructor may ask you to go through the steps of picking an appropriate statistical test. Therefore, write your null and alternative hypotheses, and then use the information in this book to pick an appropriate statistical test of your hypothesis. Also decide how you will graph your results. Check your decisions with your instructor, and then graph your results and do the appropriate statistical test if you have a sufficient sample size.

QUESTIONS FOR DISCUSSION

What are some examples of anthropomorphism?

Why should you avoid anthropomorphism? In answering this question, consider how anthropomorphism might produce misinterpretations of the behavior you are studying. Also consider how anthropomorphic thinking could actually lead a person to treat an animal inappropriately.

Which behavior patterns that you observed would you consider to be *states*? Why?

Which behavior patterns that you observed would you consider to be *events*? Why?

Pick two similar event categories and describe how you could distinguish them with verbal descriptions and sketches.

Behavior patterns sometimes occur in discrete **bouts** of the same activity separated in time by gaps during which other types of behavior occur. A bout can be of a single behavioral state that occurs continuously (such as a bout of sleeping) or can consist of a temporal cluster of the same events (such as a bout of quacking) repeated successively without interruption (Martin & Bateson 1993). Of the *states* that you observed, which occurred in clearly distinguishable bouts? What *events* did you observe that occurred in bouts? Describe ways in which you might quantify events that occur in bouts.

What was the average number of tail-wags per bout that you and your partner observed? How does this compare to values observed by others in the class? What sorts of variability did you observe in the frequency of tail-wags per bout? Describe some sources of this variability. Did tail-wags also vary in other ways, such as in the amplitude of the movement or in body postures during tail-wags?

Because events such as tail-wags occur quickly, you may not be able to get a completely accurate count of tail-wags per bout by simply watching the animals in real time. If you videotaped the animals, you could get a completely accurate count. Does this mean that videotaping is generally preferable to recording observations in real time? Discuss the advantages and disadvantages of videotaping vs observing behavior directly.

Discuss the names that you used to describe your behavior categories. Do any of these names imply anything about the action's function or the intentions of the animal? Are such names problematic or preferable? Explain.

Discuss the advantages and disadvantages of using behavior sampling.

ACKNOWLEDGMENTS

This exercise is loosely based on one originally developed by Jack Hailman and modified by H. Jane Brockmann for their undergraduate courses in animal behavior at the University of Wisconsin and the University of Florida, respectively. Thanks to Jane Brockmann for giving me the idea for this exercise, for commenting on earlier versions, and for giving me the idea of including this exercise in the book. I thank the students in my animal behavior classes of 1996, 1998, and 2000 at Hamline University and those in Jane Brockmann's 1987 and 1989 animal behavior classes at the University of Florida for testing earlier versions of this lab.

LITERATURE CITED

Altmann, J. 1974. Observational study of behavior: sampling methods. *Behaviour*, **49**, 227–267.

Hailman, J. P. & Baylis, J. R. 1991. Post-flight tail-wagging in the mallard. Journal of Field Ornithology, **62**, 226–229.

Hailman, J. P. & Dzelzkalns, J. J. I. 1974. Mallard tail-wagging: punctuation for animal communication? *American Naturalist*, **108**, 236–238.

Lehner, P. N. 1987. Design and execution of animal behavior research: an overview. *Journal of Animal Science*, **65**, 1213–1219.

Martin, P. & Bateson, P. 1993. *Measuring Behaviour: An Introductory Guide*. 2nd ed. Cambridge: Cambridge University Press.

McKinney, F. 1965. The comfort movements of Anatidae. *Behaviour*, **25**, 120–220.

WORKSHEET

For each question in this worksheet, record your observations in your data notebook then complete the worksheet from the notes in your notebook.

A. Quantifying Tail-Wags

1. In the space below, record the mean number of tail-wags per bout ±1 standard deviation, and indicate the sample size (total tail-wagging bouts used to calculate the mean).

2. Record the mean bout duration ±1 standard deviation and indicate the sample size used to calculate the mean bout duration.

3. Record the rate of tail-wagging bouts in your observation period as the number of bouts per minute.

B. Defining Behavior Categories

Attach to this worksheet a page with separate sketches of each of the five behavior patterns that constituted the different categories of self-maintenance behavior that you observed. These should be carefully drawn composite sketches based on the multiple quick sketches that you made while watching the behavior. Use arrows and/or sequential sketches to convey movements. The sketch for each behavior pattern should have a numbered figure caption that includes your name for the behavior, a 2–3-letter abbreviation of the name, and the sample size. Sketches and figure captions should be sufficiently detailed to allow a naïve reader to distinguish between the different behavior patterns.

C. Censusing Behavior with Scan Sampling

1. Fill in Table 1.1 with the results of your census of four ducks.
2. Attach a list of the names and distinguishing features of each of the four individual ducks that you censused.

Table 1.1 **Data sheet for entering results of your scan-sampling census of four ducks. On this table, fill in the names of your ducks in the I.D. spaces, and enter the abbreviations for the behavior patterns that you observed for each of these individuals during the scans for each time.**

Time (minutes)	*Duck Identity*			
	I.D.:	I.D.:	I.D.:	I.D.:
0				
2				
4				
6				

3. Fill in Table 1.1 with the results of your census. Below Table 1.1 or on a separate page, briefly describe each behavior pattern that is listed in the table, if it was not already described in Part B of this worksheet or shown in Figures 1.1 and 1.2. Each such description should follow the guidelines you used to complete Part B of this worksheet, including a sketch of the behavior pattern with its name, its abbreviation, and the sample size.

D. Sequencing Behavior

1. Attach to this worksheet a photocopy of the sequence of behavior patterns that you observed in a bout of self-maintenance behavior. If more than one sequence occurs in your observation period, attach just the *longest* sequence. DO NOT attach more than one sequence.

2. Record here the total amount of time that elapsed from the start to the finish of your sequence.

3. Did your sequence last longer than the 5-minute observation period?   Yes   No   (circle one).

E. Making Comparisons of Behavior

Work in teams of three or four to do the remainder of this worksheet.

1. In the space below, record the common name and the scientific (Latin) name of the species your group observed.

2. Record the number of individuals that your group observed.

3. Are sexes distinguishable?   Yes   No   (circle one).   If yes, explain how on an attached page.

4. Are age classes (adult, juvenile) distinguishable?   Yes   No   (circle one).   If yes, explain how on an attached page.

5. Are individuals distinguishable?   Yes   No   (circle one).   If yes, explain how on an attached page.

6. On an attached page, indicate the name and operational definition of each comparison group (independent variable) that you used. Be sure that your definitions enable you to distinguish between the groups.

7. On an attached page, indicate the name of your dependent variable, and explain how you measured this variable. If your dependent variable involved a behavior pattern (such as counting the number of occurrences of an event), then include a sketch of the behavior in its typical form to create an operational definition.

8. On an attached page, name the sampling method (e.g., *ad libitum*, focal-animal, behavior, or scan sampling) and the recording method (e.g., continuous or instantaneous) that you used, and describe how you used these to collect your data.

9. Attach a photocopy of the data you collected. These data should include values of the dependent variable collected for at least four individuals from each comparison group.

CHAPTER 2

Developing Operational Definitions and Measuring Interobserver Reliability Using House Crickets (Acheta domesticus)

TERRY GLOVER
Division of Social and Behavioral Sciences, Bloomfield College, Bloomfield NJ 07003, USA

INTRODUCTION

People have been fascinated by animal behavior for centuries. In ancient times, questions were asked by people involved in farming, fishing, herding, and hunting. Today even more kinds of questions are asked as the number of people visiting zoos and aquaria and owning pets has increased. Many people base their knowledge of animals on informal observations. Although these informal observations are a good beginning, they cannot provide information that is reliable and accurate enough to advance the science of animal behavior. Our current knowledge of interesting phenomena such as dominance hierarchies, mating behavior, and habitat preferences has been based on careful, systematic observations (Martin & Bateson 1993; Siiter 1999). Systematic observations depend on (1) clear definitions of the behavior observed and (2) an assurance that these definitions are used consistently for the all of the data collected. This laboratory exercise will introduce the use of operational definitions and a technique for measuring whether there is consistency between two observers who have observed the same animal.

Operational definitions enable researchers to translate concepts into more concrete terms so that they can be measured reliably by anyone using the definition (Ray 1997). For example, an operational definition of aggression might state that each time one animal bites another, it will count as one aggressive encounter. Controversies can result if definitions are not used in identical ways by all researchers (Carroll & Maestripieri 1998). Once a concept is operationally defined, it must be used consistently by all observers. Often the study of animal behavior, whether in the field or in the laboratory, requires more than one observer, particularly if the observations are made over an extended period. Consistency between two observers can be tested by having both observers simultaneously observe the same animal. They then compare results using some statistical procedure that will give them a measure of interobserver reliability. In this exercise, you and a lab partner will use Pearson's correlation coefficient *r* to measure interobserver reliability (Spatz 2000). If the correlation is close to +1.00, it indicates strong agreement between the observers. Martin and Bateson (1993) state that a correlation of +0.70 is the lowest acceptable level of agreement. A lower correlation indicates poor agreement. If the agreement is poor, it may be necessary to reconsider the operational definitions used.

This exercise employs the common house cricket, *Acheta domesticus*, which is native to Asia and is used extensively in North America as fishing bait and as food for reptiles. This species can survive and breed in the United States and Canada (Walker 1999). Cricket farms and local pet stores generally raise crickets in large groups. *A. domesticus* is a generalist and eats a variety of foods (Walker & Masaki 1989). Adult male crickets of several species have frequently been used in studies of territoriality, aggression, and mating behavior, especially singing (Dingle 1968; Brooks & Yasukawa 1999). Males produce cricket songs by rubbing their hind wings together. The species-specific song is used to attract females. Adult females have an ovipositor, a long, thin tube that protrudes from the tip of the abdomen. Males lack an ovipositor. When another cricket approaches a male, he uses chemoreceptors in his antennae to identify whether it is a male or a female. During mating, males transfer a jellylike spermatophore to the female. The female uses her ovipositor to lay eggs in damp soil (Loher & Dambach 1989). At warm temperatures, the eggs hatch in 7–10 days and look like small wingless crickets. House crickets molt several times during their life span of 2–3 months. Both male and female crickets have been used to study the degree to which locomotion (movement) is affected by habitat disruption (With et al. 1999). In general, crickets avoid bright light and seek enclosures or corners. In this exercise you will develop operational definitions and measure interobserver reliability while observing locomotion, contact with another cricket, and feeding.

MATERIALS

Each pair of students will need two adult crickets, a clear plastic shoebox container (31 × 17 × 8.5 centimeters) for observations, plastic wrap and tape for covering the container, a sheet of paper measuring at least 31 × 17 centimeters, a watch or classroom clock with a second hand, a small piece of food such as a Cheerio or a raisin, and a paper and pencil for recording data.

PROCEDURE

Preliminary Observation
Procedure

Approximately 24 hours prior to this observation, your instructor removed one cricket from the community container and marked its back with one drop of typing correction fluid. It was then placed in a container covered with plastic wrap. The plastic wrap has been taped to the sides and has air holes in the top. The cricket has not eaten for 24 hours, even though it has water available. You and your partner will begin by doing a 4-minute observation. During this time, on a sheet of notebook paper, record all the behaviors you observe. List each behavior you see in the order in which it occurs. For example, if the animal moves, grooms itself, and moves again, write "move, groom, move." Do not worry about using technical names for the behaviors you observe. The animal should not be aware of your presence. During the observations, adhere to the following guidelines: Do not touch or move the container; do not stand or lean over the container; minimize your movements; do not talk to your partner. Begin recording when the instructor says, "Begin" and stop recording when the instructor says, "Stop."

Questions for Discussion

Discuss the following questions first with your partner and then with the class as a whole. Did you and your partner find it easy to record the sequence of every behavior? Would it have been easier to focus on a few behaviors instead of recording every behavior?

Did your data sheet agree with your partner's? Can you quantify how much the two of you agreed?

Did you see any patterns of behavior or develop any hypotheses on the basis of your observation?

Final Observation: Sampling Techniques, Operational Definitions, Interobserver Reliability

Rationale

In answering the first two of the foregoing questions, you may have found it difficult to record all behaviors. Although recording all behaviors gives a good overview of the animal's behavioral repertoire, it can provide more information than is necessary. Research usually focuses on one or more target behaviors and frequently uses time sampling instead of recording a continuous sequence. In answering the third question, you may have found that you and your partner did not always record the same behaviors. You may have recorded grooming at the same moment your partner recorded movement. It is important for observers as well as readers of research articles to know exactly what was being recorded. Therefore, it will be necessary for you to develop, in the next part of this lab, clear and quantifiable operational definitions for movement, contact, and feeding. Once this is done, you will be able to record the animal's behavior and to measure interobserver reliability.

This part of the lab is divided into three phases. During the first phase, you will record only locomotion (movement). During the second phase, you will record locomotion and contact. During the third phase you will record locomotion, contact, and feeding. Specific details on timing and recording are given in the "Data Recording and Analysis" section.

Operational Definitions

Locomotion—How would you define locomotion so that the locomotion level of your cricket can be compared to those of the crickets other students are observing? What would you define as "one" movement? A classic way is to use an enclosed space that has been divided into a grid and to count one movement each time the animal crosses over a grid line. This open-field technique (Calhoun 1968) has been used with a variety of species such as rats (Renner & Seltzer 1991), guppies (Budaev 1997), and angelfish (Gómez-Laplaza & Morgan 1991). Discuss with fellow students how many sections your container should have. Each section should not be so small that a cricket is larger than a single section, nor should it be so large that the cricket can be active and never cross a line. Next you and your classmates must define how much of the cricket's body must cross a line in order for it to be counted as one movement. The operational definition could require as much as the entire body crossing into another section or as little as the head crossing a grid line. All members of the class must use the same operational definition. Once the class decides on the number of sections, either draw grid lines or fold your sheet of paper into the desired number of sections, and gently slide and center it under the container without disturbing the cricket.

Contact—At the end of the first observational period, your instructor will add a second cricket to the container. You are to record the number of times the original cricket contacted the newly introduced cricket. Again, the members of the class must agree on what constitutes one contact. For example, is touching with antennae sufficient or must some other part of the body be used?

Feeding—During the final recording period, feeding behavior will also be observed. A piece of food will be placed in the center of the container, and you will record the number of times the original cricket eats. Members of the class must operationally define feeding.

HYPOTHESES AND PREDICTIONS

An important question for observational studies is whether your data matched your partner's. Were your operational definitions clear enough so that each of you measured the same behaviors? Did you obtain a Pearson's correlation of 1.00? If your definitions were clear and both of you recorded identical data, you should obtain a Pearson's correlation coefficient r of 1.00. In terms of this specific exercise, did your cricket increase or decrease locomotion after the introduction of another cricket and the introduction of food? Animals usually change activity levels when new stimuli are introduced. Did your cricket contact the other cricket? Did it eat the food?

DATA RECORDING AND ANALYSES

Recording

You and your partner will each independently record your cricket's behavior on separate data sheets (Table 2.1) for a total of 18 minutes. Each time a target behavior is observed, make a mark on the data sheet. You must not talk to your partner or let your partner see your data sheet. The instructor will announce the beginning of each interval by saying "1 minute," "2 minutes," and so on until the recording session is over. At the end of the first phase, the instructor will place the second, unmarked cricket in the container. When the instructor resumes timing by saying "7 minutes," record locomotion and contact made by the original cricket. Do not record the unmarked cricket's behavior. At the end of the second phase, the instructor will introduce a piece of food. When the instructor resumes timing by saying "13 minutes," record all three target behaviors. At the end of the session, convert your tally marks into minute-by-minute totals so that you have 18 locomotion numbers, 12 cricket contact numbers, and 6 feeding numbers. Fill in the parentheses on your data sheet with your partner's minute-by-minute locomotion data.

Table 2.1 **Cricket observation data sheet.**

Name_____   Partner_____
Date_____

Minute	Locomotion
1	( )
2	( )
3	( )
4	( )
5	( )
6	( )

Minute	Locomotion	Contact
7	( )	
8	( )	
9	( )	
10	( )	
11	( )	
12	( )	

Minute	Locomotion	Contact	Feeding
13	( )		
14	( )		
15	( )		
16	( )		
17	( )		
18	( )		

After recording is completed, place your partner's locomotion totals for each minute in the parentheses, (   ).
Additional comments:

Analyses

Locomotion, Contact, and Feeding Data

The data can be presented in the form of a graph where the frequency of each behavior is plotted as a function of successive minutes. A sample is shown in Figure 2.1. It is possible for two behaviors to have the same frequency, so different symbols or lines should be used for each behavior. When drawing your graph, whether by hand or computer program, keep in mind the following tips: Separate each of the three portions of the graph with a break, because although the minutes were successive, they were not

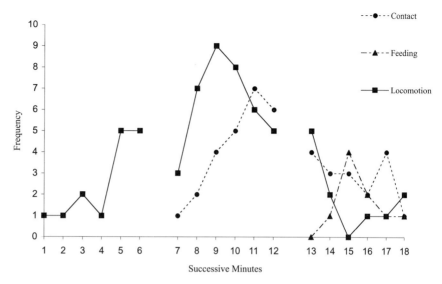

Figure 2.1 **Frequency of locomotion, contact, and feeding over successive minutes for one adult cricket.**

continuous; write a clear and complete figure caption; use a ruler to connect your points; use a pencil unless you do things perfectly the first time.

Interobserver Reliability

Do a scatter plot using your data and your partner's data for locomotion. For each minute put one point on the graph. A sample scatter plot is shown in Figure 2.2. If you saw two movements during the first minute and your partner saw three, plot the point where 2 on the *x*-axis intersects with 3 on the *y*-axis. Do this for each minute of the observation. If you and your partner record identical data, this would indicate perfect interobserver reliability. If there is perfect interobserver reliability and the same scale is used for each axis, the points will form a line going at a 45° angle from the lower left to the upper right. Perfect interobserver reliability will result in a correlation coefficient of +1.00. If you and your partner do not agree, the scatter plot will show deviations from the ideal 45° line, and the correlation will be lower than +1.00. If there is no relationship between the two sets of observations, the points will be scattered, and the correlation will be low and will approach a coefficient of 0.00. If one partner reports high numbers when the other partner reports low numbers, you will have a negative correlation. Negative correlations can range form −1.00 to 0.00, and the points on the scatter plot will show a tendency to go from the upper left down to the lower right. See Appendix C, (Introduction to Statistics), for a further discussion of correlation.

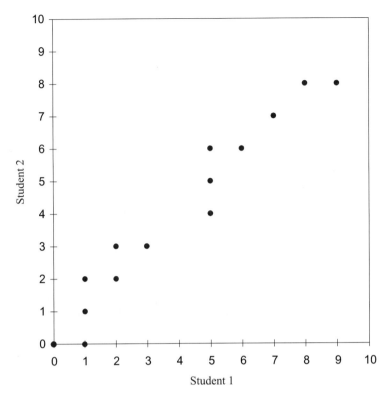

Figure 2.2 **Scatter plot of interobserver reliability for locomotion data.**

Compute a Pearson's correlation coefficient (r) using the locomotion data collected by you and by your partner. The formula for Pearson's r is included in all elementary statistics texts (such as Spatz 2000), in all statistical software packages, and in many spreadsheet programs. If your r is close to 1.00, you can feel confident that you and your partner used the same definitions and were consistent when recording the data. At this point, each of you would be ready to collect locomotion data alone and know that both of you were using the same definition. If your r is less than 0.70, you would want to modify your methods, including your operational definitions, and then retest for interobserver reliability before doing a locomotion study.

QUESTIONS FOR DISCUSSION

Were your operational definitions adequate or do they need to be clarified further?

How good was your interobserver reliability? If $r < 0.70$, do you have a possible explanation as to why?

Did you see any changes in behavior with the introduction of the second cricket or with the food? Do you think there would have been more or less activity if the crickets had been put in the individual containers right before the observations began?

Do you think the feeding measure should have been either duration of time spent eating or how long it took the cricket to begin eating (latency)? In general, how do you feel about frequency as a measure?

If you were to do this exercise over, what would you do differently?

ACKNOWLEDGMENTS

I wish to thank the Bloomfield College students who helped refine this exercise, as well as Elizabeth Glover Malcolm, Sarah Malcolm, and David S. Malcolm, who assisted in formatting figures and proofreading. I also wish to thank the editors and an anonymous reviewer for helpful comments.

LITERATURE CITED

Brooks, R. L. & Yasukawa, K. 1999. Crickets and territory defense. http://www.animalbehavior.org/ABS/Education/labs_toc.html.

Budaev, S. V. 1997. "Personality" in the guppy (*Poecilia reticulata*): a correlational study of exploratory behavior and social tendency. *Journal of Comparative Psychology*, **11**, 399–411.

Calhoun, W. H. 1968. The observation and comparison of behavior. In: *Animal Behavior in Laboratory and Field* (ed. A. W. Stokes) pp. 7–10. San Francisco: Freeman.

Carroll, K. A. & Maestripieri, D. 1998. Infant abuse and neglect in monkeys— a discussion of definitions, epidemiology, etiology, and implications for child maltreatment: reply to Cicchetti (1998) and Mason (1998). *Psychological Bulletin*, **123**, 234–237.

Dingle, H. 1968. Aggressive, territorial and sexual behavior of crickets. In: *Animal Behavior in Laboratory and Field* (ed. A. W. Stokes) pp. 89–92. San Francisco: Freeman.

Gómez-Laplaza, L. & Morgan, E. 1991. Effects of short-term isolation on locomotor activity of the angelfish (*Pterophyllum scalare*). *Journal of Comparative Psychology*, **105**, 366–375.

Loher, W. & Dambach, M. 1989. Reproductive behavior. In: *Cricket Behavior and Neurobiology* (ed. F. Huber, T. E. Moore, & W. Loher), pp. 1–42. Ithaca, NY: Cornell University Press.

Martin, P. & Bateson, P. 1993. *Measuring Behaviour: An Introductory Guide.* 2nd ed. Cambridge: Cambridge University Press.

Ray, W. J. 1997. *Methods Toward a Science of Behavior and Experience.* 5th ed. Pacific Grove, CA: Brooks/Cole.

Renner, M. J. & Seltzer, C. P. 1991. Molar characteristics of exploratory and investigatory behavior in the rat (*Rattus norvegicus*). *Journal of Comparative Psychology*, **105**, 326–339.

Siiter, R. 1999. *Introduction to Animal Behavior.* Pacific Grove, CA: Brooks/Cole.

Spatz, C. 2000. *Basic Statistics: Tales of Distributions.* 7th ed. Pacific Grove: Brooks/Cole.

Walker, T. 1999. Featured Creatures. http://ifas.ufl.edu/~insect/misc/crickets/Adomest.html

Walker, T. & Masaki, S. 1989. Natural history. In: *Cricket Behavior and Neurobiology* (ed. F. Huber, T. E. Moore & W. Loher), pp. 1–42. Ithaca, NY: Cornell University Press.

With, K. A., Cadaret, S. J. & Davis, C. 1999. Movement responses to patch structure in experimental fractal landscapes. *Ecology*, **80**, 1340–1354.

PART 2

Causation

CHAPTER 3

Courtship, Mating, and Sex Pheromones in the Mealworm Beetle (Tenebrio molitor)

ENRIQUE FONT AND ESTER DESFILIS
*Instituto Cavanilles de Biodiversidad y Biología Evolutiva, Universidad de Valencia,
Apdo. 2085, 46071—Valencia, Spain*

INTRODUCTION

Building upon the work of such pioneers as Jacob von Uexküll, Karl von Frisch, Konrad Lorenz, and Niko Tinbergen, ethologists have discovered that every animal lives in a perceptual world of its own. Indeed, much misunderstanding about animal behavior arises through our failure to realize that animals do not perceive the world around them in the same way we do. Not only do different species differ in the types of stimuli that their sense organs can detect, but in many situations, animals respond selectively to a very narrow range of stimuli. For example, male sticklebacks in breeding condition attack very crude models with red undersides even if they lack most other fishlike characteristics (Tinbergen 1952). Stimuli that elicit specific behavior patterns are termed **sign stimuli** or **key stimuli**. Sign stimuli emanating from conspecifics and used in social interactions are known as **releasers** (Tinbergen 1951). Although most studies of sign stimuli and releasers have emphasized visual stimuli, for this lab we will be primarily concerned with chemical stimuli.

ISBN 0-12-558330-3

Chemical stimuli participate in many interactions between organisms and between organisms and their environment (Dusenbery 1992). In fact, chemicals are one of the most important means of communication in many animal groups and probably also one of the oldest (Agosta 1992). However, not all interactions involving chemical stimuli are examples of communication. There are several competing definitions of **communication**, and whether or not a particular chemical stimulus is considered a signal used in communication depends on the definition adopted. Some authors, for example, have proposed restricting communication to intraspecific exchanges, a definition that would exclude all chemical interactions between members of different species. Other authors stipulate that in order for there to be communication, both sender and receiver or just the sender have to benefit from the exchange, again excluding some chemical interactions as examples of communication (Nordlund & Lewis 1976; Weldon 1980; Grier & Burk 1992).

Attempts to classify the diversity of chemical interactions have produced an intimidating list of technical terms, but here we will refer only to those most commonly used. The main distinction is between chemicals that participate in interactions between members of the same species and chemicals that participate in interactions between different species (Table 3.1). Chemicals that participate in interactions between conspecifics are termed **pheromones** (Karlson & Luscher 1959). Chemicals that are transmitted and detected between species are called **allelochemics** (Whittaker & Feeny 1971; Nordlund & Lewis 1976). Pheromones are not to be confused with the chemical stimuli known as **hormones**. Although some pheromones consist of mixtures of hormones or hormone breakdown products, hormones are chemicals that operate within a single individual, delivering a message from one part of the organism to another. Hormones regulate the internal body metabolism and are secreted by endocrine glands. In contrast, pheromones are substances or specific mixtures of substances that are released by one individual of a species and, when detected by another individual of the same species, elicit a specific behavioral or physiological response (Ben-Ari 1998). Pheromones are secreted by exocrine glands and/or body orifices associated with digestion and reproduction (Bradbury & Vehrencamp 1998).

In this laboratory exercise, you will study courtship and mating in a coleopteran, the yellow mealworm beetle (*Tenebrio molitor*). You will also perform simple experiments to investigate the role played by pheromones in sexual behavior. Pheromones used in communication between members of the opposite sex in a mating context are collectively termed **sex pheromones**. In many insect species, detection of a sex pheromone is often sufficient to initiate the species-typical courtship and mating behaviors, although the release of these behaviors can be modified by visual, tactile, acoustic, or other types of stimuli. For example, upon detecting a few

Table 3.1 **Terminology of chemicals involved in interactions between organisms and between organisms and their environment. The same chemical may function in two or three different types of interactions.**

Type of Interaction	Chemicals	Examples
Environment and organism	Chemical stimuli	CO_2, water
Interorganismal	Semiochemicals	
Intraspecific	Pheromones	
Affect receiver's behavior	Releaser (signaling) pheromones[a]	Sex attractant of female silkworm moth
Affect receiver's physiology	Primer pheromones[a]	Queen inhibitor of queen honeybee
Interspecific	Allelochemics	
Benefit to sender	Allomones	Lures, venoms, repellents
Benefit to receiver	Kairomones	Food and predator scents
Benefit to both	Synomones	Floral scents
Benefit to neither	Antimones	Chemicals released by a pathogen that cause the death of its host
Intraorganismal	Hormones	Growth hormones, sex hormones

[a]Some authors prefer the terms *releaser effect* and *primer effect* to allow for the possibility that a single pheromone has both kinds of effects.

Adapted from Whittaker & Feeny 1971; Nordlund & Lewis 1976; Hölldobler & Wilson 1990; Dusenbery 1992; Grier & Burk 1992.

hundred molecules of the female pheromone bombykol, males of the silkworm moth (*Bombyx mori*) immediately start beating their wings in a "flutter dance" and approach the source of the pheromone. As many as three different pheromones may participate in the courtship and mating of *T. molitor* (Valentine 1931; Tschinkel et al. 1967; Happ 1969; Happ & Wheeler 1969; August 1971). Of these, two are sex-specific pheromones that function as **sex attractants**: the female produces a pheromone that attracts and releases copulatory behavior from males, whereas the male emits a pheromone that stimulates locomotion in females and causes them to aggregate in the vicinity of the male. None of these pheromones has an effect on individuals of the same sex as that producing it (Happ 1969; Happ & Wheeler 1969; August 1971; Smart et al. 1980). In addition, there is evidence that male *T. molitor* may also produce an **antiaphrodisiac** pheromone that they apply to the female during copulation and that tends to discourage other males from mating with her (Happ 1969).

As part of the study of courtship and mating in *Tenebrio molitor*, you will observe a behavior known as **postcopulatory mate guarding**. Mate guarding has been reported in a variety of insects, and it can occur before, during, and/or after copulation. During postcopulatory mate guarding, the male remains near the female or in physical contact with her after copulation and attempts to prevent her from remating. Guarding presumably serves to increase a male's fitness by reducing the probability that his sperm will have to compete with the sperm of his rivals for fertilization of the female's eggs (Parker 1970, 1974, 1984; Alcock 1994; Simmons & Siva-Jothy 1998). In species in which females mate several times, sperm from two or more males compete within the reproductive tract of the female for fertilization of the eggs (called **sperm competition**; Parker 1970; see also Birkhead & Parker 1997; Birkhead & Moller 1998; Birkhead 2000). Because female *T. molitor* mate multiply, most often with different males (**polyandry**) but also with the same male (**repeated matings**), sperm competition has probably been a strong selective force in the evolution of this species (Gage 1992). The outcome of sperm competition is usually measured as the proportion of offspring sired by the second of two males to mate with a female (abbreviated as P_2). When two male *T. molitor* mate with the same female, the second male's sperm fertilizes about 90% of the eggs, but the proportion of eggs fathered by the second male declines with time since the last prior copulation (Siva-Jothy et al. 1996; Drnevich et al. 2000). Thus, by staying with the female and preventing other males from mating with her, a male increases his chances of fertilizing the female's eggs. Mate guarding is but one of several adaptations that have evolved to avoid or reduce the risk of sperm competition or to give the male a competitive edge should sperm competition take place. Other such adaptations include copulating repeatedly with the same female to swamp the sperm of rivals, removing rival sperm from the reproductive tract of the female (Gage 1992; but see Siva-Jothy et al. 1996), increasing the size of the ejaculate when other males are present (Gage & Baker 1991), and applying an antiaphrodisiac pheromone to the female to reduce her attractiveness to other males (Happ 1969).

MATERIALS

Materials needed at each laboratory station include 1 colony of *Tenebrio molitor* (the same colony can be used for several laboratory stations), at least 20 virgin male and 15 virgin female adult mealworm beetles, 1 or 2 sheets of filter paper, 5 disposable plastic petri dishes 90 millimeters in diameter, 1 pair of forceps, 1 bottle of white correction fluid and/or enamel paint for marking individuals, 5 glass rods (7 centimeters × 3 millimeters) fire-polished at both ends, 1 disposable 1.5-milliliter plastic microtube containing the female pheromone extract, 1 bottle of absolute ethanol, 2 stopwatches, 1 electronic beeper for time sampling (optional), and 1 low-power dissecting microscope or hand-held magnifier (optional).

GENERAL OBSERVATIONS

The subjects to be used in this exercise are adult mealworm beetles, *Tenebrio molitor* (Coleoptera, Tenebrionidae). The beetles are large enough that even the finer details of their behavior can be observed with the naked eye, but if one is available, a low-power dissecting microscope or a hand-held magnifier may be used for observation. Some of the experiments will require keeping track of the behavior of several interacting individuals held together in an observation arena. Watch for morphological differences that will enable you to identify the individual beetles (it helps to pair up individuals that differ in size). If necessary, you may mark beetles for individual recognition by applying 1 or 2 droplets of white correction fluid (such as Liquid Paper) or enamel paint to different parts of their elytra. Keep room lights low during this laboratory, and avoid hitting the working area, which produces vibrations that may be disturbing to the beetles.

If you have access to a culture of *Tenebrio molitor*, devote a few minutes to observing the behavior of the animals in the colony, and learn to recognize the different stages in the life cycle of the mealworm beetle. *T. molitor* undergoes a complete metamorphosis during development and has egg, larva, pupa, and adult (beetle) stages. In a colony that has been maintained for some time, all stages from egg to adult will be present at any given time. The female beetle lays several hundred bean-shaped eggs. The eggs are white, less than a millimeter in length, and very sticky. They rapidly adhere to the walls or to other objects that may be found in the culture and become coated in the powdery waste that accumulates in the bottom of the culture. Eggs hatch in about 1 week. The newborn larvae are whitish and very small; it may take several weeks before they grow large enough to be easily seen.

Mealworms are the larval stage; they are yellow with orange or brown cross bands. You may have seen them before, because they are used as bait for fishing and as food for a wide variety of small amphibians, reptiles, birds, and mammals (Martin et al. 1976; Frye 1992). Mealworms molt about a dozen times as they grow larger, the exact number of molts depending on the temperature and the type of growing medium. Newly molted larvae are easily distinguished because they are soft and white. After about 3 months, the full-grown larvae crawl to the top of the culture, become quiescent, and adopt a peculiar posture by bending their body in the shape of a letter "C." The larva then becomes a pupa. The pupae are white or light brown and do not move, but they will wiggle if disturbed. In the wild, mealworm beetles usually spend the winter as pupae and emerge as adults by the next spring. In the laboratory, where food is abundant and temperatures are high year-round, the pupae emerge as adults after 1 or 2 weeks.

The beetle that eventually emerges from the pupa is a light brown, darkening to reddish-brown and finally black after a few days. The degree of darkening can thus provide a crude estimate of the age of the beetle: Fully tanned beetles are at least 1 week old (Smart et al. 1980). The adult beetles show very little sexual dimorphism, and the sexes are difficult to distinguish without a dissecting microscope (Bhattacharya et al. 1970). The earliest matings occur at 4 days after emergence in both sexes, and egg laying is first observed when females are 8 days of age (Happ & Wheeler 1969; Happ 1970). The adults live about 2–3 months. You will probably notice that some beetles are almost twice as large as others. The size differences have nothing to do with the sex of the beetles or with their age, because the adults do not grow. After gaining some acquaintance with the different life stages of the mealworm beetle, turn to the virgin males and females.

DESCRIBING COURTSHIP AND MATING

Procedure

The first step in any ethological study is to identify and describe the action patterns that are to be studied. Action patterns are the basic units of behavior, much as atoms are the basic units of matter. For this part of the exercise you will observe and record courtship and mating behavior in full-grown beetles. The objective is to identify, describe, and quantify action patterns occurring during courtship and mating. You will be provided with two cultures, one containing virgin males, the other virgin females. Your instructor isolated virgin beetles by sexing individuals as pupae at least 1 month before today's laboratory period and raising them in separate all-male and all-female containers. Your observations today should be conducted with beetles that are at least 2 weeks old, so make sure that you use only individuals that are fully tanned.

Line your working area with a clean piece of filter paper. Put the bottom parts of 5 petri dishes upside down on the filter paper; these will be your observation arenas. Introduce a virgin male under each of the inverted petri dishes, and allow them at least 5 minutes to acclimate to the strange surroundings (use forceps to pick up the animals gently and transfer them to the observation arena). Next introduce a virgin female under one of the observation arenas, and carefully observe the ensuing sequence of events. If the male and the female do not interact after 2–3 minutes, try a different pair.

Repeat the process with additional pairs of beetles until you are familiar with courtship and mating behavior in this species (3–5 pairings are usually adequate). Try to identify discrete units or categories (that is, action patterns) in the behavior of the two sexes. Build a catalog with detailed descriptions of all the different action patterns that you have identified. When we suspect that a behavioral catalog includes most of the behaviors that an animal is capable of performing, we call it an **ethogram** (or a partial ethogram if it is restricted to one specific type of behaviors, such as those seen in courtship). Your descriptions of the action patterns should be clear and concise and should emphasize the postures and movements displayed by the animals rather than their presumed functions. Ideally, your descriptions should contain enough detail that anybody reading them has enough information to envisage the behavior perfectly (Martin & Bateson 1993; Lehner 1996). Following is a narrative account of what usually goes on during courtship and mating. Use this description as a basis for identifying action patterns and for naming the relevant body parts. However, you should strive to note and describe every behavior as you observe it.

Courtship usually starts as soon as the male touches the body of the female with his antennae (called antennation). Occasionally, a male may extend his forelegs, raising the anterior portion of his body and slowly waving his antennae from side to side. The rearing and antennal waving may indicate that the male has detected the female's pheromone (August 1971). In most cases, however, the male and the female move about the arena with no overt indication that one is aware of the other. After coming into contact with the female, the male starts tapping her body with his antennae in a rapid tattoo and rubbing her with his forelegs. Regardless of where he first touches the female, the male moves to the tip of her abdomen, tapping and rubbing her body as he moves along. At this point the female may move away, with the male following and tapping her with his antennae. When the female eventually stops, the male positions himself behind the female while rubbing the lateral margins of her elytra with his forelegs and middle legs in a wiping motion. If the female remains motionless, the male then climbs onto her back and rapidly moves forward, rubbing her sides and tapping her with his antennae. Once on top of the female, the male lowers the tip of his abdomen and extrudes his copulatory organ (aedeagus). The aedeagus, which is a light beige, bends forward and down and extends for 2–3 millimeters until it contacts the tip of the female's abdomen. If the female is receptive, she will raise the tip of her abdomen and slightly extrude her ovipositor to allow intromission. Only the dark-colored, needlelike penis at the end of the aedeagus penetrates the body of the female. When the male achieves intromission, he stops moving and gradually ceases his foreleg rubbing and antennae tapping. However, if intromission is not successful, the male continues rubbing and tapping the female while probing the end of her abdomen with his aedeagus.

Copulation usually lasts between 1 and 2 minutes, although some copulations may exceptionally last as long as 6 minutes. During copulation only the hindlegs of the male touch the substrate; the female has all legs on the substrate, and her motionless antennae are held forward in contact with the substrate (Fig. 3.1A). If the male falls off or is dislodged, he may resume courtship, although sometimes he manages to stay *in copula* even after the pair fall on their sides. Copulation ends with the male withdrawing

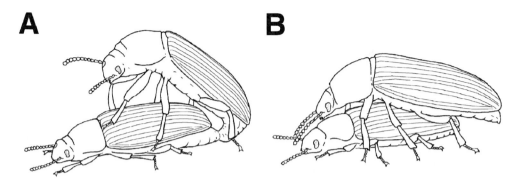

Figure 3.1 **Copulation (A) and postcopulatory mate guarding (B) in** *Tenebrio molitor.* **The male is on top.**

his copulatory organ from the female as she lowers the tip of her abdomen. During copulation the male transfers his sperm, via a spermatophore, into the female copulatory bursa. The spermatophore is long and thin, and its anterior end is partially invaginated. The invaginated anterior end of the spermatophore expands in three stages within the female's bursa before finally bursting and releasing sperm 7–10 minutes after the end of copulation. Sperm then migrates from the bursa into the female's sperm storage organ, or spermatheca, although complete sperm storage does not occur until 6 hours after copulation (Gadzama & Happ 1974; Gage & Baker 1991; Drnevich et al. 2000). After a variable period of time (from several minutes to a few hours), the female ejects the empty spermatophore, which she may then consume.

Hypotheses and Predictions

Although the emphasis in this part of the exercise is on observation and description of the behavior of (relatively) undisturbed animals, there is no reason why the empirical data that you will generate cannot be used to distinguish between competing hypotheses. In fact, not everything we might wish to know as scientists is amenable to an experiment, and there are many interesting questions about animal behavior that can be answered through observation alone.

The general research question that you will be investigating during this part of the lab involves the fundamental asymmetry that characterizes the relationships between males and females of sexually reproducing species. Because females invest more resources than do males in gamete production and other forms of parental care, sexual selection theory predicts that in most species, males should be the active and competitive sex during courtship, whereas females should be relatively passive and discriminating. This prediction, when applied to the behavior of *Tenebrio molitor*, might lead us to expect that male beetles will attempt to mate with any female that comes their way and will compete with other males for fertilization of the female's eggs.

Females, on the other hand, should be expected to exercise some kind of mate choice and allow themselves to be mated by some males but not by others.

Data Recording and Analyses

Prior to beginning the observations, you should design your own data sheets for recording data. As you observe different pairs of beetles, watch especially for any aspects of courtship and mating that you think may be suitable for quantitative analysis. For each pair of beetles you should record at least the following data: (1) latency to first contact (time elapsed from the introduction of the female until the first contact), (2) sex of the individual that makes the initial contact, (3) duration of courtship (time elapsed from first contact to intromission), and (4) duration of copulation (time elapsed from intromission to withdrawal of the male's copulatory organ). Pool all observations and calculate summary statistics (mean, standard error/standard deviation, range, coefficient of variation) for all latency and duration data. Use the pooled class data to calculate the proportion of times that individuals of each sex initiate contact and the proportion of successful copulations. In those pairs that did not copulate, was it because the male did not attempt copulation or because the female rejected the male? What useful information might females get from the male's behavior?

You may also wish to compare the behavior of males confronted with virgin and with nonvirgin females. To do this comparison, you should conduct observations of the behavior of two groups of males. Males in one group will be allowed to court virgin females, whereas those in the other group will be allowed to court nonvirgin females. Appropriate dependent variables for comparing the two groups include latency to first contact, duration of courtship, and duration of copulation. For each dependent variable, you should formulate a null hypothesis (for example, the latency to first contact is the same in both groups) as well as an alternative hypothesis (for example, males in different groups have different latency scores). You can test for statistically significant differences using Student's *t* test (parametric) for comparing means or the Mann–Whitney *U* test (nonparametric) for comparing medians. The latter should be used when working with small data sets that cannot be shown to have been drawn randomly from two populations that are normally distributed and have equal variances. Bar graphs of the dependent variables as a function of group membership would provide an appropriate graphical analysis of the data. Does the proportion of successful copulations change when males are offered nonvirgin rather than virgin females? To test the null hypothesis that the proportion of successful copulations is the same in both groups of males, you should cast the data into a 2×2 contingency table (use the actual frequencies observed rather than converting the data to percentages) and test for significance using a chi-square test of independence.

You will note during this exercise that males treat females differently from other males; that is, they are making sexual discriminations. Later you will perform simple experiments to determine what sorts of stimuli are used by males, but your observations of behavioral interactions between males and females should enable you to draw some tentative conclusions about the role of various stimuli (visual, chemical, tactile, etc.) that release male courtship behavior.

Sequential Organization of Male Courtship Behavior

If time permits, observe additional pairs of beetles and try to determine the sequence in which the different behaviors occur. Behavioral sequences can refer to a single individual (intraindividual) or to two interacting individuals (dyadic). Keeping track of the behavior of both sexes can be complicated, so you should initially restrict your observations to the male. Do not attempt to record all the behaviors that you have identified. Concentrate instead on a manageable subset (you may want to drop behaviors that are difficult to identify reliably or that are not specific to the courtship and mating context). It is also important that the behavior categories that you use be mutually exclusive (that is, one category should not include another). Use shorthand notation to record the different action patterns ("F" for follow, "T" for antennae tapping, etc.), and simply write down the codes for the different behaviors in the order in which they occur. The task is easier if you do not include successive repetitions of the same behavior (use this approach today, but realize that repetitions are sometimes important to adequately describing such sequences). It may be necessary to pool the results from different groups to achieve a sample size large enough for analysis. Sequential records from several groups of students can be summarized in the form of a "box and arrow" diagram (**kinematic graph**) to illustrate transitions between behaviors (Sustare 1978; Lehner 1996). To do this, simply draw small boxes with the different action patterns in the order in which they usually follow each other, and connect the boxes with arrows indicating the sequence of events, with numbers above arrows to indicate how many times you have observed each transition. Before collecting the data, consult your instructor about alternative recording methods and ways to analyze the sequential records that you will obtain (for suggestions see Slater 1973; Fagen & Young 1978; Martin & Bateson 1993; Lehner 1996). Think about the following questions when interpreting your analyses: Are sequences of behavior pretty much fixed and invariant (deterministic) or is there a lot of variability? How does the behavior of one sex affect the behavior of the other?

Postcopulatory Mate Guarding

After mating, most males engage in a behavior known as postcopulatory mate guarding. To study mate guarding, you may simply observe the postcopulatory behavior of the same pairs you used for the previous section.

Alternatively, you may use a new batch of beetles and let them proceed through courtship and mating to observe their postcopulatory behavior.

Mate guarding in *Tenebrio molitor* may take several forms. Usually the male remains on top of the female after withdrawing his copulatory organ (Fig. 3.1B) and/or dismounts the female and stays immediately adjacent to and in physical contact with her for a variable length of time. Sometimes a male alternates between these two forms of mate guarding. When guarding, the male frequently strokes the body of the female with his antennae and/or maxillary palps. The female remains relatively quiescent during the majority of the guarding period. If she moves away from the male, he may make rapid searching movements and follow the female until he regains contact with her. Sometimes, if the pair is left undisturbed, the male goes straight from guarding into a new round of courtship and mating. Guarding males tend to behave aggressively toward members of their own sex and to attack intruding males.

Observe the behavior of several pairs of beetles after copulation. Note and describe every action pattern that you observe. In addition, for each pair you should record the time the male spends on top of the female following copulation and, if time permits, the total duration of mate guarding (in some cases this may take over an hour). For purposes of this exercise, guarding will end whenever the female moves away from the male and he makes no attempt to resume guarding or whenever the male walks away from the female without paying further attention to her. Do all males guard their mates after copulation? Do any pairs mate more than once? If they do, measure the time interval between successive copulations. If they do not, is this because the female actively rejects the male or because the male does not attempt a second copulation?

Shortly after a male and a female have finished copulating, introduce a second male under one of the observation arenas and observe the ensuing behavior. Describe any aggressive interactions between the two males. Repeat the procedure in other arenas to observe several interactions between a guarding male and an intruding male. How effective is mate guarding? Does male size affect the outcome of guarding behavior? Does the presence of other males in the observation arena affect the form and duration of mate guarding? Use the information provided in the introduction to formulate hypotheses and predictions concerning the functional significance of male mate-guarding behavior (see also Alcock 1994).

PHEROMONE EXPERIMENT

Procedure

The objective of this part of the exercise is to study the response of male *Tenebrio molitor* to the female sex attractant pheromone. You will expose males to glass rods to which an extract of the female pheromone has been applied and will compare their response to the glass rod to their response to a conspecific female. Clean the working area and replace the filter paper if necessary. Isolate a new batch of virgin males in the observation arenas. Dip one end of a clean glass rod in the pheromone extract and allow it to air-dry for 30 seconds. Introduce the treated glass rod under one of the observation arenas and observe the behavior of the male. Try to identify and record any action patterns that you saw earlier in interactions between males and females (such as tapping with the antennae, mounting, probing with the aedeagus). If the male does not move, try a different one. (Be patient; it may be several minutes before the male contacts the end of the glass rod that carries the pheromone.) Repeat the process with the remaining observation arenas. Record the number of beetles attempting copulation with the glass rod, and measure the time elapsed between introduction of the rod and the first copulation attempt.

Hypothesis and Predictions

This laboratory exercise is designed to elicit your thoughts about the stimulus control of courtship and mating in *Tenebrio molitor* beetles. We are primarily interested in discovering the importance of chemical stimuli in releasing male courtship and mating behavior. Our main hypothesis is that in a mating context, males selectively pay attention to chemicals released by the female, while ignoring other such stimuli as size, color, shape, and texture. If this hypothesis is correct, males should court any object, no matter how unlike a real female beetle, as long as it has been impregnated with the female pheromone.

Data Recording and Analyses

Most of the simple experiments that you will be conducting for this part of the exercise will result in clear responses that require limited analysis. Think about the following questions when interpreting your results: How does the behavior of the males confronted with pheromone-impregnated glass rods differ from that of courting males? Do they exhibit any of the action patterns that you saw earlier in interactions between males and females? If so, which ones? Do males court the clean side of pheromone-impregnated glass rods? Calculate summary statistics for the time elapsed between introduction of the rod and the first copulation attempt. Use the pooled class data to calculate the proportion of males attempting copulation with the glass rods. You may also wish to compare the response of males to treated glass rods and to control rods that have been dipped in absolute ethanol (ethanol is the solvent used for extraction of the pheromone). If two groups of males are used, the null hypothesis that the proportion of males attempting copulation is the same in both groups can be tested through appropriate use of a chi-square test. If the same males respond to treated and to control rods, you should check with your instructor regarding what dependent variables and statistical tests are appropriate to use.

SMALL-GROUP EXPERIMENTS

Depending on the number of beetles and on the amount of time available, you may be asked to design and conduct your own experiment. First, brainstorm to create a list of 2–5 general questions about courtship, mating, and/or mate attraction that your group would like to explore. Then, as a group, narrow down your choices to one that you will test. After choosing your research question, design a test for this question by making appropriate modifications to the procedure used for the whole-class experiment. For example, you could compare the responses of virgin and nonvirgin males to the pheromone extract. Design a data sheet. What are the independent and the dependent variables in your experiment? Discuss how best to graph your data, and then make the graph after checking your decision with your instructor. Write null and alternative hypotheses for the specific independent and dependent variables that you tested. What statistical test will you use? Check your answers with your instructor.

QUESTIONS FOR DISCUSSION

Do your observations of courtship and mating agree with those of the rest of the class? Have all the groups of students identified the same action patterns? Are your descriptions of the different action patterns understandable to students in other groups? Have you described the behavior of the beetles as though they were little people encased in a chitinous exoskeleton (anthropomorphism), or have you simply described the behaviors that occurred? If you used anthropomorphic language, was that language just a matter of literary style (a metaphor), or was it intended as an explanation of the causes of the behavior, which you assume are common to humans and beetles?

What are the possible functions of courtship behavior in *Tenebrio molitor*? Why are males the courting sex, and what does this tell us about the evolutionary forces shaping courtship and mating patterns in this species? Do females have any control over which male fertilizes their eggs?

There are published descriptions of courtship and mating in other species of stored-product beetles (e.g., Wojcik 1969). Compare courtship and mating in *Tenebrio molitor* and in other species, and discuss reasons why courtship is species-specific.

From your observations, what conclusions can you draw about the sensory control of courtship and mating in *Tenebrio molitor*? What further experimental controls and tests would be useful in confirming or extending your conclusions (see Edvardsson & Arnqvist 2000)?

Discuss possible costs and benefits of mate-guarding behavior.

The males of some insect species guard their females before they copulate with them, whereas in other species, such as *Tenebrio molitor*, mate guarding takes place after copulation. Discuss reasons for these differences (see Thornhill & Alcock 1983; Simmons & Siva-Jothy 1998).

What advantages might accrue to females from mating with several males?

Many insects show, like *Tenebrio molitor*, a pattern of last-male sperm precedence, in which the last male to mate with a female fertilizes the majority of eggs subsequently laid by that female (Smith 1984; Lewis & Austad 1990; Simmons & Siva-Jothy 1998). Discuss possible mechanisms that may be responsible for the high levels of last-male sperm precedence observed in this species (see Gage 1992; Drnevich et al. 2000).

Does the effectiveness of the female pheromone justify considering it a releaser for courtship and mating in this species? Under what circumstances might selection favor a relationship between simple releasers and responses?

Why do males not go beyond the preliminary stages of courtship when attempting copulation with a pheromone-impregnated glass rod?

How might the release and detection of sex attractant pheromones benefit male and female mealworms?

There are two different modes of detection for pheromones and allelochemics: olfactory detection, which involves detection of airborne or waterborne chemicals from a distant source, and contact reception, which requires direct contact of the receptors with the source of the chemical stimulus (Bradbury & Vehrencamp 1998). What do your behavioral observations suggest about the mode of detection of the female pheromone in *Tenebrio molitor*? Where might the receptors be located in the male?

LITERATURE CITED

Agosta, W. C. 1992. *Chemical Communication: the Language of Pheromones.* Oxford: Freeman.

Alcock, J. 1994. Postinsemination associations between males and females in insects: the mate-guarding hypothesis. *Annual Review of Entomology*, **39**, 1–21.

August, C. J. 1971. The role of male and female pheromones in mating behaviour of *Tenebrio molitor. Journal of Insect Physiology*, **17**, 739–751.

Ben-Ari, E. T. 1998. Pheromones: what's in a name? *BioScience*, **48**, 505–511.

Bhattacharya, A. K., Ameel, J. J. & Waldbauer, G. R. 1970. A method for sexing living pupal and adult yellow mealworms. *Annals of the Entomological Society of America*, **63**, 1783.

Birkhead, T. R. 2000. *Promiscuity: an Evolutionary History of Sperm Competition.* Cambridge, MA: Harvard University Press.

Birkhead, T. R. & Parker, G. A. 1997. Sperm competition and mating systems. In: *Behavioural Ecology: an Evolutionary Approach.* 4th edn. (ed. J. R. Krebs & N. B. Davies), pp. 121–145. Oxford: Blackwell.

Birkhead, T. R. & Moller, A. P. (eds.) 1998. *Sperm Competition and Sexual Selection.* San Diego: Academic Press.

Bradbury, J. W. & Vehrencamp, S. L. 1998. *Principles of Animal Communication.* Sunderland, MA: Sinauer Associates.

Drnevich, J. M., Hayes, E. F. & Rutowski, R. L. 2000. Sperm precedence, mating interval, and a novel mechanism of paternity bias in a beetle (*Tenebrio molitor* L.). *Behavioral Ecology and Sociobiology,* **48**, 447–451.

Dusenbery, D. B. 1992. *Sensory Ecology.* New York: Freeman.

Edvardsson, M. & Arnqvist, G. 2000. Copulatory courtship and cryptic female choice in red flour beetles *Tribolium castaneum.* Proceedings of the Royal Society of London B, **267**, 559–563.

Fagen, R. M. & Young, D. Y. 1978. Temporal patterns of behaviors: durations, intervals, latencies, and sequences. In: *Quantitative Ethology* (ed. P. W. Colgan), pp. 79–114. New York: Wiley.

Frye, F. L. 1992. *Captive Invertebrates: a Guide to Their Biology and Husbandry.* Melbourne, Australia: Krieger.

Gadzama, N. M. & Happ, G. M. 1974. Fine structure and evacuation of the spermatophore of *Tenebrio molitor* L. (Coleoptera: Tenebrionidae). *Tissue Cell,* **6**, 95–108.

Gage, M. J. G. 1992. Removal of rival sperm during copulation in a beetle, *Tenebrio molitor. Animal Behaviour,* **44**, 587–589.

Gage, M. J. G. & Baker, R. R. 1991. Ejaculate size varies with sociosexual situation in an insect. *Ecological Entomology,* **16**, 331–337.

Grier, J. W. & Burk, T. 1992. *Biology of Animal Behavior.* 2nd edn. St. Louis, MO: Mosby Year Book.

Happ, G. M. 1969. Multiple sex pheromones of the mealworm beetle, *Tenebrio molitor* L. *Nature,* **222**, 180–181.

Happ, G. M. 1970. Maturation of the response of male *Tenebrio molitor* to the female sex pheromone. *Annals of the Entomological Society of America,* **63**, 1782.

Happ, G. M. & Wheeler, J. W. 1969. Bioassay, preliminary purification and effect of age, crowding and mating on the release of sex pheromone by female *Tenebrio molitor. Annals of the Entomological Society of America,* **62**, 846–851.

Hölldobler, B. & Wilson, E. O. 1990. *The Ants.* Cambridge, MA: Harvard University Press.

Karlson, P. & Luscher, M. 1959. "Pheromones": a new term for a class of biologically active substances. *Nature,* **183**, 55–56.

Lehner, P. N. 1996. *Handbook of Ethological Methods.* 2nd ed. Cambridge: Cambridge University Press.

Lewis, S. M. & Austad, S. N. 1990. Sources of intraspecific variation in sperm precedence in red flour beetles. *American Naturalist,* **135**, 351–359.

Martin, P. & Bateson, P. P. G. 1993. *Measuring Behaviour: an Introductory Guide.* 2nd ed. Cambridge: Cambridge University Press.

Martin, R. D., Rivers, J. P. W. & Cowgill, U. M. 1976. Culturing mealworms as food for animals in captivity. *International Zoo Yearbook,* **16**, 63–70.

Nordlund, D. A. & Lewis, W. J. 1976. Terminology of chemical releasing stimuli in intraspecific and interspecific interactions. *Journal of Chemical Ecology,* **2**, 211–220.

Parker, G. A. 1970. Sperm competition and its evolutionary consequences in the insects. *Biological Reviews*, **45**, 525–567.

Parker, G. A. 1974. Courtship persistence and female-guarding as male time investment strategies. *Behaviour*, **48**, 157–184.

Parker, G. A. 1984. Sperm competition and the evolution of animal mating strategies. In: *Sperm Competition and the Evolution of Animal Mating Systems* (ed. R. L. Smith), pp. 2–60. New York: Academic Press.

Simmons, L. W. & Siva-Jothy, M. T. 1998. Sperm competition in insects: mechanisms and the potential for selection. In: *Sperm Competition and Sexual Selection* (ed. T. R. Birkhead & A. P. Moller), pp. 341–434. London: Academic Press.

Siva-Jothy, M. T., Earle Blake, D., Thompson, J. & Ryder, J. J. 1996. Short and long term precedence in the beetle *Tenebrio molitor*: a test of the "adaptive sperm removal" hypothesis. *Physiological Entomology*, **21**, 313–316.

Slater, P. J. B. 1973. Describing sequences of behavior. In: *Perspectives in Ethology*, Vol. 1 (ed. P. P. G. Bateson & H. P. Klopfer), pp. 131–153. New York: Plenum Press.

Smart, L. E., Martin, A. P. & Cloudsley-Thompson, J. L. 1980. The response to pheromones of adult and newly emerged mealworm beetles (*Tenebrio molitor* L.) (Col. Tenebrionidae). *Entomologist's Monthly Magazine*, **116**, 139–145.

Smith, R. L. (Ed.). 1984. *Sperm Competition and the Evolution of Animal Mating Systems*. New York: Academic Press.

Sustare, B. D. 1978. Systems diagrams. In *Quantitative Ethology* (Ed. by P. W. Colgan), pp. 275–311. New York: John Wiley & Sons.

Thornhill, R. & Alcock, J. 1983. *The Evolution of Insect Mating Systems*. Cambridge, MA: Harvard University Press.

Tinbergen, N. 1951. *The Study of Instinct*. Oxford: Clarendon Press.

Tinbergen, N. 1952. The curious behavior of the stickleback. *Scientific American*, **187**, 22–26.

Tschinkel, W., Willson, C. & Bern, H. A. 1967. Sex pheromone of the mealworm beetle (*Tenebrio molitor*). *Journal of Experimental Zoology*, **164**, 81–86.

Valentine, J. M. 1931. The olfactory sense of the adult mealworm beetle *Tenebrio molitor* (Linn.). *Journal of Experimental Zoology*, **58**, 165–227.

Weldon, P. J. 1980. In defense of "kairomone" as a class of chemical releasing stimuli. *Journal of Chemical Ecology*, **6**, 719–725.

Whittaker, R. H. & Feeny, P. P. 1971. Allelochemics: chemical interactions between species. *Science*, **171**, 757–770.

Wojcik, D. P. 1969. Mating behavior of 8 stored-product beetles (Coleoptera: Dermestidae, Tenebrionidae, Cucujidae, and Curculionidae). *Florida Entomologist*, **52**, 171–197.

CHAPTER 4

Courtship and Mate Attraction in Parasitic Wasps

R OBERT W. M ATTHEWS AND J ANICE R. M ATTHEWS
Entomology Department, The University of Georgia, Athens, GA 30602, USA

INTRODUCTION

A thorough knowledge of the courtship behaviors of animals is important because reproduction is a critical aspect of behavior with implications for conservation, management, and control of animal populations. However, insect courtship is undeniably diverse (Thornhill & Alcock 1983). Some species exhibit almost none. Others practice complex and lengthy rituals that involve a mixture of visual, sound, chemical, and motion cues. Many sorts of courtship attractants have been documented among insects, from the visual cues of firefly flashes and the auditory cues of cricket chirps to the chemical cues, or **pheromones**, by which moths attract one another.

Melittobia digitata are little parasitic wasps also known as WOWBugs. Their tiny stingers, used only to puncture hosts, cannot penetrate human flesh. Around the world, *Melittobia* raise their young upon the larvae and pupae of other insect species in several different orders, including bees and wasps, beetles, and flies. *Melittobia* males are highly aggressive toward one another, often battling to the death for the right to court and mate with their sisters inside the darkness of the host cocoon. Females—which make up over 95% of each generation—disperse after mating to undertake a hazardous journey to search for new hosts. Up to 700 mated females may emerge from a single host cocoon. Most will perish, but the lucky few find a new insect larva or pupa upon which to lay eggs. Within 17–24 days, these eggs will mature to adults able to breed and repeat the cycle.

What form does courtship take in the bizarre little world of *Melittobia* wasps? In many parasitic Hymenoptera, mating is preceded by elaborate male courtship displays that include species-specific characteristics. Generally, the male's repertoire consists of movements involving the wings, legs, antennae, and mouthparts. These movements are performed continuously or intermittently (depending on the species) and are repeated over and over until the relatively passive female indicates her willingness to mate (**copulate**).

Continuous performance of a pattern of movements, with the female signaling her mating readiness at any random moment, is considered to represent a more or less primitive condition (Assem 1975). A great many insect species fall into this category.

In other species, different behaviors are often strung together in a predictable sequence called a **bout**. Just as a bout of a fencing match ends with success or failure for the fencer, a male wasp's courtship bout ends with the female indicating either willingness to copulate or refusal to cooperate.

Melittobia belongs in this latter group. In some species, its courtship sequences are simple, composed of a single motor pattern or a series of identical motor patterns that follow each other in rapid succession. In other species, courtship bouts incorporate a number of different motor patterns following each other. Behaviors such as wing fluttering and leg "bicycling" follow one another in a predictable sequence. If the female gives the proper signal at the end of the bout, they copulate; if not, the male often begins another courtship sequence. In fact, according to Assem (1975), a world authority on this group, the courtship of *Melittobia* is among the most complicated known in the parasitic wasps. Because of this, *Melittobia* species have been much studied (Dahms 1973; Evans & Matthews 1976; Assem et al. 1982; Gonzalez et al. 1985).

Today's inquiry begins with observing, describing, and quantifying the courtship activities of *Melittobia digitata*. Who approaches whom first? How persistent is the average *Melittobia* male? How long do courtship and copulation usually last? Does one sex predictably determine the outcome of courtship encounters?

Then we will more closely examine mate attraction, the earliest step in courtship. Does *Melittobia* courtship begin with a purely random encounter, or might the insects be signaling one another? If they do signal, what form does such communication take?

Finally, we will design individual or small-group experiments to investigate further some aspects of mate attraction. By that point you should be pretty familiar with many aspects of the fascinating courtship rituals of this parasitic wasp.

MATERIALS

Each pair of students will need 1 bioassay chamber, 1 Chemwipe® sheet cut into 1-inch (2.5-centimeter) squares, 1 deep-well projection slide with a single male *Melittobia* and a "primer" female pupa, 1 dissecting microscope with "cool" fiber-optic light, 1 fine-tipped black marker, 1 forceps, 4 gelatin capsules (size 0 or an appropriate size to fit over the ends of straws), 1 opaque paper cup or clean and empty tin can, 1 pipe cleaner or chenille stick, 1 timer (a stopwatch or wristwatch that measures seconds), 4 approximately 4-dram vials with tight cotton stoppers (2 with 10–15 virgin female *Melittobia* and 2 empty ones), 1 sheet of white unlined paper, and a dictating cassette tape recorder (optional).

DESCRIBING COURTSHIP

Hypotheses and Predictions

In this first part of the laboratory exercise, you will be examining the nature of *Melittobia* courtship. You will be watching the various courtship stages, measuring male persistence, and determining how long courtship and mating usually last. You will investigate how the interactions between males and females differ when females are virgin vs mated. Your observations should also help you generate ideas about what sensory cues male and female *Melittobia* might use to find each other.

Procedure

Read through the procedures that follow, and discuss them as a class before you begin following the protocol. If you will be working as an individual, you may wish to use a dictating cassette recorder to document your observations. If you will be working in pairs, decide how you will divide between you the tasks of observing, recording, timing, and so on. Rotate assignments; everyone is responsible for having participated in all aspects of the inquiry.

Obtain your materials and set up the fiber-optic light so that it illuminates your microscope stage. Place the vial containing the virgin female wasps on the stage, adjust the magnification, and focus and observe for a few moments. Describe their body color, wing size and form, and antennae structure.

In the deep-well projection slide, there will be either two adult *Melittobia* or an adult with a pupa. Place the slide on the microscope stage. Adjust the focus and observe for a few moments. Which is the male? How do the sexes differ? Describe the male's body color, wing length and form, and antennae structure.

If the pupa has become an adult female, remove her before proceeding. Open the lid of the deep-well slide, carefully pick up the female with the pipe cleaner, and place her in an empty vial. She will already be mated, so label this vial "mated."

Working over the white paper, turn the bottom end of the vial that contains virgin females upward and toward a light source. As the females crawl away from the cotton stopper, carefully remove it. Reach into the vial with the pipe cleaner and carefully pick up one female. Keep the bottom end of the vial upward and toward the light source to stop others from attempting to escape during this procedure, and replace the stopper in the vial promptly. Should extra females leave the vial, cover them with an empty vial until you can pick them up and return them to the rest.

First, become familiar with the general picture of *Melittobia* courtship. To begin, open the lid of the well slide. Help the female crawl off the pipe cleaner into the well with the male. Replace the lid, and place the deep-well slide back under the microscope. While one of you times the proceedings, the other should watch the wasps, verbally describing each different behavior that occurs. When the pair come together, with the male mounted on top, note the orientation of the male and female, the movements of body parts, and the sequence in which these movements are performed. Pay particular attention to patterns of alternating activity.

Once the male has climbed on the female and faces the same direction, he begins his first courting episode. We shall arbitrarily recognize the actual start of courtship as the moment when he bends his head slightly downward over hers and contacts her antennae with his (Fig. 4.1A). There follows a distinctive series of male behaviors featuring repeated rhythmic opening and closing of antennae and raising and lowering of legs. His courtship bout ends at the moment he releases contact with the female's antennae and stretches his body lengthwise, preparatory to moving backwards to attempt copulation (Fig. 4.1B). If the female is receptive, she will lower her body, tilt her head up, and change her abdominal shape to allow genital contact (Fig. 4.1C). If the female is not receptive, she does not lower her body or change her abdominal shape, and the male cannot copulate. In this latter case, after a brief period the male usually responds either by dismounting or by moving forward again and initiating a new courtship episode that begins when he again establishes antennal contact.

Like other animals placed in a new environment, the wasps may need a period of acclimatization to the chamber and exploration of each other before they will begin sexual activity. You may see a number of initial tentative contacts before the mounted male first contacts the female's antennae with his. However, if no courtship is observed within 8 minutes, record this fact on your data sheet, remove the female, and replace her with a new one. Alternatively, simply add a second virgin female to the courtship arena. If you still have no success, obtain a replacement male from your instructor and repeat the process.

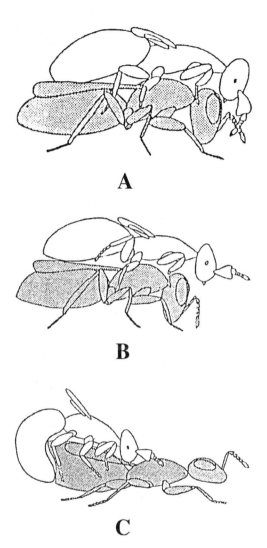

Figure 4.1 ***Melittobia*** pairs at various stages in courtship (the female's body is shaded). A. Beginning of a courtship bout with the male in forward position, head tipped down, antennae contacting those of the female. B. End of a courtship bout, with the male stretching his body after releasing contact with the female's antennae, preparatory to his backing up to attempt genital contact. C. Copulation, which begins when the receptive female tips her head upward and changes her abdomen posture to a more triangular shape that opens her genital orifice and which ends when the male withdraws from her. (Drawings courtesy of J. M. Gonzalez.)

After mating occurs, use the pipe cleaner to remove the female gently. Place her in an empty vial and plug it with cotton. Using the marker, clearly label the vial. Add all subsequent mated females to this same vial. Never return a mated female to the vial with the unmated ones!

Conduct at least three more trials with the same male to observe *Melittobia* courtship and mating in more detail. Taking turns as observer and recorder, time the courting and copulation and record these observations on the data sheet (Table 4.1).

Finally, conduct at least two similar trials, reusing now-mated females from your labeled vial. Again, time courting and copulation, and record these observations on the data sheet (Table 4.2).

Table 4.1 **Data sheet for quantifying aspects of *Melittobia* courtship between males and virgin females. The first row shows hypothetical data.**

Melittobia Pairing	*Courtship Bout Number*	*Courtship Bout Length (seconds)*	*Successful Copulation?*	*Copulation Duration (seconds)*	*Courtship + Mating (seconds)*
Example: Male 1 + virgin 1	*1*	*47*	*no*	*—*	*—*
	2	*39*	*yes*	*5*	*91*
Summary					
Number of pairs tested:	Mean number of courtship bouts per pair:	Mean ± SD for courtship bout length (seconds):	% of all pairings that were successful:	Mean ± SD for duration of copulation:	Mean ± SD for total courtship bout duration:

Table 4.2 **Data sheet for quantifying aspects of** *Melittobia* **courtship between males and mated females.**

Melittobia Pairing	Courtship Bout Number	Courtship Bout Length (seconds)	Successful Copulation?	Copulation Duration (seconds)	Courtship + Mating (seconds)
	Summary				
Number of pairs tested:	Mean number of courtship bouts per pair:	Mean ± SD for courtship bout length (seconds):	% of all pairings that were successful:	Mean ± SD copulation bout duration:	Mean ± SD for total courtship bout duration:

Data Recording and Analysis

Working together, determine the means and standard deviations for the data you and your partner recorded (Tables 4.1 and 4.2) for all components of successful and unsuccessful courtship bouts. Calculate the percentage of tested pairs that successfully mated for both virgin and previously mated females.

Record your values on the class data sheets at the front of the room. Compile the same descriptive statistics for the class data. Before you leave the laboratory today, be sure to copy these class data into your notebook or onto your computer disk. Also save your own data sheet for possible inclusion in your laboratory report.

MATE ATTRACTION: WHOLE-CLASS EXPERIMENT

Hypothesis and Predictions

Does *Melittobia* courtship begin with attraction by a signal of some sort? What form might it take? We are accustomed to thinking that the females of most species are the attractive sex. However, your *Melittobia* courtship observations probably suggest the opposite, for in the initial contact phases a female typically comes to a male, approaching him from the side and touching his abdomen with her antennae.

If males attract females, what cues might be involved? Vision seems unlikely, given that in nature, courtship occurs inside the dark cocoon of the host. What about smell, movement, or sound? To investigate, we will be using an experimental apparatus called a bioassay chamber. A **bioassay** is an analysis that uses the behavior of living organisms to measure response to a variable. Our chamber has four spokes or arms so that we can simultaneously measure females' responses to three choice conditions and a control. Its design enables us to separate a male physically, while allowing possible odor, sound, and motion to be transmitted across a permeable barrier.

Our research hypothesis will be that males of *Melittobia* attract females by smell, movement, or sound. Our independent variable will be the potential mate's condition (functional male, mated female, or squashed male). Because females are so mobile, we'll measure their reaction (our dependent variable) in terms of average number of wasps per location per minute.

Procedure

Setting up the Bioassay

The film canister bioassay chamber that we will be using (Fig. 4.2) has gelatin capsules as replaceable experimental quarters. First, separate four of the empty gelatin capsules and set aside the shorter end of each. With a felt-tip pen, number the end of each longer half from 1 to 4. Each capsule will be loaded in a different way.

Capsule 1 should remain empty. Cover its open end with a small piece of Chemwipe® tissue to act as a porous barrier between the interior of the capsule and the interior of the bioassay chamber. Holding the Chemwipe® in place with your fingers, gently push the covered end of the capsule partway over the open end of one of the tube "spokes" on the bioassay chamber, being careful not to tear the tissue.

Into capsule 2, use a pipe cleaner to introduce the experienced live male *Melittobia* from your deep-well observation slide. Cover the capsule as above, and place it on a second spoke of the bioassay chamber.

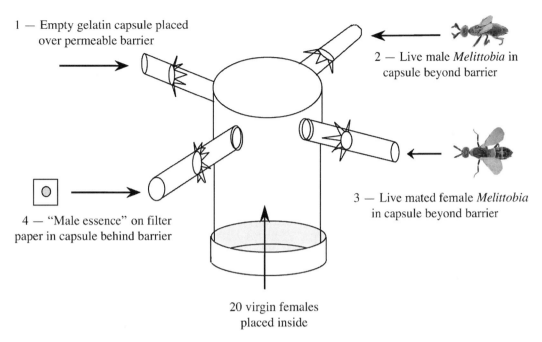

1 — Empty gelatin capsule placed over permeable barrier

2 — Live male *Melittobia* in capsule beyond barrier

3 — Live mated female *Melittobia* in capsule beyond barrier

4 — "Male essence" on filter paper in capsule behind barrier

20 virgin females placed inside

Figure 4.2 **Melittobia mate attraction can be bioassayed in a simple chamber made from a film canister, clear plastic straws, poster putty, Chemwipe® squares, and gelatin capsules.**

Follow the same procedure to place a live mated female *Melittobia* in capsule 3. Place it on a third spoke of the bioassay apparatus.

From your instructor, obtain a square containing "male essence" made by squashing a freshly dead male on filter paper. Use your forceps to place it in capsule 4. Put this on the remaining spoke of the apparatus.

Check the interior of the bioassay chamber. All the spokes should be flush with the inside of the canister so that *Melittobia* can easily enter them and should be securely sealed with the poster putty around their perimeter so that the wasps will be unable to escape. If any spokes have moved or become unsealed, fix them.

Conducting the Bioassay

Invert the chamber and open the cap at the bottom of the canister. Working together, quickly introduce 20 unmated female *Melittobia* into the chamber by removing the vial stoppers, tipping the two vials over the canister, and sharply snapping your finger against the side of each vial until all the wasps are dislodged into the canister. Quickly replace the lid, being careful not to smash any wasps between the cap and the canister. If you feel that you have lost a wasp or two, note that fact but continue the experiment. Although it is best to have 20 females in the canister, this number

Table 4.3 **Bioassay results from observing 20 virgin female *Melittobia* wasps at 1-minute intervals over a 10-minute period.**

Sample Minute	Location				
	1—Empty Tube	2—Live Male	3—Mated Female	4—Male Essence	No Choice Made, Still in Cylinder
1					
2					
3					
4					
5					
6					
7					
8					
9					
10					
Mean number of wasps observed per minute					

is not critical, and the actual number used will be confirmed at the end of the experiment.

Place the bioassay chamber, lid side down, on the table. Cover it with the paper cup and start your timer. Once a minute for the next 10 minutes, lift the cup briefly. On the data sheet (Table 4.3), record the number of females visible in each of the spokes of the bioassay chamber. (Scientists call this the "on the dot" or **instantaneous sampling** method.)

Are you seeing the same or different individuals at each count? What would be the most reliable way to answer this question? What practical problems would have to be addressed? An indirect indication can be obtained by noting how much the wasps move around once they enter the spokes.

At the end of the 10-minute recording period, remove the cup and watch any wasps in the spokes of the bioassay chamber for 2 or 3 minutes. Record what you see. Where are the wasps that are not visible in the spokes? You must account for all individuals. Open the canister over the piece of white paper, and count the wasps as you transfer them back into vials. As before, you can place the open end of a vial over any individuals that attempt to escape. Once the covered wasp begins to move up the vial, pick up the vial and place it over another wasp. When all have been gathered, plug the vial securely with a cotton stopper.

At the conclusion of the bioassay trials, remove capsule 2 from its straw, and return the male to his deep-well slide. Return the mated female from capsule 3 to the "mated females" vial. Then discard the four capsules and the "male essence."

Data Recording and Analysis

On your data sheet (Table 4.3), calculate the average number of wasps per location per minute. Then enter these averages on the class data sheet. Copy the class data sheet to use in writing your laboratory report.

For this experiment, the null hypothesis is that when virgin *Melittobia* females are given the choice of an empty capsule, one with a live male, one with a live mated female, and one with "male essence," the average number of females appearing in each location, as measured at 1-minute intervals over a 10-minute span, is not statistically different from random. The alternative hypothesis is that the average number of females appearing in each of these four locations, as measured at 1-minute intervals over a 10-minute span, is significantly different from random.

Various statistical analyses are possible. Your instructor may suggest either the chi-square goodness-of-fit test (not test of independence) or a Kruskal–Wallis test. In order for you to use the chi-square goodness-of-fit test, your class data set must be large enough for you to have expected values greater than 5; a Kruskal–Wallis test gets around this problem.

MATE ATTRACTION: SMALL-GROUP EXPERIMENTS

Hypotheses and Predictions

So far, we've only scratched the surface of *Melittobia* courtship and mate attraction. Now it is your turn to carry the research a step further. You will be working as partners or teams to investigate specific aspects of *Melittobia* mate attraction more fully.

Before doing your small-group experiment, brainstorm to create a list of 2–5 general questions about mate attraction that your group would like to explore. For example, you might decide to compare female attraction to brothers (siblings) vs unrelated (nonsibling) males, or to experienced vs inexperienced males, or to young vs older males. If virgin females responded strongly to squashed males, you might wish to study their response to squashes prepared from different male body parts. As part of the brainstorming process, be sure also to consider what outcome(s) you might expect for a particular experiment.

Then, as a group, narrow down your choices to one that you will test. Develop a working hypothesis that reflects what you think will happen.

Procedure

After choosing your research question, work together with your partner(s) to design a test for this question. You will want to use the same bioassay chambers and the same general procedures as in the whole-class experiment.

What will your variables be? How will you measure female response? What control(s) do you need to have?

You may want to use the same general approach and compare a control and three groups (such as males of three different ages). Instead of pooling data with the whole class, you will need to do repeated trials of the same experiment and to pool these means.

Alternatively, you may wish to make a smaller number of comparisons. For example, you might compare one treatment group (such as male essence) and a control and duplicate the experiment in pairs of spokes in a single chamber. This will reduce the number of trials you must run.

Data Recording and Analysis

Design a data sheet. What are the independent and dependent variables and the controls in your experiment? Write a specific null hypothesis and an alternative hypothesis for the independent and dependent variables that you tested.

What statistical test will you use? Check your choice with your instructor. As before, if you will be using a chi-square goodness-of-fit test, you will need enough replicates to generate expected values greater than 5. Alternatively, you could do a Kruskal–Wallis test. If you compared only two groups, you could also use the Mann–Whitney U test.

Presenting the Final Report

Discuss how best to present your data. Check your decision with your instructor, and then prepare appropriate tables, illustrations, and/or graphs. In your report, summarize the class data for courtship behavior and the bioassays with descriptive statistics that include means and standard deviations. Where appropriate, state the null and alternative hypotheses, and specify the independent and dependent variables used and the controls. Describe whatever statistical tests you employed, and give their result. Include your responses to the questions for discussion that follow. If requested by your instructor to do so, prepare and format your report as though for publication in a scientific journal. For help, refer to Matthews et al. (2000).

QUESTIONS FOR DISCUSSION

How long did courtship bouts and copulation usually last? How variable were these behaviors? What general stages of *Melittobia* courtship did you see? At the beginning, which sex approached the other first? What did they do? What does this suggest about the sort of cues the wasps might be using to find each other?

How many bouts of courtship did the average pair perform? On the basis of your class data, does it appear that persistent males (those performing repeated bouts) were generally successful in obtaining copulation?

Were there quantitative differences between virgin and mated females in the lengths of courtship bouts? In what ways did a mated female behave differently than a virgin female during male courtship attempts? What does this suggest about which sex controls the ultimate outcome of *Melittobia* courtship?

What correlation(s) can you draw between male and female antenna structure in *Melittobia* and courtship and mate attraction behavior?

Compare and contrast the contents of the four gelatin capsules in the mate attraction bioassay. What types of communication signals were being tested with each? What new information could capsule 4 provide that would not be evident just from capsule 2?

Which capsule contained the most females in your own trials? in the summed trials for the whole class? What evidence for pheromone usage did these experiments provide?

The protocol in the bioassay calls for introducing 20 unmated females simultaneously into the testing chamber. What fundamental premise of most statistical tests does this protocol violate? What biological variable does the protocol fail to control? What changes could you propose to address these concerns? What practical effect might these changes have on the conduct of this inquiry during a class laboratory period?

In the bioassay chamber, we used a single male in captivity to assay the behavior of 20 virgin females. Using what you know of *Melittobia* life history and behavior, predict what would happen if we were to reverse the experiment, using a single captive female and placing 20 male *Melittobia* together in the central canister of the bioassay chamber.

SUGGESTED READING

Assem, J. Van den. 1975. Temporal patterning of courtship behaviour in some parasitic Hymenoptera, with special reference to *Melittobia acasta*. *Journal of Entomology*, **50**, 137–146.

Assem, J. Van den, Bosch, H. A. J. in den & Prooy, E. 1982. *Melittobia* courtship behaviour: a comparative study of the evolution of a display. *Netherlands Journal of Zoology* **32**, 427–471.

Dahms, E. C. 1973. The courtship behaviour of *Melittobia australica* Girault, 1912 (Hymenoptera: Eulophidae). *Memoirs of the Queensland Museum*, **16**, 411–414.

Evans, D. W. & Matthews, R. W. 1976. Comparative courtship behaviour in two species of the parasitic wasp, *Melittobia* (Hymenoptera: Eulophidae). *Animal Behaviour*, **24**, 46–51.

Gonzalez, J., Matthews, R. W. & Matthews, J. R. 1985. A sex pheromone in males of *Melittobia australica* and *M. femorata* (Hymenoptera: Eulophidae). *Florida Entomologist*, **68**, 279–286.

Matthews, J. R., Bowen, J. M. & Matthews, R. W. 2000. *Successful Scientific Writing: A Step-by-Step Guide for the Biological and Medical Sciences*. 2nd ed. Cambridge: Cambridge University Press.

Thornhill, R. & Alcock, J. 1983. *The Evolution of Insect Mating Systems*. Cambridge, MA: Harvard University Press.

CHAPTER 5

Chemoreception in Lizards

C. O'NEIL KREKORIAN
San Diego State University, Department of Biology, 5500 Campanile Drive, San Diego, CA 92182-4614, USA

INTRODUCTION

Past analysis of communication systems in lizards has focused primarily on visual displays. In the last two decades, chemical communication in lizards, as well as in other vertebrates, has become an active area of research. Many species of lizards extrude their tongues while moving through their home ranges. In some cases the tongue is simply flicked into the air, whereas in other cases the tongue is touched to some object in the environment. The tongue is thought to deliver molecules to the paired vomeronasal (Jacobson's) organs in the roofs of the lizards' mouths. However, there is some evidence that taste in some species may be the chemosensory mode involved during substrate licking.

Behavioral studies suggest a variety of functions for the tongue extrusions in lizards (Simon 1983; Halpern 1992). However, there have been only a few quantitative studies on chemoreception in lizards, and only a few species—*Sceloporus jarrovi* (Defazio et al. 1977; Simon et al. 1981), *Dipsosaurus dorsalis* (Krekorian 1989; Cooper & Alberts 1990, 1991; Dussault & Krekorian 1991; Bealor & Krekorian, 2002), and *Eumeces laticeps* (Cooper & Vitt 1984; Cooper 1990)—have been examined more thoroughly. These studies convincingly show that tongue extrusion is used to obtain several types of information. Among the uses suggested for the tongue–vomeronasal system are: (1) detection of conspecifics, (2) species identification, (3) sex discrimination, (4) territoriality, (5) food seeking, (6) general exploration, and (7) predator detection.

The desert iguana (*Dipsosaurus dorsalis*), the broad-headed skink (*Eumeces laticeps*), and the iguanid lizard *Sceloporus jarrovi* frequently extrude their tongues while moving through their home ranges. Desert iguanas, for example, have been observed to extrude their tongues both in the laboratory and in the field when encountering pebbles, dried leaves, fecal material, conspecifics, food, and burrow openings (Krekorian 1989; Cooper & Alberts 1990; Pedersen 1988, 1992).

MATERIALS

The materials needed for this laboratory are 8–16 lizards, 14–16 small terraria (5–10 gallons), and eight 250-watt infrared heat lamps. Lizards can be identified by painting numbers on their backs with fingernail polish or model paint. Use sand, gravel, or paper towels to line the floor of the terraria. If possible, place 1 or 2 small rocks (70 centimeters in diameter) in each test terrarium. You will need a stopwatch or digital timer for each experimental setup.

PROCEDURE

You should work in groups of 2–4 students per experimental setup. Work quietly and avoid sudden movements of appendages (arms or hands) that might disturb the lizard. Sit approximately 60–90 centimeters from the terrarium while making your observations. Designate one person to control the timer and record the latency of the first tongue extrusion and subsequent tongue extrusions during the test period. Data from each group will be pooled for the class to provide an acceptable sample size for statistical analysis.

Begin by observing the lizards for 10 minutes. During this period, watch for tongue touches and tongue flicks. With tongue touches, the tongue contacts the substrate or some object, whereas with tongue flicks, the tongue is extruded into the air. Do the tongue extrusions occur while the lizard is stationary, after a move, or while it is moving? What objects in the terrarium are licked? Upon completion of the preliminary 10-minute period, begin the experiment. The experiment will consist of three periods: the baseline, treatment, and control periods. For each period, record the time (latency) until the first tongue extrusion and then the number of tongue flicks and the number of tongue touches ("licks") for the next 15 minutes.

The 15-minute period begins with the first tongue extrusion. For lizards that are slow to respond, enforce a maximum-latency rule of 10 minutes, and then begin the 15-minute testing period.

Baseline Period

Remove the lizard from its home cage and hold it for 1 minute before returning it to the same location in its home cage. Start your timer and observations when the lizard is placed in the terrarium, and record the following dependent variables: latency until the first tongue extrusion, the number of tongue flicks, the number of tongue touches, and the objects licked. Your 15-minute test period begins with the first tongue extrusion.

Treatment Period

Transfer the lizard from its home location to the treatment terrarium. The treatment terrarium is a clean terrarium with either gravel or sand as a substrate. Repeat the procedure used in the baseline period. Begin your observations after the transfer, and record the pertinent data.

Control Period

Return the lizard to its home cage. Begin your observations after the transfer, and record the dependent variables previously recorded in the baseline and treatment periods.

Each group of students will replicate this experiment twice using different lizards during the testing. Please handle the lizards with care and refrain from activities that will disturb or influence the behavior of the lizards.

When the replicates are completed, record your data in Table 5.1.

Table 5.1 **Data sheet for recording dependent variables for each test period.**

Lizard	Baseline (Pretest)			Treatment (Experimental)			Control (Posttest)		
	Latency (Seconds)	*Tongue Extrusion*		*Latency (Seconds)*	*Tongue Extrusion*		*Latency (Seconds)*	*Tongue Extrusion*	
		Flicks	*Touches*		*Flicks*	*Touches*		*Flicks*	*Touches*
1									
2									
3									
4									
5									
6									
7									
8									
9									

HYPOTHESES AND PREDICTIONS

The general hypothesis for the experiment is that tongue extrusion behavior varies with a lizard's familiarity with its habitat. If the treatment influences frequency of tongue extrusion in lizards, then the number of tongue extrusions in the treatment period should be greater or smaller than the numbers of tongue extrusions in the baseline and control periods. If chemoreception is used by lizards in general exploration of their environment, how will the treatment influence their behavior?

DATA RECORDING AND ANALYSES

On the chalkboard, each group should record its data on latency until first tongue extrusion and number of tongue extrusions for each testing period in a table like Table 5.1. Copy the rest of the class data onto your Table 5.1 so that you have the pooled class data. Use the pooled class data to calculate the mean latency and also the mean frequency of tongue extrusions. To test the latency data in this experiment, the null hypothesis is that the latency did not differ among the three treatment periods. The alternative hypothesis is that latency of tongue extrusion differed among the three periods. Data can be analyzed using a repeated-measures ANOVA (parametric) or the nonparametric Friedman two-way analysis of variance by ranks (Siegel 1956). The Friedman test is usually more appropriate as a result of small sample sizes, maximum assigned latencies for reluctant lizards, and data that may not be normally distributed—and because the test is simple to apply to these data.

What are the null and alternative hypotheses for frequency of tongue extrusion? What are the null and alternative hypotheses for separate frequency analysis of tongue flicks and tongue touches? Check your answers with your instructor. You can analyze these data using the same statistics you used for latency.

QUESTIONS FOR DISCUSSION

Judging on the basis of your data, what function does tongue extrusion have in this experimental context? Is the sex and/or the size of the adult lizard a factor in its response? What dependent variable appears to be the best for analysis? Do tongue touches or tongue flicks occur more frequently? Were there other changes in the behavior of the lizards when they were placed in the treatment terrarium that differed from their behavior in the baseline period or the control period? Could these possible behavioral changes be quantified and used as dependent variables to measure differences among the three test periods?

ACKNOWLEDGMENTS

I thank the editors, Bonnie J. Ploger and Ken Yasukawa, and an anonymous reviewer for their valuable comments and suggestions for improving the laboratory.

LITERATURE CITED

Bealor, M. T. & Krekorian, C. O. 2002. Chemosensory identification of lizard-eating snakes in the desert iguana, *Dipsosaurus dorsalis* (Squamata: Iguanidae). *Journal of Herpetology*, **36**, 9–15.

Cooper, W. E. 1990. Chemical detection of predators by a lizard, the broad-headed skink, *Eumeces laticeps. Journal of Herpetology*, **256**, 126–167.

Cooper, W. E. & Alberts, A. C. 1990. Responses to chemical food stimuli by an herbivorous actively foraging lizard, *Dipsosaurus dorsalis. Herpetologica*, **46**, 259–266.

Cooper, W. E. & Alberts, A. C. 1991. Tongue-flicking and biting in response to chemical food stimuli by an iguanid lizard (*Dipsosaurus dorsalis*) having sealed vomeronasal ducts: vomerolfaction may mediate these behavioral responses. *Journal of Chemical Ecology*, **17**, 135–145.

Cooper, W. E. & Vitt, L. J. 1984. Conspecific odor detection by the male broad-headed skink, *Eumeces laticeps*: effects of sex and site of odor source and of male reproductive condition. *Journal of Experimental Zoology*, **230**, 199–209.

DeFazio, A., Simon, A. C., Middendorf, G. A., & Romano, D. 1977. Iguanid substrate licking: a response to novel situations in *Sceloporous jarrovi. Copeia*, **1977**, 706–709.

Dussault, M. H. & Krekorian, C. O. 1991. Conspecific discrimination by chemoreception in the desert iguana, *Dipsosaurus dorsalis. Herpetologica*, **47**, 82–88.

Halpern, M. 1992. Nasal chemical senses in reptiles: structure and function. In: *Biology of the Reptilia*, Vol. 18 (ed. C. Gans & D. Crews), pp. 423–523. Chicago: University of Chicago Press.

Krekorian, C. O. 1989. Field and laboratory observations on chemoreception in the desert iguana, *Dipsosaurus dorsalis. Journal of Herpetology*, **23**, 267–273.

Pedersen, J. M. 1988. Laboratory observations on the function of tongue extrusions in the desert iguana, *Dipsosaurus dorsalis. Journal of Comparative Psychology*, **102**, 1–4.

Pedersen, J. M. 1992. Field observations on the role of tongue extrusion in the social behavior of the desert iguana (*Dipsosaurus dorsalis*). *Journal of Comparative Psychology*, **106**, 287–294.

Siegel, S. 1956. *Nonparametric Statistics for the Behavioral Sciences*. New York: McGraw-Hill.

Simon, C. A. 1983. A review of lizard chemoreception. In: *Lizard Ecology: Studies of a Model Organism* (ed. R. B. Huey, E. R. Pianka & T. W. Schoener), pp. 119–133. Cambridge, MA: Harvard University Press.

Simon, C. A., Gravelle, K., Bissinger, B. E., Eiss, I. & Ruibal, R. 1981. The role of chemoreception in the iguanid lizard *Sceloporus jarrovi. Animal Behaviour*, **29**, 46–54.

CHAPTER 6

Behavioral Thermoregulation in Field Populations of Amphibian Larvae

HOWARD H. WHITEMAN[1] AND NANCY L. BUSCHHAUS[2]
[1]*Department of Biological Sciences, Murray State University, Murray, KY 42071, USA*
[2]*Department of Biological Sciences, University of Tennessee at Martin, Martin, TN 38238, USA*

INTRODUCTION

Most physiological processes and many developmental traits are extremely sensitive to environmental change and perform optimally within a narrow range of conditions. Small changes in factors such as temperature, pH, and salt concentration can have an enormous impact on the efficiency of biological systems. As a result, organisms have evolved mechanisms to respond to fluctuations in their environment and maintain internal conditions at an optimal level. This regulation of internal conditions, which is termed **homeostasis**, is a fundamental principle of **physiological ecology**, the study of how organisms respond to and cope with their abiotic environment (Begon et al. 1990; Ricklefs & Miller 1999).

Responses to changes in the environment can be made through physiological and/or behavioral means, and different types of organisms may exhibit different adaptations to a common problem. However, regardless of the mechanism(s), it is important to remember that controlling the internal environment in the face of external fluctuations requires the expenditure of energy. All organisms must meet certain basic minimum requirements of necessary resources (such as food, water, and oxygen) to survive; any particular strategy may place constraints on where and how these resources can be obtained (Begon et al. 1990; Ricklefs & Miller 1999).

Thermoregulation, the ability to maintain optimal temperature, can be critical to the success of organisms because of its importance to growth, development, and cellular function. Organisms can be categorized in terms of the way they thermoregulate. For example, **endotherms** regulate their temperature primarily via the production of heat from metabolic reactions within their own bodies; **ectotherms** rely primarily on external sources of heat. In animals, this is generally a distinction between the birds and mammals in the first case, and the invertebrates, fishes, amphibians, and reptiles in the second case. However, exceptions to this general rule occur in some species, when endotherms temporarily lower their body temperature or when ectothermic organisms increase their body temperature by metabolic means for short periods of time. Irrespective of whether an organism is endothermic or ectothermic, maintaining optimal temperature always entails an energetic cost (Ricklefs & Miller 1999).

Thermoregulation and other physiological processes are influenced by body size—specifically by the ratio of surface area (SA) to volume (V). Because the ratio of SA to V (SA/V) decreases as body size increases, smaller organisms have the highest SA/V. Increased SA/V means that smaller organisms are more affected by thermoregulatory problems than are larger organisms, because they have more surface exposed to the outer environment as a function of their body size (Calder 1984). Imagine the comparison between a 5-pound bag of ice cubes and a 5-pound solid block of ice. Which will melt faster? Although both have the same volume, the ice cubes have greater surface area—and thus a higher SA/V ratio—and will melt first because of the increased exposure. A similar situation will occur if we do not control for the volume of the ice. Imagine an ice cube and a block of ice with four times the cube's volume. In the same time period, a greater percentage of the small cube than of the large block will melt because of the greater SA/V ratio in the small cube. These same conditions affect organisms of small size.

Behavioral Thermoregulation and Amphibian Larvae

Behavioral thermoregulation is regulation of temperature by altering one's behavior. Although both endotherms and ectotherms behaviorally thermoregulate, it is especially important in ectotherms because they cannot regulate their body temperature through physiological means as well as endotherms can (Ricklefs & Miller 1999). Examples of behavioral thermoregulation include endothermic dogs panting to increase heat loss on hot days; ectothermic lizards basking on warm rocks before foraging to help obtain sprint speeds for capturing prey, and after foraging to aid digestion; and endothermic hummingbirds migrating to the warmest areas on a daily basis to minimize temperature loss and energy expenditure (Begon et al. 1990; Calder & Calder 1992).

During this field exercise, we will study behavioral thermoregulation in amphibian larvae. Amphibians are ectothermic, and their larvae (termed tadpoles in frogs and toads-larvae in salamanders) require a narrow range of temperatures for optimal growth, development, and metamorphosis. To maintain this narrow range, amphibians often behaviorally thermoregulate (Brattstrom 1962; Lillywhite 1971; Lillywhite et al. 1973; Heath 1975). Quick growth and development are especially important in ephemeral ponds, where amphibian larvae have only a limited time to reach the minimum size suitable for metamorphosis (Wilbur & Collins 1973). In addition, maximizing growth through optimal use of temperature within the pond should be under strong selection, because body size at metamorphosis is correlated with adult fitness in a number of amphibian species (Scott 1994; Semlitsch et al. 1998). In habitats where there are short growing seasons and/or extreme variation in ambient daily temperature (such as high-elevation ponds and desert pools), maintaining optimal temperature is even more important (Lillywhite et al. 1973; Heath 1975). However, there are potential costs to behavioral thermoregulation, including increased exposure to predators during migrations to warm areas and the lost opportunity cost if prey are most abundant in areas of the pond not used for thermoregulation. We will observe the spatial and temporal distribution of amphibian larvae in relation to water temperature within a pond to determine whether they behaviorally thermoregulate.

MATERIALS

Equipment and Supplies

Each class will need a thermometer tied to an adequate length of string to reach the deepest parts of the pond. One thermometer can be used to measure all areas, but having multiple thermometers (at least one deep and one shallow) makes the process much faster and more accurate. The class will also need flags and string for marking at least six 1- 2-m^2 sampling quadrats (at least 40 flags), or wooden quadrats (made of thin wooden strips) that can be anchored to the bottom. Extra flags will be needed to mark deep-water temperature-recording sites. In murky ponds, you can use a D-shaped sampling net to capture amphibians rather than observing them directly. Hip or chest waders may be necessary in some habitats if you are measuring deep-water temperatures. Flashlights or head lamps will be needed if night-time observations are performed. Students should also have data sheets, pencils, and a watch.

Organisms

Amphibian larvae and/or adults of various sizes and species can be used for this exercise. Common species include pond-breeding salamanders of the genus *Ambystoma* (such as tiger, mole, spotted, and marbled salamanders) and frogs of the genus *Rana* (such as leopard frogs, green frogs, and bullfrogs) and the genus *Pseudacris* (chorus frogs and spring peepers).

PROCEDURE

At least one day prior to observations, your instructor placed a minimum of six 1- 2-m^2 quadrats parallel to the shoreline in shallow-water areas around the perimeter of the study pond(s). You will observe and estimate densities of amphibian larvae during at least three time periods throughout the day, including morning, noon, afternoon, and (if possible) evening or night. Larval densities will be estimated by counting the number of larvae within quadrats. At each observation period, you should approach quadrats very slowly, carefully, and silently to minimize disturbance of amphibian larvae and to ensure an accurate count. During daytime, you should be especially careful not to cast a shadow over the quadrat. If you have distinctly different size classes of larvae and/or if adults are also available, you should separate the data by recording different size classes and/or adult numbers. Adults may be recorded as the number observed in shallow water within a few meters of each quadrat sampling site rather than as the number within each quadrat, because they are often found at much lower densities than larvae, and few may actually enter quadrats.

After all observations have been taken for a given period, water temperature will be recorded with a thermometer at each quadrat site as well as at a deep-water area. Deep-water areas should be several meters away from the shallow-water site and at least 0.5 meter deep, and each site should be flagged to allow repeated samples of the same area. You can obtain shallow temperatures from shore and temperatures in deep sites by wading. You should wade slowly and carefully to avoid underwater obstructions and to minimize disturbance of amphibians.

Because of the temporal nature of this project, you may be asked to volunteer to collect data during the later time periods (2–4 students per time period).

HYPOTHESES AND PREDICTIONS

Given the foregoing description, formulate a specific biological hypothesis for each of the following questions. For each hypothesis, list at least one prediction that will occur if the hypothesis is correct. (1) How will the difference in pond temperature between shallow and deep sites change across the day? (2) How will temperature influence the behavior of amphibians in the pond? (3) How will body size affect amphibian behavioral thermoregulation?

DATA RECORDING AND ANALYSIS

You should create a data sheet for each pond with the following columns: time of sample, quadrat number, number of larvae observed of the focal species (different columns for each species, size class, or adult forms if appropriate), quadrat water temperature, and deep-water temperature. After data collection, the difference in water temperature between matched sites should be determined for each sampling period by subtracting deep-water temperature from the temperature recorded in shallow water.

Note that the predictions from your biological hypotheses can be used to form both null and alternative statistical hypotheses. For example, one possible null hypothesis is that the number of amphibian larvae does not vary with temperature differences between shallow and deep areas. An alternative hypothesis would be that larval numbers vary with water temperature. You should create similar null and alternative hypotheses for each of the questions specified, using your biological hypotheses and predictions as a guide.

Data will be analyzed using one-way analysis of variance (ANOVA) or the nonparametric equivalent (Kruskal–Wallis test) to determine whether temperature and larval densities differ across time in shallow areas, using each quadrat as a replicate sample. You will need to pool your data with other members of the class unless each group recorded data for six or more quadrats at each sampling time. To aid in your interpretation of the results, plot the mean value ±1 SE of each response variable (e.g., larval density, difference in water temperature) for each sample period. How larval densities in shallow water vary as a function of the difference in water temperature between deep and shallow sites can also be analyzed using linear regression (or the nonparametric Spearman Rank correlation). Making scatter plots with regression lines will enhance your understanding of this analysis. If different size classes of larvae or both larvae and adults are being compared, an analysis of covariance (ANCOVA) may be used to test for differences in slope and intercept between regression lines, with regression lines plotted on the same figure. The ANCOVA provides a test of how body size influences the thermoregulatory abilities of organisms of different size.

QUESTIONS FOR DISCUSSION

Why is temperature an important resource for amphibians?

Did the amphibians in your observations show evidence of behavioral thermoregulation? How do you know?

Does the ability of amphibians to thermoregulate depend on body size? How do you know?

What kinds of trade-offs (such as energetic or mortality costs) might amphibian larvae experience by behaviorally thermoregulating?

What other factors besides temperature might explain the changes in behavior seen during this exercise? How might you test for these factors through observation? through experimentation?

ACKNOWLEDGMENTS

We thank Rick Williams for encouragement during the early development of this lab at the Rocky Mountain Biological Laboratory, and Darrin Good and the editors for constructive criticism.

LITERATURE CITED

Begon, M., Harper, J. L. & Townsend, C. R. 1990. *Ecology: Individuals, Populations, and Communities.* Cambridge, MA: Blackwell Scientific.

Brattstrom, B. H. 1962. Thermal control of aggregation behavior in tadpoles. *Herpetologica,* **18,** 38–46.

Calder, W. A. 1984. *Size, Function, and Life History.* Cambridge, MA: Harvard University Press.

Calder, W. A. & Calder, L. L. 1992. Broad-tailed hummingbird. In: *The Birds of North America,* No. 16 (ed. A. Poole, P. Stettenheim, & F. Gill), pp. 1–16. The Academy of Natural Sciences, Philadelphia.

Heath, A. G. 1975. Behavioral thermoregulation in high-altitude tiger salamanders, *Ambystoma tigrinum. Herpetologica,* **31,** 84–93.

Lillywhite, H. B. 1971. Temperature selection by the bullfrog, *Rana catesbeiana. Comparative Biochemistry and Physiology,* **40A,** 213–227.

Lillywhite, H. B., Licht, P. & Chelgren, P. 1973. The role of behavioral thermoregulation in the growth energetics of the toad, *Bufo boreas. Ecology,* **54,** 375–383.

Ricklefs, R. E. & Miller, G. L. 1999. *Ecology.* New York: Freeman.

Scott, D. E. 1994. The effect of larval density on adult demographic traits in *Ambystoma opacum. Ecology,* **75,** 1383–1396.

Semlitsch, R. D., Scott, D. E. & Pechmann, J. H. K. 1988. Time and size at metamorphosis related to adult fitness in *Ambystoma talpoideum. Ecology,* **69,** 184–192.

Wilbur, H. M. & Collins, J. P. 1973. Ecological aspects of amphibian metamorphosis. *Science,* **182,** 1305–1314.

CHAPTER 7

Temperature Dependence of the Electric Organ Discharge in Weakly Electric Fish

GÜNTHER K. H. ZUPANC[1,2], JONATHAN R. BANKS[1], GERHARD ENGLER[1], AND ROBERT C. BEASON[3]

[1]School of Biological Sciences, University of Manchester, 3.614 Stopford Building, Oxford Road, Manchester M13 9PT, UK

[2]School of Engineering and Science, International University Bremen, Campus Ring 1, D-28759 Bremen, Germany

[3]Department of Biology, University of Louisiana at Monroe, 700 University Avenue, Monroe, LA 71209, USA

INTRODUCTION

Some species of cartilaginous and bony fish have developed specialized sensory systems through which they can sense electric signals originating from physical or biological sources, even if these signals are of such low amplitude that we can detect them only with sensitive and highly sophisticated equipment. Many of these fish have, in the course of evolution, developed their own specialized organs that produce electric discharges, known by the name of **electric organ discharge** or **EOD** (for reviews see Bullock & Heiligenberg 1986; Zupanc 1988a; Zupanc & Banks 1998). On the basis of the voltage of these discharges, electric fish are divided into two groups—**strongly electric** and **weakly electric**. Strongly electric fish generate voltages between 20 volts and 600 volts. Representatives are the marine electric ray (*Torpedo*) and the freshwater electric catfish (*Malapterurus*). The discharges are used to stun prey, and some of these strongly electric fish can be dangerous to humans. By contrast, the signals produced by weakly electric fish are far too low in amplitude to be detected unless we amplify them. In 1951 Hans Lissmann was the first to succeed in making the EODs of weakly electric fish visible on an oscilloscope and audible by means of loudspeakers (Lissmann 1951). Given that the voltage usually does not exceed 1 volt (measured through electrodes placed directly at the head and tail of the fish), it is not surprising that these signals remained unknown for so long. The well-known electric eel (*Electrophorus electricus*) incorporates features found in both strongly and weakly electric fish. This South American freshwater species has three different electric organs. The **main organ** and **Hunter's organ** produce extremely powerful electric shocks, whereas **Sachs' organ** generates only weak electric discharges.

The weakly electric fish comprise two orders of teleost fish: the **Gymnotiformes**, which live in freshwater in South and Central America, and the **Mormyriformes**, which are native to African freshwaters. All mormyriforms except one species (*Gymnarchus niloticus*) produce electric pulses that are followed by relatively long intervals of "silence". The pulses are very short, lasting from 0.3 to several milliseconds. Because of their discharge pattern, fish of this group are referred to as **pulse-type fish**. In contrast, most gymnotiforms, as well as *Gymnarchus niloticus*, produce discharges that resemble continuous waves when displayed on an oscilloscope screen. Because of the appearance of the signals on the oscilloscope screen, the fish that make up this group are referred to as **wave-type fish**.

The discharge rate of the **electric organ** is under control of the **pacemaker nucleus**, a neuronal oscillator in the brainstem (for review see Dye & Meyer 1986). The pacemaker nucleus projects, via relay cells and spinal electromotoneurons, to the electric organ, which, in most electric fishes, is derived from muscle tissue. In the family Apteronotidae, however, the axonal endings of the spinal electromotoneurons themselves compose the electric organ (for review see Bass 1986). Weakly electric fish are able to perceive the discharges produced by their own electric organ, as well as discharges generated by other fish, using **electroreceptors** located in the skin (for review see Zakon 1986). These signals are centrally processed by specialized areas in the brain of the fish. Various parameters of the EOD (such as frequency and waveform) subserve several behavioral functions, including orientation in the environment (**electrolocation**; for reviews see Heiligenberg 1977, 1991) and communication with conspecifics (**electrocommunication**; for reviews see Hopkins 1988, 1999; Zupanc & Maler 1997). Whereas the range for detection of objects by analysis of the self-generated electric field is restricted to a few centimeters, the electro-communication range may extend up to 1 meter (Knudsen 1975).

In wave-type gymnotiform fish, the EOD typically is extremely stable in both frequency and waveform under constant environmental conditions, even over several hours. On the other hand, the frequency (in some species also the waveform) of the discharges is often **sexually dimorphic**. The information contained in the different frequency ranges occupied by males and females is likely to be used for sexual mate recognition. However, it is possible for unambiguous sex recognition to be severely impeded by factors resulting in modulations of the EOD. The frequency of the EOD, for example, can be altered by changes in ambient water temperature (Coates et al. 1954; Enger & Szabo 1968; Boudinot 1970; Feng 1976; Zupanc 1988b, 1988c). You will examine this effect and its behavioral significance in the following exercise, which will demonstrate that behavior is not an isolated entity. Rather, an understanding of a behavioral system must include knowledge about ecological factors that influence a behavioral pattern and about the physiological mechanisms that mediate such changes.

MATERIALS

The following is a list of materials needed for the lab exercise to be run by one student group. The components are assembled as shown in Figure 7.1. If not enough time is available, the set-up will have been prepared by your instructor prior to your arrival. In this case, we recommend that you discuss the design of the set-up with your instructor before the start of the experiments.

> 1 Wave-type gymnotiform fish, such as knifefish (*Eigenmannia sp.*), brown ghost (*Apteronotus leptorhynchus*), or black ghost (*Apteronotus albifrons*).
> 1 Plastic aquarium (volume approximately 25 liters).
> 1 Standard laboratory sink with taps for hot and cold water. Alternatively, a larger aquarium (volume approximately 100 liters) with provision for hot- and cold-water inputs may be used. The sink or aquarium must be large enough to contain the smaller tank and its supports.
> 4 Ceramic flower pots (or similar dense materials) to support the smaller tank.
> 1 Opaque (preferably grey or black) tube shelter for the fish made from plastic waste-water pipe that is 150–250 millimeters in length and has an inner diameter of approximately 30 millimeters. This tube must be at least 50 millimeters longer than the total length of the experimental fish used.

Fine-mesh netting, such as nylon mosquito netting, and rubber bands to close the ends of the plastic pipe.

> 1 External water pump system (such as an Eheim filter pump 2211) to circulate the water in the smaller tank. Its capacity should be approximately 300 liters per hour.

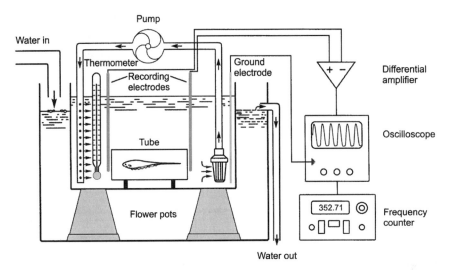

Figure 7.1 **Setup for studying the effect of water temperature on the EOD frequency of wave-type weakly electric fish. A small tank with the fish is placed on inverted flower pots in a larger aquarium (or a sink). Water of the desired temperature flows continuously into this larger tank, the level of which is regulated by the outflow. This arrangement enables the experimenter to control the temperature of the water in the smaller tank. Different temperature layers in the smaller tank are avoided by continuous circulation of the water via a pump. The fish is kept in a plastic tube slightly larger than the fish's total length. The ends of this tube are closed with plastic netting. Recording electrodes are put as close as possible at either end of this tube. In addition, a ground electrode is placed in the smaller aquarium. The temperature is measured by a mercury or electronic thermometer. The signal from the fish is fed, via a differential amplifier, into an oscilloscope and a frequency counter.**

1 Mercury or electronic thermometer with an accuracy and resolution of at least 0.2°C.

1 Pair of recording electrodes and 1 ground electrode.

Wires and connectors for connecting the electrodes to the amplifier, oscilloscope, and frequency counter.

1 Differential amplifier having a switchable gain of ×10 and ×100 and a frequency range of 10 hertz–10 kilohertz.

1 Digital storage oscilloscope with the facility to output stored traces to a plotter or a printer.

1 Printer/plotter for making printouts of oscillograms.

1 Frequency counter (optional).

1 Audio amplifier with loudspeaker.

Appropriate clamps, stands, and clips to support the thermometer and the electrodes.

As an alternative to the described recording set-up, a computer-based integrated data recording and analysis system, such as PowerLab (formerly MacLab), can be employed. Also, several rather inexpensive digital acquisition and analysis programs, such as Canary or Cool Edit Pro and Cool Edit 2000, have proved useful. These software packages, which are designed primarily for analysis of acoustic signals, make use of the analog input to the sound card for data acquisition.

PROCEDURE

Transfer of the Fish from Its Home Tank to the Experimental Aquarium

After having assembled the components listed in the "Materials" section according to Figure 7.1, fill the smaller aquarium with water from the home tank of the fish to be used in the experiment. Make sure that the pump system is working properly. If necessary, remove air bubbles from the hoses connecting the tank with the pump. Then transfer the experimental fish from its home tank to the experimental aquarium. This is best achieved by using the shelter tubes (as described in the list of Materials) in the home tank. The fish readily spend most of their time in these tubes. Close the ends of the tube with the fine-mesh netting, cover them with your palms, and transfer the tube filled with water and containing the fish into the experimental tank. This procedure avoids any direct contact with the fish; handling them with fish nets may cause skin abrasions, which can result in fungal infections.

Recording of the Fish's Electric Organ Discharges and Demonstration of the Constancy of the Discharge Frequency under Stable Environmental Conditions

Record the EOD by placing the recording electrodes at either end of the tube housing the fish. The ground electrode should be positioned approximately halfway between these two recording electrodes. Turn on the audio system. The EOD signal should be audible and visible on the oscilloscope display. You may need to adjust the gain of both the differential amplifier and the audio amplifier to obtain a clear and complete signal. Note the constancy of both frequency (pitch) and amplitude (intensity) of the signal. Movements of the fish may result in changes in its position relative to the recording electrodes—and thus in modulations of the recorded amplitude. This variability, which is particularly pronounced when the fish turns around, is not caused by changes in the amplitude of the EOD signal produced!

Set the time base of the oscilloscope so as to display 2–3 EOD cycles on the screen. The length of one EOD cycle corresponds to the time between two adjacent positive peaks. Make one printout of the recording for each student. Indicate on the printout the duration of one cycle, *T*. Add the printout to your worksheet.

From the printout (or the oscilloscope screen) determine the frequency of the recorded EOD. The frequency (*f*), in hertz (Hz), is the inverse of the duration of one cycle (*T*s), in seconds, of the EOD:

$$f = 1/T.$$

Change the time base of the oscilloscope so as to display 20–30 EOD cycles on the screen. Again, make one printout for each student. This recording should demonstrate the constancy of the frequency and the amplitude of the EOD, as well as its waveform, under constant environmental conditions.

Investigation of the Effect of Variation of Water Temperature on the Frequency of the Electric Organ Discharge

Set the time base of the oscilloscope to display approximately 10 EOD cycles on the screen. Measure the water temperature with the thermometer. At the same time, determine the EOD frequency with the oscilloscope by measuring the time over at least 10 EOD cycles (or with the frequency counter, if one is available).

Start increasing the water temperature in the holding tank by adding hot tap water to the surrounding sink (or the larger aquarium) in which the smaller tank is placed. (Why is hot or cold water not directly added to the aquarium?) Measure the EOD frequency for 0.5°C changes in ambient water temperature. As soon as the temperature in the holding tank has reached 30°C, turn off the hot tap and empty the sink (or larger aquarium). Fill the sink (or larger aquarium) with cold water, and gradually reduce the water temperature in the aquarium. MAKE ABSOLUTELY SURE THAT THE TEMPERATURE IN THE FISH TANK DOES NOT RISE ABOVE 30°C OR SINK BELOW 20°C!

Continuously measure the EOD frequency while the water temperature is lowered. Reverse the temperature once more as soon as the tank has reached 20°C by emptying the sink (or larger aquarium) again and adding hot water. Terminate the experiment by turning off the hot-water tap after the fish tank has reached the original starting temperature. At this point, the fish should be removed from the experimental set-up and returned to its home tank. Avoid a possible temperature shock by slowly adjusting the water temperature of the experimental tank to the temperature of the water in the home tank.

HYPOTHESES AND PREDICTIONS

What might happen, in principle, when you examine the effect of ambient temperature on the EOD frequency of the fish? What do you predict will happen?

DATA RECORDING AND ANALYSES

Recording of Data

Enter the data from the temperature experiment on your fish onto data sheets that provide space for (1) the number and the time of each recording, (2) the temperature measured, and (3) the frequency value obtained by calculation from the time of 10 EOD cycles or directly from the frequency counter. A sample data sheet is provided in Table 7.1.

Table 7.1 Sample data sheet for recording temperature and frequency measurements.

Number of Recording	Time	Temperature (°C)	Duration of 10 EOD Cycles (ms)	Duration of 1 EOD Cycle (s)	Frequency (Hz)

Plotting of the Frequency of the Electric Organ Discharge as a Function of Water Temperature

Plot the recorded EOD frequencies as a function of ambient water temperature. Indicate the starting frequency with an asterisk. Distinguish, by using different symbols, between frequency values obtained when raising the water temperature and those obtained when decreasing the water temperature. If it is visible, indicate an **hysteresis effect**. Hysteresis is the effect observed when a biological or physical system produces different output values for a given input value, depending on whether the input variable increases or decreases during the experiment.

Calculation of the Q_{10} value

The Q_{10} value is used to characterize the temperature dependence of biological, chemical, or physical processes. It indicates how much faster a reaction takes place if the temperature is increased by $10°C$. For enzymatic reactions, Q_{10} values between 2 and 4 are characteristic; physical reactions, on the other hand, exhibit a less pronounced temperature dependence, with Q_{10} values between 1.1 and 1.4.

For the experiment conducted in the lab exercise, calculate the Q_{10} value for the range between the two extreme temperatures measured when decreasing *or* increasing the water temperature. If this range, as defined by the lower temperature t_1 and the upper temperature t_2, is different from $10°C$, the Q_{10} value for the corresponding frequency values f_1 and f_2 can be calculated by employing the following formula:

$$\log Q_{10} = [10/(t_2 - t_1)]\log(f_2/f_1).$$

Correlation and Regression Analysis

Using the frequency values obtained from your fish after increasing or decreasing the water temperature, calculate the Spearman rank correlation coefficient and perform a regression analysis. Either of these two types of analysis examines the relationship between temperature and EOD frequency, but the regression analysis also gives you information about the extent to which the EOD frequency changes when the temperature increases or decreases. In general terms, a regression analysis provides you with a linear equation of the form $y = mx + b$. Geometrically, this equation defines a straight line, with m representing the slope of the line. In our example, m indicates how much the EOD frequency changes when the water temperature is altered by $1°C$. Both correlation analysis and regression analysis are available in many spreadsheet programs, such as Excel (Microsoft Corporation) and in common statistics software packages, such as SPSS (SPSS Inc.).

In our analysis, you will be testing the hypothesis that the EOD frequency of your fish depends on the temperature of the water. What is the null hypothesis?

Note that after performing such an analysis on only one fish, you can draw conclusions only about this particular individual. To be able to generalize about the relationship between temperature and EOD frequency in the whole species, you would need to do statistical analysis on multiple fish.

QUESTIONS FOR DISCUSSION

In both *Eigenmannia* and *Apteronotus*, the EOD is sexually dimorphic. Male *Apteronotus leptorhynchus* typically discharge, within the species-specific range (approximately 500–1000 hertz at a temperature of 27°C), at higher frequencies than females. In *Eigenmannia* sp., this **sexual dimorphism** is reversed. In this species, females discharge, within the species-specific range (240–600 hertz at a temperature of 25°C), at higher frequencies than males. It is likely that this information is used for recognition of potential sexual mates and for electrocommunication in each genus. As demonstrated in the lab exercise, however, a "male" EOD can be transformed into a "female" EOD (and vice versa) by changing the ambient temperature. How, then, can the signals produced by the sender still be unambiguous for a potential receiver? (Similar **sender–receiver matching** problems occur in other communication systems, such as the acoustic communication of crickets; see Doherty 1985.) Which behavioral and physiological mechanisms may compensate for the temperature dependence of the EOD frequency?

ACKNOWLEDGMENTS

Part of the work described in this laboratory exercise has been funded by grant 34/S09223 from the Biotechnology and Biological Sciences Research Council to Günther K. H. Zupanc. We thank the editors, Bonnie J. Ploger and Ken Yasukawa, as well as the two anonymous reviewers, for their comments on earlier versions of the manuscript.

LITERATURE CITED

Bass, A. H. 1986. Electric organs revisited: evolution of a vertebrate communication and orientation organ. In: *Electroreception* (ed. T. H. Bullock and W. Heiligenberg), pp. 13–70. New York: Wiley.

Boudinot, M. 1970. The effect of decreasing and increasing temperature on the frequency of the electric organ discharge in *Eigenmannia* sp. *Comparative Biochemistry and Physiology*, **37**, 601–603.

Bullock, T. H. & Heiligenberg, W. 1986. *Electroreception.* New York: Wiley.

Coates, C. W., Altamirano, M. & Grundfest, H. 1954. Activity in electrogenic organs of knifefishes. *Science,* **120,** 845–846.

Doherty, J. A. 1985. Temperature coupling and "trade-off" phenomena in the acoustic communication system of the cricket, *Gryllus bimaculatus* De Geer (Gryllidae). *Journal of Experimental Biology,* **114,** 17–35.

Dye, J. C. & Meyer, J. H. 1986. Central control of the electric organ discharge in weakly electric fish. In: *Electroreception* (ed. T. H. Bullock and W. Heiligenberg), pp. 71–102. New York: Wiley.

Enger, P. S. & Szabo, T. 1968. Effect of temperature on the discharge rates of the electric organ of some gymnotids. *Comparative Biochemistry and Physiology,* **27,** 625–627.

Feng, A. S. 1976. The effect of temperature on a social behavior of weakly electric fish *Eigenmannia virescens. Comparative Biochemistry and Physiology A,* **55,** 99–102.

Heiligenberg, W. 1977. *Principles of Electrolocation and Jamming Avoidance in Electric Fish: A Neuroethological Approach.* Berlin/Heidelberg/New York: Springer-Verlag.

Heiligenberg, W. 1991. *Neural Nets in Electric Fish.* Cambridge, MA: M.I.T. Press.

Hopkins, C. D. 1988. Neuroethology of electric communication. *Annual Review of Neuroscience,* **11,** 497–535.

Hopkins, C. D. 1999. Design features for electric communication. *Journal of Experimental Biology,* **202,** 1217–1228.

Knudsen, E. I. 1975. Spatial aspects of the electric fields generated by weakly electric fish. *Journal of Comparative Physiology A,* **99,** 103–118.

Lissmann, H. W. 1951. Continuous electrical signals from the tail of a fish, *Gymnarchus niloticus* Cuv. *Nature,* **167,** 201–202.

Zakon, H. H. 1986. The electroreceptive periphery. In: *Electroreception* (ed. T. H. Bullock and W. Heiligenberg), pp. 103–156. New York: Wiley.

Zupanc, G. K. H. 1988a. *Fish and Their Behavior.* 2nd ed. Melle: Tetra-Press.

Zupanc, G. K. H. 1988b. Das Experiment: Temperatureinflüsse auf das Verhalten von schwachelektrischen Fischen. *Biologie in unserer Zeit,* **18,** 25–30.

Zupanc, G. K. H. 1988c. Temperatur und Verhalten: Physiologische Versuche an schwachelektrischen Fischen. In: *Praktische Verhaltensbiologie* (ed. G. K. H. Zupanc), pp. 166–181. Berlin: Paul Parey.

Zupanc, G. K. H. & Banks, J. R. 1998. Electric fish: animals with a sixth sense. *Biological Sciences Review,* **11,** 23–27.

Zupanc, G. K. H. & Maler, L. 1997. Neuronal control of behavioral plasticity: the prepacemaker nucleus of weakly electric gymnotiform fish. *Journal of Comparative Physiology A,* **180,** 99–111.

CHAPTER 8

Observing and Analyzing Human Nonverbal Communication

PENNY L. BERNSTEIN
Biological Sciences, Kent State University Stark Campus, 6000 Frank Avenue, Canton, OH 44720, USA

INTRODUCTION

When people think about human communication, they usually think about talking. Vocalizations are so pervasive in our everyday life that we tend to overlook the ongoing silent communication signals that surround us. Yet these **nonverbal signals** are crucial to our daily lives; conscious of them or not, we are quite good at giving them and responding to them appropriately. These signals often support and reinforce our talking, but they can also act as independent sources of communication. **Smiles** comprise a set of nonverbal **graded signals**, running through a continuous set of changes from a simple closed-mouth smile all the way to a full, open-mouthed grin (Figs. 8.1 and 8.2; and for overviews see Smith 1977; Ekman 1992; Ekman & Friesen 1982; Ekman et al. 1988; Fridlund 1994; Russell 1994; for discussion of the evolution of smiles see van Hooff 1972; Chevalier-Skolnikoff 1982; Preuschoft 1992; Preuschoft & Preuschoft 1994; Preuschoft & van Hooff 1997). People use this set of signals abundantly every day, in many different situations. And recipients respond in what seem to be

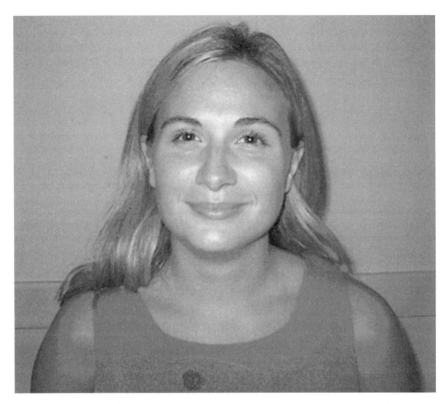

Figure 8.1 **Closed-mouth smile. Lips are pulled back, mouth is closed, no teeth are showing. This posed photo lacks the normal social context.**

organized, meaningful ways. This exercise is designed to help you explore human nonverbal communication by observing smiles in a variety of contexts and by using your observations to analyze the role these signals may play in human interaction.

The emphasis in this exercise will be on developing a **message analysis** of the signals. This approach, best exemplified by the work of Smith (1977, and as reviewed in 1997, 1998), attempts to assess the kinds of information made available by signals (messages) that are then used by recipients, in context, as they formulate their responses (meaning). Alternative approaches have emphasized ecological and coevolutionary aspects of signaling (e.g., Dawkins & Krebs 1978; Krebs & Dawkins 1984; Ryan & Rand 1993; Ryan & Wilczynski 1988; Endler 1993; Zahavi & Zahavi 1997), structural/ motivational rules (Morton 1982; Morton & Page 1992), management hypotheses (Owings & Morton 1998), and economic or game theory approaches, which focus on the evolutionary payoffs of signal interactions that could be predicted from consideration of the benefits and costs accruing to both signaler and receiver (Bradbury & Vehrencamp 1998, 2000).

Figure 8.2 **Open-mouth smile. Lips are pulled back and up, mouth is open, teeth are showing. In a full grin, there would be space between upper and lower teeth. This posed photo lacks the normal social context.**

These approaches are important in their own right, and you might choose to explore any of these communication perspectives further or to compare them in ongoing independent or small-group studies. Our objective in this exercise is to practice observing and to couch our observations in an inter-actional framework, using information as our organizing factor—that is, a message analysis. You will be working on a whole-class project and discussion that involve smiling. If time permits, this will be followed by small-group projects in which you will develop your own experiments involving other nonverbal signals.

MATERIALS

You will be observing humans as they interact in natural settings. You will need paper and pencil to write down your observations. Your instructor may provide data sheets, or you may be free to take your own notes. You will be relying on your ability to focus and observe and to take detailed notes.

PROCEDURE

The procedures that follow are for a whole-class project on smiling. After you complete this assignment, your instructor may have you work in small groups to design your own project to examine other signals, such as tongue-showing or mirroring and synchrony.

WHOLE-CLASS EXPERIMENT

Smile Signal Form

Smiles form a graded set. In all smiles, the lips are pulled back and somewhat compressed. They may be pulled back very little or a lot; they may also be raised toward the ears. The mouth may or may not be opened, and if open, it may be open slightly or open widely (with a full range of variation in between). Teeth may or may not be visible. Generally, the set runs from closed-mouth smile (lips stretched somewhat but not parted, no teeth showing) to a wide-open grin (lips stretched back and up, lips open, teeth showing), with variations between the two extremes, including the closed grin and the half-smile. To simplify data collection, you will concentrate on observing the obvious extremes, the closed-mouth smile and the wide-open grin (Figs. 8.1 and 8.2).

Observation Procedure

Data for this exercise are gathered outside of the classroom, either during class time or on your own time, with each student working on his or her own or in small groups, depending on the instructor's directions. You can look for smiles in any situation, although some situations are better than others (because of the way the signals work and the information they provide), and hints about good situations might be given or discussed in class before you begin observing.

You will be observing 3, 5, or 10 smiles, each from a different individual, as directed by your instructor. These will not be enough cases for you to develop a message hypothesis by yourself. However, cases will be pooled and discussed in class, and this should provide enough data for patterns to begin to emerge in signal use and in the messages these signals may provide. Your instructor may direct you to have all of your cases be of the same kind of smile (such as all closed-mouth smiles or all wide-open grins) or a mixed set. But *the cases observed by the class as a whole should be mixed* so that the class can determine whether and how patterns vary between the two ends of the continuum. To avoid sampling problems, you should observe a different individual for each smile (see the "Data Recording and Analysis" section).

In this type of observational procedure, it is critical that you be an *unobtrusive, uninvolved observer.* You will need to find a place to observe where you will not be playing any role in the interactions as they unfold and will not be readily noticed by the people you are observing.

Observations involve describing *behaviors* and the *contexts* in which they occur. Be sure to write down observations, not interpretations—that is, describe rather than infer. (For example, if you think someone was happy, *describe* the behavior and contextual aspects that you *observed* that led you to this interpretation). Further, it is critical to focus on the behaviors of the communicator before, during, and after the signal is given; the responses that occur; and the context. Here is an example: "Two people are talking in low voices in a cafeteria. One then takes the other person's hand, gazes at the person, kisses the person, gives a wide-open grin (lips stretched back and up, mouth open, teeth showing) while still holding the person's hand, and initiates a long conversation while continuing to hold hands and look directly at the other person; the other person listens, talks, and smiles as well." This description illustrates the sort of detail necessary for analyzing a signal. Some sample descriptions are provided in the sample data sheet given in Table 8.1. This sort of description might sound complicated, but you will find it is easier to do it than to read about it. The observing process is described in greater detail in the next section.

Data Sheet

When you observe a smile, it is critical that you write down the following pieces of data. (1) *Form.* Record what the signal looked like: lips closed or open, pulled back or not and generally how far, teeth showing or not, etc. If you define your terms ahead of time, you might be able to write down specific terms instead of a general description (such as closed or open). (2) *Communicator behavior.* Record what the communicator was doing before, during, and after the signal. Be as specific and descriptive as possible of the smile, of when it occurred, and of the variety of other behaviors with it, such as body, head, and eye orientation. It is especially important to keep track of when the smile actually occurred and of the behavior after the smile ends. Try to keep in mind the following kinds of questions: Was there an interaction before, during, or after the smile? Did the communicator initiate interaction as they smiled or soon after, did someone else, or did interaction end? Was he or she looking at a person or away from the person before and during the smile? Was the body facing the person or away? Did he or she continue to interact after the smile or leave the interaction? (3) *Context.* Record what else was going on besides the signal and communicator behavior. What were the responses? Where did the interaction take place? What else was happening? Who was involved? Who else was there? Did the people know each other or not? What were the relationships? the ages? the sexes? See the sample description quoted

.... 6.1 Sample data sheet.

Form	Behavior			Context	Duration
	Before	*During*	*After*		
Closed	Looks at girlfriend She is talking	Nods up & down, Closed smile as looks at girlfriend (lips back, not open, small)	Looks down Looks at hands as fiddles with napkin She stops talking, they sit in silence	Boy and girlfriend She is talking loudly complaining about another friend	1 minute
Closed	Hurrying toward class Looks at approaching person	Gives smile (lips back, not open), nods head and continues walk at same rapid pace	Continues down hall and into room	In hall just before next class period	10 seconds
Closed	Kid looks at mother Asks what is for dinner	Gives smile (small, closed) as mom tells him they will have liver for dinner	Thanks mother and turns and leaves room	My brother, dinner time, is hungry, does not like liver	30 seconds
Open	Looks at girlfriend She is talking	Nods up and down (lips back, open, teeth show)	Talks animatedly as girlfiend listens Continue for over 5 minutes and observer moves on	Discussing plans for dinner date	>6 minutes
Open	Approaching friend in hallway	Stops and begins talking to friend, friend stops, Engage in conversation as smile continues (lips back far, open, teeth show)	Says goodbye and moves on down hall	Two girl-friends meet in hallway not at a class time	4 minutes
Open	Two friends sitting together in conversation	Smile while talking, and then as listen to response, back and forth (lips back far, grin, open, teeth)	Continues until observer leaves	Gestures are slow conversation not rapid Do not seem in hurry	>6 minutes

in the "Observation Procedure" section for the kind of detail you should be trying to include. (4) *Duration.* You may also wish to time the interactions that occur, if possible. See Table 8.1 for examples.

In-Class Discussion

In class you will discuss the variety of smile cases collected by all students, looking carefully at the communicator behaviors before, during, and after the signal is given and at the contexts. As a class, you will uncover a pattern of usage for this signal—that is, a pattern of communicator behaviors that are associated with the signal. In message/meaning analysis, this pattern of communicator behavior will reflect what information a recipient could predict from the signal and is referred to as the *message* (Smith 1977, 1997, 1998). Message theory holds that messages remain the same for a given signal across many contexts; they are consistent sources of information. The recipient responses, in context, are considered the *meanings*. Because contexts and responses vary, meanings may change in different situations. You will be focusing on the message. Further detail can be found in the "Data Analysis" section.

SMALL-GROUP EXPERIMENT

Signal Forms

You will first need to define the signals you are going to study. Tongue-showing and mirroring and synchrony are recommended because other observers have already defined and studied them and, in the case of tongue-showing, have tested the message hypothesis (for tongue-showing see Smith et al. 1974; Smith 1977; Dolgin & Sabini 1982; for mirroring and synchrony see Condon & Ogston 1967; Condon 1976; Kendon 1970; see Weitz 1979 for an overview), which will enable you to compare your findings with a known set. Other signals already researched include spacing behaviors (proxemics; see Hall 1966, 1968 and in Weitz 1979 for early work) and hand gestures (kinesics; see Birdwhistell 1970 and in Weitz 1979 for his pioneering studies).

Tongue-Showing (TS)

In the most common form of this display, an individual extends the tip of the tongue out between closed lips (the tongue tip shows) and holds it in that position briefly (Fig. 8.3). Both the actual position of the tongue and how much of it shows are quite variable. However, the individual does *not* lick the lips, wiggle the tongue, or move the tongue in and out (these are other sorts of tongue behaviors). The tip may be held at the center of the lips, or at either side of the mouth, at the corners. Variations include keeping

Figure 8.3 **Tongue-showing. Note that this person is trying to hold the phone with her shoulder while stretching at an awkward angle to reach something. A social interruption at this point might disrupt the delicate balance she is trying to maintain.**

the tongue inside and pushed up against the inside of the cheek (balled up, so it can be noticed but is hidden) and the special case of sticking out the tongue (the "nyah, nyah" position). Although you may think tongue-showing is a rare signal, it is much more common than people realize. It occurs most commonly in situations in which individuals are concentrating and in which an interruption could be a problem (such as when lifting a heavy object). But it occurs in many other situations as well.

Mirroring and Synchrony (M&S)

William Condon and Adam Kendon pioneered research into these behaviors (see Weitz 1979 for classic papers). These behaviors apparently have to do with social agreement and are considered part of "formalized interactions" (Smith 1977). They are easiest to observe during conversations. If individuals are in social agreement, they will usually mirror one another's behavior and synchronize their movements. For example, if one person picks up a cup and drinks, the other person will, too, within a few seconds; if one crosses the arms or legs, the other will, too, within a few seconds;

Figure 8.4 **Mirroring and synchrony. Note that both individuals are leaning forward in their chairs at about the same angle, both have their arms on the table and at similar angles, and both have their heads at similar angles. They are engaged in an ongoing conversation. They know each other well and are concluding a deal in which the bookstore manager will buy items from the salesman for the campus bookstore. They illustrate social agreement.**

if one leans forward, so will the other, etc. (Fig. 8.4). Even at a distance, we can often tell whether two or more people are together just by noticing whether they look alike (mirror: arms and/or legs are crossed in similar fashion, for example) and act alike (synchronize: one moves, and the other soon does, too). When individuals are in social disagreement, they often show it by failing to mirror or synchronize (Fig. 8.5). Again, this can often be spotted at a distance, and it helps us know, for example, whether people in a crowd are together or are simply standing near one another without any social relationship.

Observation Procedure

Observation procedures are the same as those you used for the previous whole-class activity. However, you will need to gather more cases, enough for you to analyze without the rest of the class. You will need to look at as many different situations as possible, sample the range of behaviors and contexts, and gather a number of cases; 30–50 would be better than 10–20.

Also, you will be generating your own hypotheses and designing specific methods to test them, which might vary to some extent from those used for smiling. For example, given the foregoing information,

Figure 8.5 **Off-mirroring and synchrony. The conversation has ended. The man is packing up his folder, and the store manager is looking toward the bookstore to see whether the workers are still busy. Note that their body postures, head angles, and arms and hand gestures are no longer mirroring or in synchrony. They are no longer in social agreement.**

you might predict that mirroring and synchrony behaviors would be different between friends in a class vs between strangers. You might test this by stating explicit independent and dependent variables. For example, you might examine "slouching in the seat during lecture" as your dependent variable with respect to your independent variable, friends vs nonfriends or strangers. You would predict that friends would be slouched at the same time more often than nonfriends, or that if a person began to slouch, then her or his friend would also slouch, within a few seconds, but a nonfriend would not. Or you might predict that a class of students that was very attentive to a professor's lecture would all mirror one another in behavior and in timing of movements (and perhaps mirror the professor as well), whereas a class that was not interested in a lecture would show a less uniform, more haphazard set of body postures and less synchrony in the timing of movements.

Data Sheet

If you continue to pursue a message analysis, your data sheet should be the same as it was for the smiling investigation. It is important to describe the form of the display; the behavior of the communicator before, during, and after the signal; and the context in which the signal interaction takes place, including the responses of recipients. You may wish to time certain behaviors, on the basis of your hypothesis. Your descriptions will include information about whether or not interaction occurs, what responses occur, etc. for later analysis.

HYPOTHESES AND PREDICTIONS

Whole-Class Experiment
Smile

The biological hypothesis that you will be testing is that smiles function within friendly interactions. To test this hypothesis, we can examine its predictions, such as the prediction that someone who smiles is likely to interact, and the interaction is more likely to be friendly in nature (not involving aggression) than unfriendly (involving aggression in some sense). Further, because smiles form a continuous, graded series, we can hypothesize that the size of the smile (how far back the lips are pulled and whether the mouth is open or closed) provides additional information about the likelihood of initiating interaction and continuing an interaction once it is begun. This second hypothesis predicts that although closed-mouth smiles correlate with a readiness to interact, an individual giving this small-sized signal is less likely than individuals giving wide-open grins to initiate further interaction or continue an interaction. Individuals with big smiles are more likely to engage in interaction and actually to initiate further interaction and continue interaction.

Small-Group Experiment
Tongue-Showing, Mirroring and Synchrony

For these signals, you should form your own hypotheses, on the basis of the information given here, readings, and pilot observations. *Hint*: Tongue-showing is often found in situations in which social interaction can be troublesome. Mirroring and synchrony signals work one way in friendly groups or situations and opposite ways in unfriendly groups or situations.

DATA RECORDING AND ANALYSIS

Whole-Class Experiment

The procedure for collecting and recording data is outlined in the "Observation Procedure" section. For analyzing data, the first step involves compiling data from the whole-class discussion. You will need to add new cases to your data sheets as they are discussed—that is, add information about form; behavior before, during, and after; and context (see Table 8.1 for sample descriptions). Once this is completed, you can begin to test the data.

Analysis of the whole-class data requires us to examine two alternative statistical hypotheses in each case (note that these are different from the biological hypotheses for the function of smiling). The first is traditionally called the statistical null hypothesis (H_0). The first such hypothesis in which we are interested is that smiling of any kind is not associated with interaction. If this null hypothesis is correct, smiling should be equally likely when there is no interaction and when an interaction is occurring. The alternative hypothesis (the alternative to the null hypothesis, H_a) is that smiling is associated with interaction. If this alternative is true, then smiling should be more common during interactions than when individuals are not interacting. A second null hypothesis states that smiles are not associated with any particular *kind* of interaction. If this null hypothesis is correct, smiling should be equally likely in both friendly and unfriendly interactions (we will not worry too much about formal definitions of these terms for this exercise but will use the everyday sense of these words). The alternative hypothesis is that smiling is associated with friendly interaction rather than unfriendly. If this alternative is true, then smiling should be more common in friendly interactions than in unfriendly ones, and, more specifically, the individual who smiles should initiate or participate in a friendly interaction rather than an unfriendly one. A third null hypothesis is that the two forms at the opposite extremes of the graded continuum of smiles, closed-mouth and open-mouth, are not associated with different probabilities of initiating and continuing interaction; the alternative hypothesis is that there are differences in interaction detail between the two forms.

To test these hypotheses, raw data (see Table 8.1 for sample cases) are examined for patterns of communicator behavior in context. For example, when an individual gave a closed smile, was she already in an interaction or not, did she initiate an interaction when she smiled, did she continue or maintain an interaction? The raw data can be recast into an analysis sheet (see the sample data analysis given in Table 8.2). Data could be grouped by smile type (independent variable) and examined for four main dependent variables: interaction (Was there an interaction context, and did interaction occur in conjunction with the smile or not?), type of interaction (friendly or unfriendly), initiation (Did the signaler initiate further interaction during or after the smile or not, or did someone else initiate further interaction or not?), and continuation of interaction (Did the signaler continue the interaction or not, or did another individual continue the interaction or not?). If the class timed durations of interaction, these results should also be recorded. Duration provides a more objective measure of continuation of interaction and thus offers another way to test the last hypothesis.

Means generated for smiles during interaction may be tested with a binomial or sign test (Siegel 1956; Siegel & Castellan 1988). However, you should be aware that your sample may be biased toward interaction situations. It is difficult to find smiles in noninteractive situations; by looking for smile situations, you may find yourself noticing only interaction situations because

Table 8.2 Sample data analysis.

Smile Type (Form)	Interaction (Yes/No)	Initiation (Signaler vs Other)	Continuation (Signaler vs Other)	Duration
closed1	yes	other	none	60 seconds
closed2	yes—greeting	sig	none	10 seconds
closed3	yes	sig—before	other	30 seconds
open1	yes	other then sig	both	>360 seconds
open2	yes	sig	both	240 seconds
open3	yes	unknown	both	>360 seconds

Interact

Smile Type	Yes	No
closed	6	0
open		

Binomial or sign test
$P < 0.05$

Friendly	Unfriendly
3	3

t test
$P < 0.05$

Initiator

Smile Type	Signaler	Other or Unknown
closed	2	1
open	2	1

Chi-square test of independence
$P < 0.05$

Continue Interaction

Smile Type	Yes	No
closed	1	2
open	3	0

Chi-square test of independence
$P < 0.05$

Duration of Interaction

Smile Type	Average
closed	33.3 seconds
open	320 seconds

Mann—Whitney U test
$P < 0.05$

that is where smiles are most likely to occur. Means generated by interaction type (friendly vs unfriendly) may be tested with a chi-square test for goodness of fit (Dunn 2001). A chi-square test for independence may be used to compare initiation of further interaction by smile type (Dunn 2001). The same test can be used to compare continuation of interaction by smile type. If duration data were collected, a Mann–Whitney U test may be employed to test differences in duration of social interaction following closed-mouth vs open-mouth smiles (Dunn 2001).

Small-Group Experiment

After reading the lab material, reviewing the references, making some test observations, and predicting what the message of tongue-showing or mirroring/synchrony might be, you will need to generate specific hypotheses (H_0 and H_a) to test with your data. Decide what statistical tests might be appropriate.

QUESTIONS FOR DISCUSSION

Whole-Class Experiment

Based on your data, what information (messages) do you think smiling provides?

Why might it be important for a communicator to provide interaction information with smiles?

Why might it be important for a signaler to be able to use a graded set of signals to provide more finely tuned information about interaction?

Was there any evidence to suggest that there might be a gender difference in frequency of smiling?

Did you find any evidence to suggest that smiles might *not* necessarily always be associated with friendly interaction?

What sorts of limits might there be to signaling using a graded set of signals?

Do smiles function primarily for the communicator, for the recipient or for both? Must both benefit in the same way, or to the same extent? At what point or under what conditions would you expect recipients to cease responding to this display?

What question would you ask next? Can you design an experiment to test your message hypothesis? Would you suggest the use of more technical equipment than direct observation and paper and pencil recording, and if so, why?

What role might deception play in communication using smiles? Did you detect any evidence of deception?

Small-Group Experiment

Based on your data, what information set (messages) do you think your signal provides?

Why might it be important for a communicator to provide the information that you found was associated with the signals you observed?

If you observed mirroring and synchrony, how was this signal set different from signals such as smiling? Can it be done alone or by one signaler to one receiver, or does it necessarily involve more than one individual?

Did the signal you examined function primarily for the communicator, for the recipient, or for both? Must both benefit in the same way, or to the same extent? At what point or under what conditions would you expect recipients to cease responding to this display?

What question would you ask next? Can you design a further experiment that would test your message hypothesis?

ACKNOWLEDGMENTS

I thank W. John Smith for training me in an inquiry teaching style before there was such a term, and W. John Smith, Kim Dolgin, Anna Lieblich, and Julia Chase for sharing with me their work on tongue-showing behavior. I also thank the many students, both majors and nonmajors, whose data collection and analyses over the years have demonstrated the consistency of the messages of smiles, tongue-showing, and mirroring and synchrony.

LITERATURE CITED

Birdwhistell, R. L. 1970. *Kinesics and Context.* Philadelphia: University of Pennsylvania Press.

Bradbury, J. W. & Vehrencamp, S. L. 1998. *Principles of Animal Communication.* Sunderland, MA: Sinauer.

Bradbury, J. W. & Vehrencamp, S. L. 2000. Economic models of animal communication. *Animal Behaviour,* **59,** 259–268.

Chevalier-Skolnikoff, S. 1982. A cognitive analysis of facial behavior in Old World monkeys, apes, and human beings. In: *Primate Communication* (ed. C. T. Snowdon, H. Brown, & M. R. Peterson), pp. 303–368. Cambridge: Cambridge University Press.

Condon, W. S. 1976. An analysis of behavioral organization. *Sign Language Studies,* **13,** 285–318.

Condon, W. S. & Ogston, W. D. 1967. A segmentation of behavior. *Journal of Psychiatric Research,* **5,** 221–235.

Dawkins, R. & Krebs, J. 1978. Animal signals: information or manipulation? In: *Behavioural Ecology* (ed. J. R. Krebs & N. B. Davies), pp. 282–309. Oxford: Blackwell Scientific.

Dolgin, K. & Sabini, J. 1982. Experimental manipulation of a human non-verbal display: the tongue-show affects an observer's willingness to interact. *Animal Behaviour*, **30**, 935–936.

Dunn, D. 2001. *Statistics and Data Analysis for the Behavioral Sciences.* New York: McGraw-Hill Higher Education.

Ekman, P. 1992. Facial expressions of emotion: an old controversy and new findings. *Philosophical Transactions of the Royal Society of London, B*, **335**, 63–70.

Ekman, P. & Friesen, W. V. 1982. Felt, false and miserable smiles. *Journal of Nonverbal Behavior* **6**(4), 238–252.

Ekman, P., Friesen, W. V. & O'Sullivan, M. 1988. Smiles while lying. *Journal of Personality and Social Psychology*, **54**, 414–420.

Endler, J. A. 1993. Some general comments on the evolution and design of animal communication systems. *Philosophical Transactions of the Royal Society of London, B*, **340**, 215–225.

Fridlund, A. 1994. *Human Facial Expressions: An Evolutionary Perspective.* New York: Academic Press.

Hall, E. T. 1966. *The Hidden Dimension.* New York: Doubleday.

Hall, E. T. 1968. Proxemics. *Current Anthropology*, **9**, 83–95, 106–108.

Hooff, J. A. R. A. M. van. 1972. A comparative approach to the phylogeny of laughter and smiling. In: *Noverbal Communication* (ed. R. A. Hinde), pp. 209–241. Cambridge: Cambridge University Press.

Kendon, A. 1970. Movement coordination in social interaction: some examples described. *Acta Psychologica*, **32**, 100–125.

Krebs, J. R. & Dawkins, R. 1984. Animal signals: mind-reading and manipulation. In: *Behavioural Ecology: An Evolutionary Approach* (ed. J. R. Krebs & N. B. Davies), pp. 380–402. Oxford: Blackwell Scientific.

Morton, E. S. 1982. Grading, discreteness, redundancy and motivational-structural rules. In: *Evolution and Ecology of Acoustic Communication in Birds* (ed. D. E. Kroodsma & E. H. Miller), pp. 183–212. New York: Academic Press.

Morton, E. S. & Page, J. 1992. *Animal Talk: Science and the Voices of Nature.* New York: Random House.

Owings, D. H. & Morton, E. S. 1998. *Animal Vocal Communication: A New Approach.* Cambridge: Cambridge University Press.

Preuschoft, S. 1992. "Laughter" and "smile" in Barbary macaques (*Macaca sylvanus*). *Ethology* **91**, 220–236.

Preuschoft, S. & Preuschoft, H. 1994. Primate nonverbal communication: our communicative heritage. In: *Origins of Semiosis: Sign Evolution in Nature and Culture* (ed. W. Noth), pp. 61–102. Berlin: Mouton de Gruyter.

Preuschoft, S. & van Hooff, J. A. R. A. M. 1997. The social function of "smile" and "laughter": variations across primate species and societies. In: *Nonverbal Communication: Where Nature Meets Culture* (ed. U. Segerstrale & P. Molnar), pp. 171–190. Mahwah, NJ: Lawrence Erlbaum.

Russell, J. A. 1994. Is there universal recognition of emotion from facial expression? A review of cross-cultural studies. *Psychological Bulletin*, **115**, 102–141.

Ryan, M. J. & Rand, A. S. 1993. Sexual selection and signal evolution: the ghost of biases past. *Proceedings of the Royal Society, London, B*, **340**, 187–195.

Ryan, M. J. & Wilczynski, W. 1988. Coevolution of sender and receiver: effect on local mate preference in cricket frogs. *Science*, **240**, 1786–1788.

Siegel, S. 1956. *Nonparametric Statistics for the Behavioral Sciences.* New York: McGraw-Hill.

Siegel, S. & Castellan, Jr., N. J. 1988. *Nonparametric Statistics for the Behavioral Sciences.* 2nd ed. New York: McGraw-Hill.

Smith, W. J. 1977. *The Behavior of Communicating.* Cambridge, MA: Harvard University Press.

Smith, W. J. 1997. The behavior of communicating, after twenty years. In: *Perspectives in Ethology* (ed. D. H. Owings, M. D. Beecher & N. S. Thompson), **12**, pp. 7–53. New York: Plenum Press.

Smith, W. J. 1998. Cognitive implications of an information-sharing model of animal communication. In: *Animal Cognition in Nature: The Convergence of Psychology and Biology in Laboratory and Field* (ed. R. P. Balda, I. M. Pepperberg & A. C. Kamil), pp. 227–243. New York: Academic Press.

Smith, W. J., Chase, J. & Lieblich, A. K. 1974. Tongue showing: a facial display of humans and other primate species. *Semiotica*, **11**, 201–246.

Weitz, S. 1979. *Nonverbal Communication.* 2nd ed. Oxford: Oxford University Press.

Zahavi, A. & Zahavi, A. 1997. *The Handicap Principle.* Oxford: Oxford University Press.

CHAPTER 9

Foraging Behavior of Ants, or Picnics: An Ant's-Eye View

SYLVIA L. HALKIN
Department of Biological Sciences, Central Connecticut State University,
New Britain, CT 06050-4010, USA

INTRODUCTION

The foraging decisions of many species seem to be based on maximizing energy intake per unit of time (Cuthill & Houston 1997). However, acquisition of essential nutrients (Dadd 1985; Belovsky 1978) and/or consumption of a varied diet (Krebs & Avery 1984) may also be important, as may risk of being preyed on while feeding (Gotceitas 1990). Competitive interactions with other foragers may hinder access to preferred foods, or cooperative communication between foragers may facilitate access (Hölldobler & Wilson 1990). In this lab we will be exploring the foraging behavior of ants. Do ants simply maximize caloric intake, or do they also display preferences for protein, carbohydrates, or fat? How do the water content of food and its texture affect food choices? Do ants find a big piece of food more quickly if a trail of smaller pieces leads to the big one? Will they travel farther to obtain more preferred foods? How does one ant lead another to a source of food? Do food preferences vary with the species of ant, or between colonies of the same species? To try to answer some of these questions, we will find nest entrances of local ants, set up food sources nearby, and monitor feeding visits by ants.

There are hundreds of different species of ants in the United States and over 8800 ant species in the world (Hölldobler & Wilson 1990). In a typical ant colony, most of the individuals are sterile female workers; often there is just one fertile queen, whose primary activity is laying eggs. The workers collect food for the colony; feed their mother, the queen, and feed and care for their young siblings; enlarge the colony as needed; and defend the colony. Ant colonies generally contain tens to thousands of workers (depending on the species), so the ants that you see out foraging during this exercise will be only a very small proportion of the workers living in their colonies. Once a year, colonies that have reached sufficient size produce large numbers of winged reproductive males and females. Their emergence from the colony is timed to coincide with the emergence of reproductives from other colonies of the same species (Hölldobler & Wilson 1990). Males and females find one another and mate. Males die shortly after mating, but the female loses her wings and goes to excavate a chamber and lay eggs that will become the workers for a new colony, with her as queen.

Ants, like all insects with complete metamorphosis, develop from eggs into successively larger stages of larvae. The largest larval stage becomes a pupa, from which the adult emerges at the size it will have for the rest of its life. In some species of ants, workers perform different tasks at different ages. Workers may also be anatomically specialized for different roles. For example, there may be workers of different sizes that collect different sizes of food items; or there may be workers, referred to as soldiers, that have very large heads and mouthparts and defend the colony entrance. Ant workers live from weeks to years (depending on the species); ant queens that have successfully established colonies may live for decades (Hölldobler & Wilson 1990)! When the queen dies, so does the entire colony, because no new workers can be born. This type of social system, in which colonies consist of many sterile female workers caring for the offspring of one or a few fertile queens, is typical of many bees and wasps as well as of ants. Its evolution is believed to have been favored by the genetics of sex determination in these species, which causes female workers to share more genes with the sisters that they help to raise than they would share with their own offspring (Hamilton 1964).

Ants, like people, use protein to build muscle and other structural body parts and also to construct enzymes to catalyze the chemical reactions in their cells. The quickly growing and developing larvae of insects require more protein than adults (Dadd 1985). Fat and carbohydrates provide energy. Fat is the most space-efficient way to store energy and therefore by weight is the richest energy source to consume; however, the energy in carbohydrates is more readily available. Water plays a key role in many metabolic processes, but excess water adds weight to food. Salt functions in helping to maintain water balance, and vitamins serve as coenzymes. On the basis of this information, what predictions would you make about food preferences of ants?

Workers in most species of ants locate food either on solitary searches or by following a pheromone trail laid down by other colony members (Hölldobler & Wilson 1990). Trails are laid to food sources too large for an individual worker to ingest or carry by herself. Food may be scarce, and competition for food may be severe, so there is a real benefit to recruiting help in quickly transporting food into the colony. Trail-laying ants drag the tip of the gaster along the ground; trail-following ants follow the trail using their antennae.

MATERIALS

Data sheets (one of each kind, for each student, plus two extra data sheets on nonforaging ants for each group) and ant key sheets (Table 9.2) and nutrient content lists (Table 9.1). Small quantities of common picnic foods. For each group of 2–4 students: 4 paper or plastic cups (or 8-ounce yogurt containers); a serrated knife (plastic picnic knives work fine); 4 paper towels; a small ruler marked in millimeters; 2 clipboards; 2 pencils with erasers; a stopwatch or wristwatch with a second hand; a magnifying glass or hand lens (either 5× or 10× magnification works well); 2 small glass or plastic vials with screw tops, snap-on caps, or corks; a small (9–12 centimeters in diameter) finger bowl and crushed ice to fill it; a blunt probe (or pair of entomological forceps); and a dissecting microscope.

PROCEDURE

In the Lab Room

Discuss with your classmates what factors may be important to ants in choosing among potential food sources. Consider not only characteristics of foods themselves but also where foods are located relative to the ant colony and to other food sources. After this class discussion, work with your lab group (2–4 students, as specified by your instructor) to design a controlled experiment or set of experiments. For each experiment, decide which factor or factors you would like to test and what foods will best enable you to test these factors. Use the data sheets and Table 9.1 (or package labels of foods provided in the lab) to help plan your experiment. Plan to use four food items, some (or all) of which may be the same, depending on the experiment that you design. Plan what size and shape to make your pieces of food, and plan where you will place them (note that in the time allotted, ants may

Table 9.1 **Nutrient contents of some common picnic foods.**[a]

Food	Protein (%)	Fat (%)	Carbohydrates (%)	Water (%)	Sodium (mg/g)
			Percentage by Weight of		
American cheese	21%	32%	<4%	39%	14.5
Cheddar cheese	25%	32%	<4%	37%	6.3
Hard–boiled egg	12%	10%	2%	75%	1.4
Butter, salted	<1%	81%	<1%	16%	8.3
Apple (with peel)	<1%	<1%	15%	84%	0.007
Orange	<1%	<1%	11%	87%	trace
Banana	1%	<1%	23%	74%	0.009
Grapes	<1%	<1%	18%	81%	0.02
White bread	8%	4%	48%	37%	5.2
Beef hot dogs	12%	28%	2%	55%	11.2
Turkey hot dogs	14%	18%	2%	63%	12.2
Ham lunch meat	18%	11%	4%	65%	13.1
Peanut butter	25%	50%	22%	1%	4.7
Jelly (e.g., grape)	<1%	<1%	72%	28%	0.2
Potato chips	7%	36%	54%	2%	4.8
Corn chips	7%	32%	57%	1%	6.1
Marshmallows	4%	<4%	82%	16%	0.9
Milk chocolate	7%	32%	61%	1%	0.8
Romaine lettuce	2%	<1%	2%	95%	0.07
Iceberg lettuce	1%	<1%	2%	96%	0.09
Carrots	1%	<1%	10%	88%	0.3
Dill pickles	<1%	<1%	5%	92%	14.3

[a]Calculated from values in Whitney & Rolfes (1993). For prepared foods purchased for this lab, see package labels for information. Sodium reflects salt content.

have trouble finding foods placed more than 10 centimeters from the nest entrance). You will want to be sure that there is enough of each food item to last 30 minutes at a hungry colony, but not so much that ants can easily hide under it and be difficult to count and observe. Thus food items should be at least $1/3$ cm^3 and no more than 1–2 cm^3.

Try to control as many as possible of the factors that you don't mean to test. To do this, plan to make your food items, and the way that you present them, as nearly identical as possible in every way except for the factor or factors that you are trying to test. Note that if you are using different types of food, these are extremely likely to differ in more than one way. Make a list of all of the ways in which they differ, and realize that all of these factors may affect the outcome of your experiment.

You will probably have time to complete one or two experiments in lab today. Once you have designed the first experiment, you can plan to repeat it with or without modifications. (For example, placement of food items can be rotated from the initial locations to test for unanticipated position effects, effects of lingering food or ant trail odors, and the effect of previous forager experience; more-preferred foods may be placed farther away than less-preferred foods to see whether preferences switch; trails of "crumbs" can lead to a bigger piece of food to see whether ants discover it more quickly than a piece of food of the same type and size at the same distance without the trail of crumbs).

Check your plans with your instructor, or present them to the rest of the class for discussion, before you begin to prepare your food items. Be sure to explain how you have controlled for factors that you do not mean to test and how you will interpret results that could be explained by more than one factor. Note that you will make observations of nonforaging ants only if the ants that you watch do not eat any of the food that you put around their nest and you are consequently unable to collect foraging data.

Once your plans have been approved, prepare enough pieces of food for all the experiments that you will have time to conduct, plus a few extras for unanticipated disasters. Use the knives provided to cut off pieces of the foods that you would like to use, and carry each type of food back to your place in one of the cups or other containers provided; do any further preparation on a paper towel at your place, and then put your food pieces into the cup or container to carry them outside.

Take two clipboards for your group's data sheets (one for Forager Count Data Sheets, and the other for Forager Behavior Data Sheets), a ruler, a magnifying glass, four small vials, and something to write with. Be sure that you have as many copies of each foraging data sheet as the number of experiments you intend to conduct. Also bring several data sheets for Observations of Nonforaging Ants, for just in case you need them. (All the data sheets appear at the end of this chapter.) If nobody in your group has a watch that can measure seconds, borrow one or take a stopwatch if these are available.

Outside

Look in lawns and sidewalk cracks to find ant nest entrances. Look for ant mounds that have a few ants already wandering around them and that are at least 5 ant mound diameters from the closest other ant mound; this will reduce the confusion of competitive interactions between ants from different nests. If ants with painful stings live in your area, be sure to avoid them (ask your instructor for guidance in identifying these ants). Once you have found an ant mound with at least several ants outside the entrance, set up your experiment, record the start time and number of ants present at each food item at 0 minutes, and draw on the Forager Count Data Sheet the locations and sizes of each food item. Use the ruler to measure the distance of each food item from the nest entrance; to avoid disturbing the ants while making these measurements, hold the ruler slightly above the ant mound and ground. One group member can be responsible for timing the 3 minutes between count periods. All students should participate in counts so that counts for different food items occur as close together in time as possible. Between these counts, use the magnifying glass to observe the behavior of individual foragers, and fill in the table on the Forager Behavior Data Sheet. Work in pairs with one person observing and the other writing down the observations, and trade roles partway through the experiment so that everyone has a chance to make observations. When using the magnifying glass, be careful not to focus the sun's rays on your ant.

Between experiments at the same ant nest, be sure to clean up any leftover food from the previous experiment; if ants are on this food, gently knock them off (a finger flick will often suffice), or move the food with ants to 1–3 meters away from the nest entrance (the ants should have no trouble finding their way home but won't disrupt your next experiment).

If you have waited for 20 minutes and no ants have visited any of your food items, start to work on a data sheet on nonforaging ants (leave your food items in place until you have completed as many of these data sheets as you have time to complete).

After your last experiment at an ant nest, trap one or two live workers in the vials (by holding the opening of the vial against the ground in front of a running ant, with the end held up at about a 30° angle and with light allowed to shine in through the vial walls and end). Bring these ants back to the lab to identify under the dissecting microscope; at the end of the lab, you will release them back at their nest entrance.

Back in the Lab

When you return to the lab, fill a small finger bowl with crushed ice, and cover the ice with a moist (not wet) paper towel. Then release the ant(s) onto the paper towel. Within a few seconds, the ants should stop moving around, without curling up as they do when dead, which would make it very hard to observe them. Place the finger bowl under the dissecting microscope to observe the ants; they can be repositioned with gentle use of the blunt probe or entomological forceps, if necessary. Try to find all of the anatomical features shown in Figure 9.1 that are present on your ants. Use the dichotomous key to try to identify your ants; in a dichotomous key, you proceed through a series of pairs of choices until you arrive at a species identification. For example, if you have a little black ant, *Monomorium minimum*, the steps in identification are 1a → 1b → 3a → 3b (at each step, if the characteristics listed in the "a" description do not match your specimen, go on to the characteristics listed in the "b" alternative). The ants listed in the key are common across much of the United States, but it is certainly possible that you have a species that is not included. If you believe this is the case, say which of the species (or genera) listed your ants most closely resemble, and then describe how they differ from that species or genus. Check your identification with your instructor,

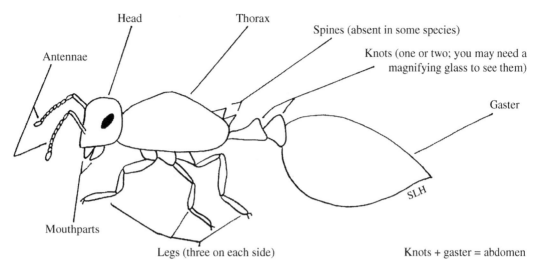

Figure 9.1 **Generalized worker ant anatomy.**

who may also want you to list the order of steps you followed in identifying your ant with the key. Your instructor may also provide an additional or alternative key for you to use.

Once you have identified your ants, return them to the vials in which you captured them and tabulate your data. On the Forager Count Data Sheet, this will involve summing the numbers of foragers counted at each food item, over all 11 counts (of course, many individual ants will have been counted multiple times, but this comparison should still give an accurate relative measure of food preference, given that multiple (or prolonged) visits by the same individual are weighted as strongly as single visits by different individuals). Fill out the top of the Forager Behavior Data Sheet.

Be sure to return your ants to their colonies at the end of the lab.

Table 9.2 **A dichotomous key to workers of ant species commonly found outdoors in New England.**

1.	a.	Single knot present	2
	b.	Double knot present	3
2.	a.	Color black, or black with a red thorax; thorax rounded; 8–14 mm long	**Carpenter Ant,** *Camponotus* **sp.**
	b.	Color light brown to dark brown; about 4 mm long; common at picnics!	**Cornfield Ant,** *Lasius* **sp.**
3.	a.	Color dark brown; two very short spines present before knots; fine sculptured lines parallel to long body axis, on head thorax; 3–4 mm long	**Pavement Ant,** *Tetramorium caespitum*
	b.	Color black; no spines present before knots; body shiny, without sculptured lines on head or thorax; about 2 mm long	**Little Black Ant,** *Monomorium minimum*

HYPOTHESES AND PREDICTIONS

In what ways do your food items vary? For each of these ways, whether it is a factor that you are testing or a factor that you do not mean to test but cannot fully control, predict whether that factor will add to or detract from a food item's preferability. For example, you may predict that foods with higher fat content will be preferred over foods with lower fat content, because they contain more energy per unit volume. For each factor, rank the four food items you plan to use in predicted order of preference. Then, taking all of these factors into account (you may give some more weight than others), make a prediction about the ants' order of preference for the food items.

DATA RECORDING AND ANALYSES

See the end of this chapter for the three data sheets. Note that the Forager Count Data Sheet should be completed in the field (except for the Total line) and that the summary information from this data sheet will be used to answer the questions at the top of the Forager Behavior Data Sheet. Between 3-minute counts, make more detailed observations of single ants, using the Forager Behavior Data Sheet. See Figure 9.1 for a drawing of an ant with names of different body parts.

Make a graph to show how the number of foragers changed over time at different food items. Indicate the number of foragers on the vertical axis (from 0 at the origin up to the maximum number of foragers observed in a single count at one food item) and the time of the count on the horizontal axis (0 minutes = time at which food was first placed, subsequent labels at 3-minute intervals through 30 minutes). Graph four separate lines, each tracking the forager count at a different food item. All four lines should start at the origin, and they may well overlap for part of their extent and/or cross one another; distinguish them by using different colors or patterns (such as lines that are continuous vs dotted vs dashed vs alternating dashes and dots).

Perform a chi-square one-sample test for goodness of fit to statistically compare the four food items in terms of the maximum number of foragers seen simultaneously feeding on that food item. The null hypothesis for this test is that each food item attracted the same maximum number of foragers (1/4 of the total obtained by adding together the maximum number of foragers for each of the four food items). The alternative hypothesis is that the maximum numbers of foragers were different for the different food items. If the maximum numbers of foragers seen at all four food items sum to <20 (making a chi-square test invalid), or if the maximum numbers of foragers are not very different for the different food items, conduct a chi-square one-sample test of goodness of fit to statistically compare each food item's summed ant counts from all 11 observation periods. The null hypothesis for this test is that each food item had the same total number of ants counted during the 11 observation periods (1/4 of the total of all the ants counted at all the food items). The alternative hypothesis is that the total numbers of ants counted during the 11 observation periods were different for the different food items.

QUESTIONS FOR DISCUSSION

You have a number of different measures that might be counted as indicators of food preference: the order in which different food items were discovered, the maximum number of foragers seen simultaneously at a food item, and a food item's sum of ant counts from each of the 11 observation periods. Can you think of other appropriate measures? Do all of these measures give the same rankings for the four food items that you used? If not, attempt to explain why not, and explain which ranking you believe best reflects the ants' food preferences. For your graph, attempt to explain any differences between the lines for the different food items. Do higher-ranked items have lines with trajectories different from those of lower-ranked items?

Which (if any) of your predicted rankings is the best match for your actual data? Which factor or factors do your data show are important in determining ant food preferences, and which can be ruled out? Are the differences in preference for your different food items statistically significant at the 0.05 α level?

Compare your data to those of other groups. Were the same factors preferred in food choices of all colonies of the same species of ants, or were there differences between colonies? Were the same factors favored by different species of ants? Explain how the food preferences you observed may help to fulfill the ants' nutritional needs at the time of year when you did the experiment.

If you made observations of nonforaging ants, what did these ants appear to be doing? Were their activities coordinated with those of other ants from the same colony? How did behavior vary between individuals? Did you notice any morphological differences between ants doing different things, and if so, did these morphological differences appear to be adaptive for the behavior you observed? Speculate about why some ants don't forage even when food is readily available near the colony entrance. Is the behavior that they are performing more important, or are they morphologically or experientially unsuited for foraging? If no ants ate any of your food, was the lack of foraging behavior restricted to the colony you observed, or did it include all colonies of your ant's species, or some or all colonies of all species? Did another group's ants eat the foods that were ignored at your colony? If so, were their ants of the same species as yours or of a different species?

ACKNOWLEDGMENTS

I thank Michelle Scott for instructions on how to handle and subdue ants for observation; Nathan Sanders, Bonnie Ploger, and Ken Yasukawa for their thoughtful reviews of this exercise; and Theresa Singer for assistance in planning how to demonstrate this lab in a new part of the country. Valerie Kruse and Mary Waters wrote the main body of the key to ant species, and Mary Salerno provided useful advice. The students in my classes at Central Connecticut State University, Carleton College, and Moravian College have contributed to the development of this lab by collecting data over the course of its many iterations and by showing me what works and what does not.

LITERATURE CITED

Belovsky, G. E. 1978. Diet optimization in a generalist herbivore: the moose. *Theoretical Population Biology,* **14**, 105–134.

Cuthill, I. C. & Houston, A. I. 1997. Managing time and energy. In: *Behavioural Ecology: An Evolutionary Approach* (ed. J. R. Krebs & N. B. Davies), pp. 97–120. Oxford: Blackwell Science.

Dadd, R. H. 1985. Nutrition: Organisms. In: *Comprehensive Insect Physiology Biochemistry and Pharmacology,* Vol. 4: *Regulation: Digestion, Nutrition, Excretion* (ed. G. A. Kerkut & L. I. Gilbert), pp. 313–390. Oxford: Pergamon Press.

Gotceitas, V. 1990. Foraging and predator avoidance: a test of a patch choice model with juvenile bluegill sunfish. *Oecologia,* **83**, 346–351.

Hamilton, W. D. 1964. The genetical evolution of social behaviour, I, II. *Journal of Theoretical Biology,* **7**, 1–52.

Hölldobler, B. & Wilson, E. O. 1990. *The Ants.* Cambridge, MA: Harvard University Press.

Krebs, J. R. & Avery, M. I. 1984. Chick growth and prey quality in the European bee-eater (*Merops apiaster*). *Oecologia,* **64**, 363–368.

Whitney, E. N. & Rolfes, S. R. 1993. *Understanding Nutrition.* Appendix H, pp. H-0–H-65. Minneapolis, MN: West.

ADDITIONAL SUGGESTED READING

Wilson, E. O. 1987. The little things that run the world. *Conservation Biology,* **1**, 344–346.

Wilson, E. O. 1990. Empire of the ants. *Discover,* **March 1990**, 44–50.

Wilson, E. O. 1991. Ants. *Wings,* **16**, 3–13.

FORAGER COUNT DATA SHEET

Observers: _____

Ant species (English and Latin names): _____

Show on the diagram below the placement, sizes, shapes, and types of food items you used. The "o" shows the location of the nest entrance. Record the actual measured distance from each food item to the nest entrance beside a line drawn from that food item to the nest entrance; be sure to include units of measurement.

o

Start time: _____. Use the table below to identify food items (on the lines provided) and to record the number of foragers at each food item for each observation time. If you need to move food items to count ants beneath them, describe on the back of this page any differences in how or whether different food items were moved.

Observation Time (Minutes)	Food Item 1:	Food Item 2:	Food Item 3:	Food Item 4:
0				
3				
6				
9				
12				
15				
18				
21				
24				
27				
30				
TOTAL				

FORAGER BEHAVIOR DATA SHEET

Observers: _____

Ant species (English and Latin names): _____

Which food item was discovered   1st? _____

2nd? _____

3rd: _____

4th? _____

Which food item had the most foragers at once? _____

Which food item had the smallest maximum number of foragers? _____

Ant number	At which food item was ant observed?	Which body part contacted food first?	How many seconds after this to first contact with mouthparts?	Did ant follow a trail to food (using antennae)?	Did ant make a trail to food (dragging gaster)?	Did ant make a trail away from food?
1						
2						
3						
4						
5						
6						
7						
8						
9						
10						
11						
12						
13						
14						
15						

OBSERVATIONS OF NONFORAGING ANTS

Observers:_____

Ant species (English and Latin names):_____

If ants are coming and going to and from the entrance to their nest but seem to be ignoring food, what do they seem to be doing? Possibilities include (but are not limited to) patrolling the area to guard the nest and removing material to enlarge the underground nest. In some species the excavated material is carefully constructed into a mound that functions in controlling air circulation, temperature, and humidity of the nest and may serve as living quarters for part of the colony.

In the center of the space below, draw a life-sized entrance to the nest, and draw at the same scale the paths taken by 10 different ants, noting locations at which they perform any behavior other than walking and standing still. If they are carrying anything, note what it is, and if they lay it down, note where. Describe any interactions with other ants; if these include physical contact, name the body parts that they use to contact one another. You will probably have time to complete at least two of these sets of 10 observations during a 30-minute period; use a separate data sheet for each 10 ants whose paths you draw. Any group member who is not drawing ant paths should assist by keeping track of the ant's path and behavior while her or his partner looks away to draw; each group member should take a turn at drawing the paths taken by several ants.

CHAPTER 10

Hummingbird Foraging Patterns: Experiments Using Artificial Flowers

ALASTAIR INMAN
Department of Biology, Knox College, Galesburg, IL 61401, USA

INTRODUCTION

Hummingbirds (order Apodiformes, family Trochilidae) are the smallest of all North American birds; an adult ruby-throated hummingbird weighs only about 3 grams (Johnsgard 1983). Their small size and high mass-specific metabolic rate mean that hummingbirds must forage as frequently as 8 times per hour, consuming half their weight in sugar each day (Stokes & Stokes 1989). This makes them an ideal species for studying foraging patterns.

Unlike most prey, which try to elude their predators, flowers stand to benefit when foraged upon by hummingbirds, because a visit by a hummingbird allows the flower to receive pollen from conspecifics and pass on its own pollen. Thus we might expect to see adaptations on the part of flower species to make them easy for hummingbirds to find. At the same time, finding flowers is clearly beneficial to the hummingbirds, so we should expect the birds to have evolved effective methods for finding flowers. This is a classic example of **coevolution**, in which two species evolve in response to each other.

In this lab, you will make fake flowers that enable you to easily monitor the time it takes for them to be visited by a hummingbird. In this way, you can investigate what kinds of flower attributes enhance detection and visits by hummingbirds.

MATERIALS

Each group will need red, yellow, and blue flagging tape, several microcentrifuge tubes, elastic bands, 1 micropipetter, 1 bottle of 20% sucrose solution (20 grams sucrose added to 80 milliliters distilled water), several wooden stakes (1.0–1.5 meters high), 1 rubber mallet, opaque plastic garbage bags, 1 watch, 1 clipboard, 1 data sheet, 1 compass, and 3 blank cards.

HYPOTHESIS AND PREDICTIONS

You will begin by running a whole-class experiment that will ask whether hummingbirds are more attracted to flowers of a particular color. Anecdotal stories of hummingbirds being attracted to red objects, such as fire extinguishers, are supported by a number of studies that find hummingbirds prefer red flowers over flowers of other colors (Melendez-Ackerman et al. 1997; Stiles 1976). However, the degree to which the preference for red in these studies was influenced by the birds' previous experience with the natural flowers that were used is unclear. As a class, you will use artificial flowers to test the prediction (of the biological hypothesis that hummingbirds prefer red flowers) that hummingbirds will forage from red flowers sooner than they will from either yellow flowers or blue flowers. The advantage of using artificial flowers is that we can be certain that they are novel flower types to the hummingbirds. You will test this prediction by placing single artificial flowers of three different colors (red, yellow, and blue) in the field and monitoring how long it takes until each flower is visited by a hummingbird.

PROCEDURE

As a class you will need to construct the fake flowers that will be used for the first experiment. In order for you to combine the results obtained by different groups, it is important that all flowers be constructed in the same way, so follow the instructions carefully, and compare your flowers with those made by other members of the class to make sure that they are not

Table 10.1 **Sample data sheet. All times are expressed in hours: minutes. Your data sheet should have the same column headings, except that visit times may differ. A "√" indicates full tube, and an "X" indicates an empty tube.**

Trio	Stake	Location Notes[a]	Color	Start Time	0:15	0:30	0:45	1:00	1:15	1:30	1:45	2:00
1	1		Blue	2:13	√	√	√	X	X	X	X	X
1	2		Red	2:14	√	√	√	√	√	X	X	X
1	3		Yellow	2:14	√	√	X	X	X	X	X	X
2	1		Red	2:25	X	X	X	X	X	X	X	X
2	2		Yellow	2:25	√	√	√	√	X	X	X	X
2	3		Blue	2:26	√	√	X	X	X	X	X	X

[a]Use this area to record notes about the location of each stake to help you find it again.

constructed differently. Your instructor will tell you how many flowers of each color each group will test. Each flower is made with a microcentrifuge tube and two pieces of flagging tape 8 centimeter's long. Begin by cutting enough 8-centimeter pieces of the appropriately colored flagging tape to make all the flowers your group will use. To make a flower, rip off the top of a microcentrifuge tube and then pierce the closed end of the tube through the center of one of the pieces of flagging tape, sliding the flagging tape up the tube until the tape sits snugly up around the tube's opening. Pierce the tube through a second piece of (same color) flagging tape, and once again push the flagging tape up the tube until the tape is snug against the opening of the tube, with this second piece of tape running perpendicular to the first (they should form a cross).

Once you have completed all your flowers, make a data sheet with column headings that are the same as those on the sample data sheet (Table 10.1). Next, proceed as a class to the field area where you will run the experiment. To ensure that groups do not place flowers too close to each other, your instructor will assign each group an area in which to place its flowers. Take all your equipment and proceed to the area you have been assigned.

Flowers will be placed in the field on "stems" in the form of wooden stakes, with one flower mounted on each stake. In placing your stakes around your area, you want to keep the stakes as far apart from each other as you can but, at the same time, to ensure that the stakes are placed in locations that are relatively similar in terms of the surrounding vegetation. To do this, place the stakes out in sets of three, using one flower of each the three colors in each trio of stakes. Find three sites with similar surrounding vegetation that are at least 30 meters apart from each other (and from all other fake flowers). Use the mallet to drive a wooden stake into the ground at each location. As you place each stake, note on your data sheet the location of the stake to make sure that you can find it again. If necessary, use your compass to determine the direction of prominent landmarks, and then estimate the distance to these landmarks.

Table 10.2 **Sample class data sheet of the latency until first visit for each flower. Latencies have been calculated from the data shown in Table 10.1 and are presented in minutes.**

Trio	Red Flower	Blue Flower	Yellow Flower
1	90	60	45
2	15	45	75
.	.	.	.
.	.	.	.
.	.	.	.

Once all three stakes in a set have been placed in the ground, randomly assign the three different flower colors to the three stakes. You can use any method to do this, but one simple technique is to write the three colors (red, yellow, and blue) on one side of three blank cards. Shuffle the cards, hold them upside down, and have a group member draw one card each time you arrive at a stake. Use an elastic band to attach a flower of the chosen color to the stake, 5 centimeters from the top. Be sure to attach the flower in such a way that the flagging-tape "petals" don't get folded down against the microcentrifuge tube or the stake. Once the flower has been set up, place a garbage bag over the stake in such a way that it completely covers the flower. Continue on to the other stakes of this trio, attaching flowers and covering the stakes with garbage bags.

Using the same procedure, continue setting up sets of three flowers until you have placed all your flowers in your area. Remember to cover each stake with a garbage bag as you set it up.

Once you have set up all of your flowers, you are ready to fill the flowers with sugar water and begin monitoring them for hummingbird visits. Return to each stake, remove the garbage bag, and fill the microcentrifuge tube with 20 microliters of sugar water. Insert the micropipetter all the way to the bottom of the tube when adding the sugar water so that it forms a single drop, clearly visible at the bottom of the tube. As you fill each tube, record the time in the appropriate space on your data sheet. Move among all of your stakes, filling each tube and recording the times on your data sheet.

To monitor visits by hummingbirds, you will check each flower at regular intervals. How frequently you check will depend on the number of flowers you have set up and on the distance between them. If possible, visit each flower at least once every 15 minutes. When a hummingbird visits a flower, it will drink the sugar water out of the tube, leaving it empty. To check each flower, look to see whether the sugar water is still present in the end of the tube. If it is, place a check mark in the appropriate space in your data table (as illustrated in Table 10.1). If the tube is empty, place an "X" in the appropriate space in your data table. Continue checking each flower at regular intervals for as long as you have been told to do so by your instructor. When the experiment is completed, gather up all your equipment and return to the lab.

DATA RECORDING AND ANALYSIS

Combining the data from all groups makes it possible to analyze the data statistically. Enter the latency until first visit for each flower onto the class's combined-data table. The latency until first visit is simply the time from the experiment start time until the first scan when the nectar had been taken (the earliest scan marked with an "X"). Table 10.2 provides an example of such a data table for the class experiment.

Begin the analysis of the class results by graphing the average latency until first visit for flowers of each of the three colors. Include error bars of 1 standard error.

Statistical tests enable us to determine whether observed differences between the flower colors in the time taken until flowers are visited are larger than would be expected by chance alone. The statistical null hypothesis (H_0) is that birds do not visit red flowers sooner than blue or yellow flowers; the alternative hypothesis (H_a) is that the birds visit the red flowers significantly sooner than the flowers of other colors. The data we have collected are not parametric (that is, their distribution does not approximate a bell-shaped curve), so we will have to analyze them using nonparametric statistics. An appropriate statistic for comparing the means of three groups where the data have been collected in matched samples is the Friedman test. Your data are "matched samples" because you placed one flower of each color in each of the areas where you set up a group of three stakes.

SMALL-GROUP EXPERIMENTS

This same basic technique with fake flowers can be used to ask many different questions about the factors that affect the foraging patterns of hummingbirds. As a group, brainstorm to come up with a number of testable hypotheses. Think about what kind of prediction you might make from each hypothesis. You might, for example, ask whether the height of a flower affects how quickly hummingbirds detect it, and whether flowers are more likely to be visited if they are part of a group or if they are on their own.

Choose one of your hypotheses and design an experiment to test it. Decide what the null and alternative hypotheses are for your experiment, as well as how you will analyze your data. Be sure to discuss your ideas and plans for your experiment with your instructor before beginning.

QUESTIONS FOR DISCUSSION

In running the class experiment, you placed all three stakes of one trio in the ground and then randomly determined which stake would hold the flower of each color. Why was this important? On the basis of the results of your experiments, what kind of selective pressures do you think hummingbird-pollinated flowers might be subject to as a result of the behavior of the birds. Why do you think the birds have evolved the foraging patterns you observed? Are these patterns adaptive for the birds? What evidence would you seek if you wanted to show that the foraging patterns of hummingbirds and the display patterns of the flowers they visit have actually evolved through coevolution?

ACKNOWLEDGMENTS

Thanks to Bonnie Ploger, Ken Yasukawa, Anne Houtman, and Harry Tiebout for helpful comments on an earlier version of this paper.

LITERATURE CITED

Johnsgard, P. A. 1983. *The Hummingbirds of North America.* Washington, DC: Smithsonian Institution Press.
Stokes, D. & Stokes, L. 1989. *The Hummingbird Book.* New York: Little, Brown.
Melendez-Ackerman, E. O., Campbell, D. R. & Waser, N. M. 1997. Hummingbird behavior and mechanisms of selection on flower color in *Ipomopsis. Ecology,* **78,** 2532–2541.
Stiles, G. F. 1976. Taste preferences, color preferences and flower choice in hummingbirds. *Condor,* **44,** 189–204.

ADDITIONAL SUGGESTED READING

Alcock J. 1998. *Animal Behavior.* 6th ed. Sunderland, MA: Sinauer, pp. 341–381.
Ricklefs, R. E. 1997. *The Economy of Nature.* 4th ed. New York: Freeman, pp. 474–495.

CHAPTER 11

Honey Bee Foraging Behavior

MONICA RAVERET RICHTER AND JASMIN M. KERAMATY
*Department of Biology, Skidmore College, 815 N. Broadway, Saratoga Springs,
NY 12866, USA*

INTRODUCTION

The intricate, highly ordered social organization (Seeley 1995) that characterizes colonies of the honey bee *Apis mellifera* (Hymenoptera: Apidae) has long fascinated biologists and nonbiologists alike (Wheeler 1923). Observations on these insects have provided important insights on topics ranging from the behavioral ecology of foraging (Seeley 1985) to the organization and evolution of social behavior (Wilson 1971; Michener 1974). Most residents of honey bee colonies are nonreproductive females that raise their mother's offspring, and many authors (Gould & Gould 1988; Seeley 1995) view not the individual bee but rather the integrated colony as the unit on which natural selection acts.

Communication makes possible the coordinated functioning of colonies. Chemical communication is well developed and is used in recognizing hivemates, in alarm, and in assembly, as well as in maintaining reproductive dominance of the queen (Wilson 1971; Michener 1974). Most impressive to the human observer is the dance language of honey bees, through which dancers alert hivemates to the presence and location of nectar and pollen sources and recruit them to forage on a particular resource (von Frisch 1967). The circular round dance indicates that food is located near the colony, and aspects of the dance, including the speed and rate of directional reversal of the dancing bee, correspond to the profitability of the resource (Waddington 1982; Raveret Richter & Waddington 1993). In the figure-8-shaped waggle dance, the number of waggle runs per unit time indicates the distance of a resource from the colony, and the direction of these runs indicates the direction of the food from the colony, relative to the position of the sun. The odor of the resource clings to a dancer's body and is perceived by her hivemates, and floral nectar that she regurgitates to bees after performing her dance provides a direct indication of the concentration of the nectar that she has collected (von Frisch 1967).

Honey bees gather nectar as a source of energy and gather pollen as a protein-rich food source for developing young (Wilson 1971). In this laboratory, we will focus on the foraging behavior of bees collecting carbohydrate resources. Flowers present a diverse array of potential resources to bees, and flowers differ in characteristics such as color, odor, and shape (Faegri & van der Pijl 1966), as well as in the concentration and amount of nectar that they provide to foragers. The distribution, abundance, and quality of these resources are constantly changing (Gould & Gould 1988). How does a forager choose among available resources? Are particular colors, odors, or shapes favored by foragers?

Optimal foraging theory examines how animals, when presented with choices among foods that have different energetic profitabilities, make choices among the available options. The theory assumes (1) that the behavior exhibited by animals has persisted because it enhances the animal's fitness and has thus been favored by natural selection and (2) that maximizing the rate of intake of energy enhances fitness (Begon et al. 1990). If honey bees forage optimally, one would predict that, given a choice, they would forage at the most "profitable" resource—the one giving them either the greatest energy gain per unit time or the greatest gain per unit of foraging cost. Do honey bees forage in a manner consistent with the predictions of this theory?

In this laboratory, you will learn about the foraging biology of bees and, working in groups, design and conduct an experiment to determine whether and how particular floral attributes influence the flower choice of individual bees. You will present your research hypothesis to your classmates for discussion and critique. You will then test your hypothesis in the field. Finally, you will analyze and interpret your experimental results, indicating whether you find evidence that foraging bees have preferences among available resources.

MATERIALS

For your experiments, the following supplies will be available: blue, yellow, red, green, and black tape (for making floral designs); razors or scissors for cutting the tape; orange, anise, and peppermint extracts for scenting the solutions in your experimental feeders (use 0.5 microliter of extract for every 1 milliliter of solution); instruments for measuring quantities of solution and scent; 4-ounce baby food jars and petri dishes (for constructing feeders); microcapillary tubes (these work like tiny bee-sized straws and extend from the flower designs on the surface of a petri dish into the sugar or honey solutions within the dish, providing bees with access to the solutions); small triangular files for filing notches into baby food jars and for cutting segments of microcapillary tubing; a hand drill for making holes in petri dish lids; a round file for finishing these holes; and 20%, 30%, 40%, and 50% sucrose (table sugar) or honey solutions. Check with your instructor to find which (sucrose or honey) the bees at your study site are accustomed to feeding on, and use it in your experiments. In the field, you will carry paint and marking tools for marking individual foragers.

HYPOTHESES AND PREDICTIONS

Generation of Hypotheses and Predictions

Consider all of the factors that might influence a foraging bee's choice of resources. You may find it helpful, as a group or as a class, to make a list of these and to discuss them. Then, working in groups, use your knowledge about the characteristics of flowers, the capabilities of honey bees and, if you wish, ideas about optimal foraging to generate two *simple* hypotheses about the foraging behavior of bees. For each of your hypotheses, what do you predict would be the resource choice of foragers if the hypothesis were correct? Choose one of your hypotheses and design a simple experiment—one that uses the materials provided by your instructor and can be completed in the field during your next laboratory—to test it.

Consider the following as you discuss and generate your hypotheses and predictions.

Bees visit flowers to collect nectar and pollen. How might bees discriminate between flowers and nonflowers? How might they discriminate among flowers? Why might it be important for bees to be able to make these sorts of discriminations? Use the flowers on display to help you to generate ideas. In your experiment, you might choose to consider how a floral characteristic influences the bees' choice of feeders (artificial flowers).

One floral attribute that often escapes the attention of humans, but is of great potential importance to bees, is the quantity and quality of nectar, the bees' energy source. How much is present, and how concentrated (sweet) is it? Flowers vary in these nectar characteristics both within and among species. How might the quantity and sweetness of nectar influence a bee's choice of flowers, and why?

Optimal foraging theory examines how animals presented with choices among foods that have different energetic profitabilities make choices among their options. According to this theory, bees presented with a choice are predicted to forage at the resource that provides either the greatest energy gain per unit time or the greatest gain per unit of foraging cost. In your experiment, you might choose to consider whether the food choice of bees is influenced by the concentration of the sugar solution in a feeder.

Background on Your Experimental Subjects

Bees can be trained to visit feeders that contain sweet solutions. Your instructor has trained bees to forage at one or more field sites, and the bees are accustomed to finding food at these sites during your laboratory period. A 40% honey/water solution or scented sucrose solution has been presented daily in baby food jar feeders. These jars were filled with 100 milliliters of solution and inverted onto petri plates marked with simple floral patterns created with either blue or yellow tape—ask your instructor to provide the details for the bees that your class will observe in the field.

Bees can "remember" past foraging experience, and this experience may influence subsequent foraging behavior. Individual honey bees can also learn to associate one cue with another (for example, a blue flower with a particular concentration of sugar), can learn the position of a resource, and can learn to expect food at a particular time at a specific location. Recall what you know about honey bee dance—what information about resources can a dancing bee give to other bees in her colony? Remember that not all of the information that a dancing bee provides is directly encoded in the pattern, speed, and directionality of the dance. Also recall that dancing bees *can't* give other bees information about visually perceived characteristics of the flowers that they visit (in other words, the dancer can't communicate the message "The blue flowers are the most profitable").

When you design your experiments, consider the past experience of the foraging bees, the capacity of bees to learn to associate one floral characteristic with another, and the information that bees are communicating to their hivemates.

Presentation of Your Hypothesis and Protocol

As a group, with all members of the group participating, present your hypothesis and experimental protocol to the class for comments and critique in a brief oral report. In your presentation, address each of the following issues. (1) What is the rationale for your experiment? (What is the goal of your investigation, and why have you chosen to conduct this particular test?) (2) Formally present your null and alternative hypotheses. (3) How will you gather your data? Present a diagram of your experimental apparatus, and explain how it works. (4) How will you analyze your data? Ask your lab instructor for assistance in selecting an appropriate analysis.

(5) What would the different possible experimental outcomes tell you about honey bee foraging?

PROCEDURE

Summary of Laboratory Activities

Before the First Lab Meeting

Read this entire laboratory, and complete any additional readings assigned by your instructor.

Week 1

Observe flowers and demonstrations on honey bee morphology, life history, and dance behavior, as available in the laboratory. Then, working in groups, discuss the factors that might influence resource choice by honey bees, and design a testable hypothesis about resource choice. Present your hypothesis to the class for comments and critique in an oral report. Refine your hypothesis on the basis of the input that you receive from your class-mates and instructor, and discuss the final version with your lab instructor.

Build the feeders for your experiment; they should be ready to go for your next lab. Place your equipment in a paper bag labeled with your names and lab section. As a group, complete one copy of the group worksheet at the end of this chapter. Also prepare a sample data collection sheet. Hand both your group worksheet and your data collection sheet to your instructor before you leave.

Week 2

It is essential that you dress appropriately for this lab so that the bees will be attracted to the feeders, not to you. Wear light-colored clothing (white, beige, and khaki are great), avoiding blue, black, yellow, and floral prints. Tie back your hair, and wear a hat, shoes or hiking boots (*not* sandals), socks, a loose-fitting (but not baggy) long-sleeved shirt, and pants, that you will tuck into your socks. Do not use perfume or scented products such as soap, lotion, or shampoo. Immediately prior to lab, clean any food residue (especially sweets and fruit) from your skin and clothing.

At the beginning of the lab, you will receive your group worksheet back from your instructor. Read your instructor's comments, and make any necessary last-minute changes in your experimental protocol. Then assemble your experimental apparatus and prepare it for transport to the field.

Obtain your sucrose or honey solutions from your instructor, and if your experimental protocol requires, scent them with *0.5 microliter of scent per milliliter of solution*. Use a micropipette to dispense the scent into the sucrose or honey solution, and gently swirl the solution to distribute the scent. Transport the sweet (and, if required, scented) solutions to your field site in a covered plastic bee-proof box. Baby food jar feeders may be filled in the lab, capped tightly, and transported full to the field. Microcapillary tube feeders will need to be filled in the field and should be transported empty, in a container separate from that containing the sucrose or honey solution with which they will be filled. Following the detailed instructions in the "Equipment" section appropriate for the type of feeder that you are using, set up your experimental apparatus in the field and conduct your experiment. When you have finished gathering your data, collect your equipment and return to the lab, where you will clean and put away all of your experimental apparatus. Make one photocopy of your data for each student in your group and one for your instructor; hand in the instructor's copy before you leave.

Equipment

Preparation and Use of Baby Food Jar Feeders

Baby food jar feeders, consisting of notched jars filled with sweet solution and inverted onto a petri plate ornamented with a flower design, are best used in experiments where the exact design of the flower is not crucial; these include tests for discrimination of color, odor, and sucrose solution concentration. The flower design should reach to the edges of the petri dish because the center portion of the design will be obscured by the jar. See the examples provided by your instructor.

To assemble each feeder, you will need a 4-ounce baby food jar with 8–12 evenly spaced notches filed in the rim, a petri dish 9 centimeters in diameter, colored tape, and supplies for tracing and cutting your petal design. To construct flower petals, cut petal-shaped stencils from an index card. Stick colored tape on an appropriate cutting surface (such as a piece of clean glass) and trace the petal shapes onto the tape with a pencil. Use a mat knife or safety razor to cut the petals and try to make them as uniform as possible. A metal ruler works well for guiding the mat knife when cutting out petals with straight edges. If necessary, scissors may be used in this step, but they are not as quick and accurate. Affix the petals onto the inside of the lid of the petri dish. Don't worry; the tape will not peel off when it gets wet.

Transport solutions to the field in tightly sealed baby food jars placed in a sealed plastic box. Take care to transport your solutions with finesse so that they do not spill. Spilled solution may interfere with your experiment by attracting bees away from your feeder array.

To assemble your feeders in the field, remove the lid from each baby food jar; place the petri dish lid with the appropriate petal color and design on top of the open jar; and, holding them in close contact, invert the whole apparatus. The sugar solution will leak out slowly into the petri dish through the holes filed in the rim of the jar, and honey bees will feed around the rim of the jar. During your experiment, store baby food jar lids in the sealed plastic transport containers. As you set up and conduct your experiment in the field, soak up any spills promptly with cloth or paper towels, and rinse the area of the spill with water; store used towels in a sealed container until transport back to the lab.

Preparation and Use of Microcapillary Tube Feeders

On a microcapillary feeder, the flower pattern is fully visible (see the examples in your laboratory), and a cluster of microcapillary tubes is centered within each flower design (made with colored tape) on a petri dish lid. This feeder can be used to test the attractiveness of more intricate flower designs.

To construct a microcapillary tube feeder, follow a pattern from one of the feeder templates provided by your instructor, or make a template of your own design. A triangular arrangement of three holes at the center of each flower design works well for dispensing solution from the center of an artificial flower. Invert the petri dish lid on your template and mark the location of each hole on the lid with a water-soluble pen. Then place the inverted petri dish lid on a piece of scrap wood so that the plate is supported when you are drilling. This will protect your work surface and reduce the chance of cracking the petri dish. Protect your eyes with safety goggles, and use a hand-held drill with a 5/64-inch drill bit to drill holes through the lid of the a petri dish. To perfect your technique, practice drilling several holes on a "scrap" petri dish lid before beginning work on your feeder. When you have finished drilling your feeder, attach petals on the outer surface of the petri dish lid in an appropriate flower design.

To prepare microcapillary tubes for your feeder, measure the depth of the petri dish (with its lid on) and mark this distance on a piece of paper. To cut a microcapillary tube, line up the tube with the mark that you made on the paper. Place your fingernail or fingertips at the cutting mark to guide a 3-inch triangular file—it has a tendency to slide around. Using a gentle sawing motion, file the tube and score the glass 2/3 of the way around the tube circumference. Then, holding the tube adjacent to the scored portion, snap it away from you. This will create a fairly clean break. Each tube that is placed in the feeder should have the factory-finished end exposed at the top of the feeder, where the bees will be feeding; the end that you snapped should be in the solution.

After you have finished cutting your first tube, test it in the feeder to make sure it fits snugly into its hole (you can use a small round file to work the drilled holes to the right size) and is the appropriate length; it should extend above the surface of the petri dish lid about 1 millimeter to reduce the chances of its falling into the solution. Some trial and error will probably be necessary in this step. Have a receptacle available for broken glass, and remember to wear safety goggles.

When you use these feeders, carry the empty feeders into the field and fill them at your study site from a stock bottle of sucrose or honey solution. Take care to pour neatly and to avoid spilling solution over the sides of the dish (if you spill any, bees may feed at the edge of your dish rather than at your flowers). If necessary, use the head of a small insect pin to wick the solution up the microcapillary tubes in the field. Carry extra microcapillary tube pieces of the appropriate size into the field in a snap-top vial so that if any of your tubes fall into the solution, you can quickly replace them without having to open the feeder. As you set up and conduct your experiment in the field, soak up any spills promptly with cloth or paper towels and rinse the area of the spill with water. Store used towels and the stock bottle of soultion in a sealed container until transport back to the lab.

DATA RECORDING AND ANALYSIS

Because each group of students has a unique experimental design, each group will need to consult with the instructor to determine an analytical method appropriate for its analysis. Most groups will use a chi-square goodness-of-fit test (illustrated with a honey bee example on pp. 152–155 of Brown & Downhower 1988). You are testing to see whether your data differ from what you would expect if bees are equally likely to visit all of the feeders; detection of a significant difference from this expectation is evidence for a preference among feeders by foragers.

Your data collection sheet should be designed for the collection of census data at regular intervals. It might look something like the example in Table 11.1.

A tally of the numbers of bees feeding should be recorded in the appropriate spaces at each census time. It is also very important to record temperature, precipitation, and wind and sun conditions during the experiment.

Table 11.1 **Sample data collection sheet.**

Time (Minutes)	Color				Comments
	Blue	Yellow	Red	Black	
0					
3					
6					
9					
12					
etc.					

In addition, the feeders often get covered with bees, which makes it difficult for an approaching bee to make a foraging decision based on a color preference. Observe and record all of this information during data collection, and discuss these observations when you interpret your data.

Because foragers make repeated visits to feeders, summing up all of the visits to feeders over the duration of your experiment violates the assumptions of the chi-square statistical test. It is preferable to continue to collect data for at least 20 minutes after you have achieved a high visitation rate at your experimental site (but not so high a rate that the number of visitors obscures the variable whose influence on foraging you are testing) and then randomly to select the data from one census of bee visitation (for example, your tally of the number of bees feeding at each feeder at your 9-minute census) as a "snapshot" of the bees' preference and analyze those data.

QUESTIONS FOR DISCUSSION

How do your results compare to your predictions? What are all the possible explanations for the pattern of visitation that you observed, and how might you discriminate among them? How do your results compare to the findings of other authors who have observed the foraging behavior of bees? Are there any ways in which you might strengthen your experimental approach or your analysis in order to be better able to answer your research question? Propose a follow-up study that would enable you to test your research hypothesis more effectively or that would enable you to test a related idea about honey bee foraging behavior.

ACKNOWLEDGMENTS

MRR would like to thank Keith Waddington and colleagues from the Bee Laboratory at the University of Miami for teaching her much of what she knows about foraging honey bees; a Whitehall Foundation Grant to KW made this collaboration possible. MRR's experiences as a teaching assistant in Cornell University's introductory biology course, in which students made indoor observations of honey bee colonies, planted the seeds for the development of this field laboratory. Hannah Thomas, Kelly Baker, and Caitlin Richter helped critique and field-test portions of this lab. Students in Skidmore's population biology course and in the Summer Life Science Institute for Girls, teaching associates (Betsy Stevens, Sue Van Hook, Laurie Freeman, and Karen Kellogg), and lab assistants have helped this lab to evolve through both their comments and their creative and enthusiastic participation. Bonnie Ploger and Ken Yasukawa commented on drafts of this manuscript, much to its benefit. John Myers assembled observation hives, and Rick Green prepared and stocked the hives for our 1999 and 2001 laboratories; their efforts greatly enriched our students' experience in the lab. Presentation of this workshop at the Animal Behavior Society Meeting was supported by a Skidmore College Student/Faculty Collaborative Research Grant, funded by the W. M. Keck Foundation, to MRR and JMK, and by a Travel to Read Grant from Skidmore College to MRR.

LITERATURE CITED

Begon, M., Harper, J. L. & Townsend, C. R. 1990. *Ecology: Individuals, Populations and Communities.* 2nd ed. Boston: Blackwell Scientific.

Brown, L. & Downhower, J. F. 1988. *Analyses in Behavioral Ecology: A Manual for Lab and Field.* Sunderland, MA: Sinauer.

Faegri, K. & van der Pijl, L. 1966. *The Principles of Pollination Ecology.* Oxford: Pergammon.

von Frisch, K. 1967. *The Dance Language and Orientation of Bees.* Cambridge, MA: Harvard University Press.

Gould, J. L. & Gould, C. G. 1988. *The Honey Bee.* New York: Scientific American Library.

Michener, C. D. 1974. *The Social Behavior of the Bees.* Cambridge, MA: Harvard University Press.

Raveret Richter, M. & Waddington, K. D. 1993. Past foraging experience influences honey bee dance behaviour. *Animal Behaviour,* **46**, 123–128.

Seeley, T. D. 1985. *Honeybee Ecology.* Princeton, NJ: Princeton University Press.

Seeley, T. D. 1995. *The Wisdom of the Hive.* Cambridge, MA: Harvard University Press.

Waddington, K. D. 1982. Honey bee foraging, profitability and round dance correlates. *Journal of Comparative Physiology,* **148**, 297–301.

Wheeler, W. M. 1923. *Social Life Among the Insects.* New York: Harcourt.

Wilson, E. O. 1971. *The Insect Societies.* Cambridge, MA: The Belknap Press of Harvard University Press.

ADDITIONAL SUGGESTED READING

Barth, F. G. 1985. *Insects and Flowers: The Biology of a Partnership.* Princeton, N-J: Princeton University Press.

Bitterman, M. E. 1976. Incentive contrast in honeybees. *Science,* **192**, 380–382.

Butler, C. G. 1974. *The World of the Honeybee.* London: Collins Clear-Type Press.

von Frisch, K. 1950. *Bees: Their Vision, Chemical Senses and Language.* Ithaca, NY: Cornell University Press.

von Frisch, K. 1974. Decoding the language of the bee. *Science,* **185**, 663–669.

Gould, J. L. 1979. Do honey bees know what they are doing? *Natural History,* **88**, 66–75.

Gould, J. L. 1986. Pattern learning by honey bees. *Animal Behaviour*, **34**, 990–997.

Gould, J. L. 1987. Honey bees store learned flower-landing behaviour according to time of day. *Animal Behaviour*, **35**, 1579–1581.

Heinrich, B. 1976. The foraging specializations of individual bumblebees. *Ecological Monographs*, **46**, 105–128.

Menzel, R. & Erber, J. 1978. Learning and memory in bees. *Scientific American*, **239**, 102–110.

Siegel, S. 1956. *Nonparametric Statistics for the Behavioral Sciences.* New York: McGraw-Hill.

Thorp, R. W., Briggs, D. L., Estes, J. R. & Erickson, E. H. 1975. Nectar fluorescence under ultraviolet irradiation. *Science*, **189**, 476–478.

Waddington, K. D. 1985. Cost-intake information used in foraging. *Journal of Insect Physiology*, **31**, 891–897.

Waller, G. D. & Bachman, W. W. 1981. Use of honey-sac load and dance characteristics of worker honeybees to determine their sugar preferences. *Journal of Apicultural Research*, **20**, 23–27.

Date: _____

Lab Section: _____

Group Worksheet

Names of group members:

_____

Sucrose or honey solution concentrations and scents needed for your test:

_____

1. Present a carefully worded statement of your null hypothesis (H_0) and the corresponding alternative hypotheses (H_a).

2. Clearly describe (use illustrations if necessary) the design of the feeders you will construct. How many of these feeders will you build? List all of the equipment that you will need to build your feeders and perform your test.

3. Carefully describe the methods that you will use to gather and analyze your data. Attach a data collection sheet showing what data you will collect and how you will record it.

4. List all of the possible results for your experiment. What would each of these results tell you about honey bees? (Use additional space if necessary.)

CHAPTER 12

Individual Constancy to Color by Foraging Honey Bees

PEGGY S. M. HILL AND HARRINGTON WELLS
Faculty of Biological Sciences, University of Tulsa, Tulsa, OK 74104, USA

INTRODUCTION

Flower constancy, wherein foragers actually bypass rewarding flowers to confine their visits to a single plant species (Waser 1983), was known to Aristotle (ca 340 B.C.). This behavior is exhibited by a variety of insects (see Goulson & Wright 1998; Chittka et al. 1999), but floral constancy is perhaps best documented in the honey bee, *Apis mellifera* L. (e.g., Ribbands 1953; Heinrich 1975; Wells & Wells 1983; Waser 1986; Real 1991; Hill et al. 1997, 2001; Chittka et al. 1999). Further, **individual constancy** (Wells & Wells 1983) to flowers is spontaneous in Italian honey bees (Hill et al. 1997). By this we mean that individuals choose a flower on the basis of a suite of cues on their first visit to the patch from their hive, and they continue to visit the same flower type on repeated visits from the hive, irrespective of the alternatives available and irrespective of what their hivemates are doing. Honey bees show a high degree of constancy to both floral odor and floral color (Wells & Wells 1983, 1984, 1986).

Flower constancy raises a host of ecological, evolutionary, and physiological questions. For example, why would an insect benefit from focusing on a single plant species? Are there costs to switching from one plant species to another? What selection pressures would favor flower constancy? Can you imagine selection against flower constancy? Do any benefits accrue to the plant when a pollinator restricts foraging visits to a single species?

In order for a honey bee to restrict visits to a flower of a specific color, the bee must be able to differentiate among colors. Honey bees have **trichromatic vision,** or true color vision with three primary color receptors (Mazokhin-Porshniakov 1969). Darwin (1895) knew that honey bees are capable of differentiating among the hues of colors produced by flowers of the same species, and von Frisch (1914/1915) did the classic definitive experiments to show that honey bees can be trained to discriminate among colors. We know that the honey bee visible spectrum includes ultraviolet light, which humans cannot see, but a long-standing concept that they cannot see red, on the other end of the human visible spectrum, is currently being challenged (Chittka & Waser 1997). Consider that the honey bee can discriminate among colors as well as humans, but they do not perceive the "flower" in the same way that you do (Vorobyev et al. 1997).

The experimental model system you will use in this exercise will enable you to explore honey bee fidelity to flowers of a particular color by using a **dimorphic** (two-color) artificial flower patch similar in design to that used by Wells & Wells (1983, 1986) and Hill et al. (1997). Because energy (food) is a basic requirement for all life forms, and assuming that populations continue to increase until food supplies become limiting (Malthus 1798), natural selection is expected to promote efficiency in foragers. This idea is the nucleus of the **energy maximization** foraging theory: evolution should select for behaviors that maximize net energy gain from foraging (energy collected as food minus energy spent foraging). Use this theory as a guide as you design experiments to predict the behavior of foraging honey bees.

MATERIALS

Each group of four students will require the following materials, which will be provided by the instructor: honey bees (*Apis mellifera*) trained to visit a feeding dish out of doors in an open area; materials to assemble an artificial flower patch (2–36 individual flowers: see Fig. 12.1): 4-millimeter-thick Plexiglas squares (30 × 30 millimeters) that have been painted on the underside with Testor's model paints (blue: #1208 and yellow: #1214) and that have 2 drilled holes, flower "pedicels" of 6-millimeter dowel rods that have been cut into 90-millimeter pieces to fit into the holes on the undersides of the Plexiglas squares, and a pegboard base; nectar rewards (unscented sucrose of 1 molar and 2 molar solutions, approximately 250 milliliters of each); nectar dispensers; and 3 data sheets for recording

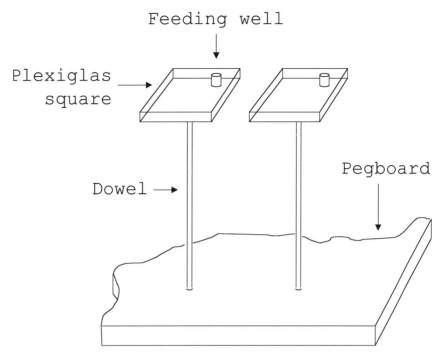

Figure 12.1 Artificial flower patch made from Plexiglas squares mounted on dowel rods and supported by a pegboard base. Shallow depressions in the surface hold "nectar" rewards. The undersides of the Plexiglas are painted with Testor's paints.

honey bee visits. (A sample data sheet is included as Table 12.1, but you may want to design your own. If you keep a standard lab notebook for all your experiments, be sure also to record the kinds of data requested on the sample.)

PROCEDURE

In the first part of the lab, you will assemble artificial flower patches consisting of multiple blue and yellow artificial flowers arranged randomly as to color in a matrix of rows and columns on a pegboard base. In the field, you will conduct a three-part experiment in which you will record which flower colors are visited by individual bees. You will mark individual bees so that they can be recognized by your group and followed through all three sections of the experiment. In the first part of the experiment, blue and yellow flowers will have the same nectar reward. In the second part, one flower color will be enriched while the reward of the other color will be unchanged. In the third part, the rewards will be reversed so that the other color will become enriched. The data will reveal whether individual bees show a preference for one of the flower colors (constancy),

Table 12.1 **Sample data sheet.**

Date_____   Color dimorphism_____

Time: From_____ To_____ Rewards_____

Additional remarks: _____

Bee	*Color Visited*	*Totals*

whether the bees from the same colony behave similarly with respect to color choice, and whether individuals maintain a preference (constancy) as the rewards are varied over the three sections of the experiment.

Working with Honey Bees

The honey bees that you observe will be from either a **feral** (wild) colony or a commercial hive, and prior to the lab period, all will have been trained by your instructor to visit a scented sucrose solution in a feeding dish in an open area outside. You will be working with an unscented reward solution that will limit the number of recruits on your experimental apparatus.

Most experiments documented in the literature have been conducted with Italian bees (*A. m. ligustica*), which are the most common subspecies in North America. Most of the feral colonies are Italian bees that have escaped from domestic hives. Unless you live in an area where Africanized bees are found, honey bees tend to be docile and to focus on the food instead of the observers! Still, you should exercise all possible caution. Movement that startles a bird or squirrel will probably also affect a honey bee. Maintaining a calm observational manner is essential when working with any wild animals, and honey bees are no exception. If you know you are allergic to bee stings, notify your instructor before the beginning of lab. Even if you do not know that you have such an allergy, assume that you do, and avoid stings.

You can usually avoid being stung by not wearing, on the days of the experiments, perfume that might make you resemble a flower! Generally avoid all scented personal care products on the day of the experiment. If you have longer hair, wear it pulled back. You should also wear shoes rather than sandals so that you will not be stung if you step on a bee in the grass. Students should follow all requests of their instructor for dress appropriate to the day's activities.

Prior to the Experiment

Use of an artificial flower patch of the Wells et al. (1981, 1992) design will enable you to control for morphology, distribution, and abundance of flowers, as well as for quality and quantity of nectar rewards, reward frequencies, odor, and number of foragers. The flowers (see Fig. 12.1) are Plexiglas squares 4 millimeters thick that measure 30×30 millimeters and are painted on the undersurface with Testor's model paints (blue: #1208; yellow: #1214). Flowers vary only with respect to color. Each flower should be mounted on a 90-millimeter pedicel of 6-millimeter doweling, fixed by inserting the doweling into the hole on the underside of the Plexiglas square. On the upper surface, each flower has a small "nectary" located in one corner, created by drilling a small hole 3 millimeters deep and 2 millimeters in diameter. This is where you will later load the nectar reward. Flowers should be organized in a Cartesian coordinate grid, spaced 75 millimeters apart in rows and columns, on a piece of pegboard. Place an equal number of yellow and blue flowers (from 2 to 36 flowers, depending on availability of materials) onto the pegboard in a random fashion. Your instructor will inform you at this time of any deviations from the experimental design described here. Align the nectaries so that they are all in the same relative position (for example, upper right-hand corners as you face the patch).

At the site selected outside, you will fill the nectaries with a predetermined quality and quantity of reward and then, as they are emptied, refill them with the same quantity and quality of reward. Measure the volumes of reward (with a pippetteman or syringe or glass pipette) that will allow delivery of consistent volumes over the course of the experiment. Before you carry your flower patch to the designated location, make sure that you can manipulate the measuring apparatus provided by your instructor.

Because you will be filling the flowers with sucrose rewards as they are emptied, consider how you can minimize your impact on the bees' behavior. If time permits in the days prior to the experiment, or, at minimum, just before you begin recording data, watch individuals at the training feeders carefully to see how your movements affect them, if at all. Take these observations into consideration as you plan your experiment. Also, consider how many individuals you will need to include in the experiment to obtain an adequate sample. Consider that an individual honey bee tends to continue foraging until she fills her crop with 60 microliters of reward (Hill et al. 1997), so the number of visits expected are linked to reward volumes.

Conducting the Experiment

Place your assembled flower patch near the training feeder and fill the nectaries with the reward solution. Record on your data sheet the odor used to train your bees to the site. (If several groups are working in the same area, more than one training feeder will have been used.) After the instructor has transferred bees to your flower patch, the feeder will be sealed in an airtight container. (If the scented dish is left nearby, bees will find it and will continue to visit that dish rather than the experimental patch.) As the bees arrive on the experimental patch, mark them with daubs of Testor's paints on the thorax and/or abdomen in a fashion that can be recognized on subsequent visits. You will be using the same individual bees throughout this experiment. You may use a toothpick, a pin sticking through a pencil's eraser, or even a pine needle as your "paint brush," and you may devise your own pattern of dots to use in marking the bees. Just be sure you are able to identify bees by this marking so that you can make accurate notations on the data sheet (e.g., orange ab = orange daub on the abdomen). Record visits of each individual bee to the patch by noting the color of the flower on which it lands and drinks. If a honey bee is drinking instead of just touching down briefly on a flower, it "bobs" its abdomen up and down. Record only visits when bees actually drink from the nectary. Continue observations for 45 minutes for each section of the experiment, making sure to record the time on the data sheet.

Conduct the experiment in three sections. First, fill all flowers with a standard reward. With the help of your instructor, you will already have determined what volume and quality of reward to use. Check with the instructor at this time if you are not sure how much of which reward to use first. (Remember that one bee typically forages until she fills her crop with 60 microliters of solution, so the lower volumes of reward will result in more visits and yield more data!) The second section of the experiment requires enrichment of the reward on one floral color. For example, you might enrich blue flowers to a 2 molar sucrose reward from a standard 1 molar sucrose reward, while continuing to fill the yellow flowers with the standard (1 molar sucrose). For the third section, switch the rewards so that the floral morph with the enriched reward (2 molar sucrose) is provisioned with the standard (1 molar sucrose), and vice versa. Between sections of the experiment, quickly remove the rewards and replace them with the new rewards, or replace all flowers with a new set of flowers. Use a new data sheet for each section of the experiment, making sure to record any changes you make. Begin the next section of the experiment as soon as the rewards are loaded into the nectaries.

HYPOTHESES AND PREDICTIONS

What do you predict will happen when honey bees are offered a patch of yellow or blue flowers where rewards are identical? Will all honey bees visit all flowers? If one of the floral morphs holds an enriched reward, such as twice the energy of the other morph, what do you predict foraging honey bees will do? When the reverse experiment is conducted, and the flower of the other color holds the enriched reward, what will the foragers do? Will they switch to the higher-energy reward? Is there an advantage to switching? On the basis of these general questions and others you have posed for this exercise, choose one hypothesis to test, and state clearly the null and alternative hypotheses before you begin any experiments. Choose an appropriate statistical test to use in interpreting your results. Check your decisions with your instructor before collecting data. Test as many hypotheses generated by your questions as time allows.

DATA RECORDING AND ANALYSES

Record individual honey bee visits to the artificial flower patch. The data sheet should include the following: date, beginning and ending times of the experiment, any unusual climatological data (e.g., very hot, 40°C), the colors being used in the choice test, the molarity of the rewards used (quality), the volume of the rewards used (quantity). Data will be the color of the flower visited each time the bee lands on a flower and drinks the reward. Remember that honey bees "bob" their abdomens as they imbibe a reward, so it is easy to tell whether they are actually "drinking" or merely "landing" temporarily before flying to another flower.

Because you will have a large number of visits from each individual bee to your group's flower patch, analyze your data separately from those of other groups. Construct from the raw data sheet a table of data summarizing the number of visits of each bee to each of the alternative colors. It is easy to convert the summarized data to bar graphs depicting total visits to blue and total visits to yellow flowers in each of the sections. It might also be of interest to plot percent visits of each individual forager to each color morph in each of the three sections. However, a more sophisticated analysis with the replicated goodness–of–fit, or G, statistic (Sokal & Rohlf 1981, pp. 721–731; or 1995, pp. 715–724) is more appropriate for the kinds of data you will obtain with this experimental design. The G statistic allows for testing data for homogeneity of foraging behavior among the bees in the experiment and for random visitation to floral morphs by individuals, as well as by the bees as a group. Check your decisions with your instructor before beginning to collect data.

QUESTIONS FOR DISCUSSION

Did all bees visit all alternative flowers in the first section when rewards were equal?

Did you observe constancy of all bees to one of the color choices, or were visits random?

When the blue flowers held an enriched reward, did individuals change their foraging behavior?

Did behavior change when the yellow flowers held the enriched reward?

Do you consider the individual behaviors of this race of honey bees to be an example of energy maximization? Are these behaviors in some other way optimal for the colony, or for the individuals?

What selection pressures might have led to this type of behavior?

LITERATURE CITED

Aristotle. ca 340 B.C. *Historia Animalium* (trans. A. L. Peck., 1965). Cambridge, MA: Harvard University Press.

Chittka, L. & Waser, N. 1997. Why red flowers are not invisible for bees. *Israel Journal of Plant Science*, **45**, 169–183.

Chittka, L., Thomson, J. D. & Waser, N. M. 1999. Flower constancy, insect psychology, and plant evolution. *Naturwissenschaften*, **86**, 361–377.

Darwin, C. 1895. *The Effects of Cross and Self Fertilisation in the Vegetable Kingdom.* New York: Appleton.

von Frisch, K. 1914/1915. Der Farbensinn und Formensinn der Biene. *Zoologische Jahrbuch (Physiologie)*, **35**, 1–188.

Goulson, D. & Wright, N. P. 1998. Flower constancy in the hoverflies *Episyrphus balteatus* (Degeer) and *Syrphus ribesii* (L.) (Syrphidae). *Behavioral Ecology*, **9**, 213–219.

Heinrich, B. 1975. Energetics of pollination. *Annual Review of Ecology and Systematics*, **6**, 139–170.

Hill, P. S. M., Hollis, J. & Wells, H. 2001. Foraging decisions in nectarivores: unexpected interactions between flower constancy and energetic rewards. *Animal Behaviour*, **62**, 729–737.

Hill, P. S. M., Wells, P. H. & Wells, H. 1997. Spontaneous flower constancy and learning in honey bees as a function of colour. *Animal Behaviour*, **54**, 615–627.

Malthus, R. T. 1798. *An Essay on the Principle of Population As It Affects the Future Improvement of Society.* London: Johnson.

Mazokhin-Porshniakov, G. A. 1969. *Insect Vision.* New York: Plenum Press.

Real, L. A. 1991. Animal choice behavior and the evolution of cognitive architecture. *Science*, **253**, 980–985.

Ribbands, C. R. 1953. *The Behavior and Social Life of Honey Bees.* London: Bee Research Association.

Sokal, R. R. & Rohlf, F. J. 1981. *Biometry: The Principles and Practice of Statistics in Biological Research.* 2nd ed. New York: Freeman.

Sokal, R. R. & Rohlf, F. J. 1995. *Biometry: The Principles and Practice of Statistics in Biological Research.* 3rd ed. New York: Freeman.

Vorobyev, M., Gumbert, A., Kunze, J., Giurfa, M. & Menzel, R. 1997. Flowers through insect eyes. *Israel Journal of Plant Science*, **45**, 93–101.

Waser, N. M. 1983. The adaptive nature of floral traits: ideas and evidence. In: *Pollination Biology* (ed. L. Real) New York: Academic Press, pp. 241–285.

Waser, N. M. 1986. Flower constancy: definition, cause and measurement. *American Naturalist*, **127**, 593–603.

Wells, H., Hill, P. S. & Wells, P. H. 1992. Nectarivore foraging ecology: rewards differing in sugar types. *Ecological Entomology*, **9**, 467–473.

Wells, H. & Wells, P. H. 1983. Honey bee foraging ecology: optimal diet, minimal uncertainty or individual constancy? *Journal of Animal Ecology*, **52**, 829–836.

Wells, H. & Wells, P. H. 1986. Optimal diet, minimal uncertainty and individual constancy in the foraging of honey bees, *Apis mellifera*. *Journal of Animal Ecology*, **55**, 375–384.

Wells, H., Wells, P. H. & Smith, D. M. 1981. Honey bee responses to reward size and colour in an artificial flower patch. *Journal of Apicultural Research*, **20**, 172–179.

Wells, P. H. & Wells, H. 1984. Can honey bees change foraging patterns? *Ecological Entomology*, **9**, 467–473.

PART 3

Development

CHAPTER 13

Dog Training Laboratory: Applied Animal Behavior

LYNN L. GILLIE[1] AND GEORGE H. WARING[2]
[1]*Division of Mathematics and Natural Science, Elmira College, One Park Place, Elmira, NY 14901, USA*
[2]*Department of Zoology, Southern Illinois University at Carbondale, Carbondale, IL 62901, USA*

INTRODUCTION

Oftentimes the behavior of animals can be utilized or modified to accomplish a goal we as humans deem beneficial. This is the essence of applied animal behavior (e.g., Craig 1981; Markowitz 1982; Monaghan 1984; Fraser 1990; Grier & Burk 1992). Using knowledge about animal behavior, we can, for example, train animals to do useful work, shift them away from undesirable behaviors, or use their keen perception to sense things we cannot. As an introduction to applied animal behavior, this exercise focuses specifically on dog training. Dogs traditionally have been companions as well as working partners to humans (Fox & Bekoff 1975; Kilgour & Dalton 1984; Taggart 1986). They have helped humans hunt game, defend resources, herd livestock, guide the blind, conduct search-and-rescue operations, pull sleds, and detect contraband and the bombs of terrorists.

Dogs are receptive to training, are easy to handle, and seem to enjoy work. They learn readily throughout their lives.

A basic understanding of learning theory is beneficial to trainers. Dog training utilizes principles of **operant conditioning**, or **instrumental learning**, in which reinforcement plays a role in the learning process as the animal gives a desired behavioral response (e.g. Pryor 1984; Taggart 1986; Klein & Mowrer 1989; McConnell 1996). **Reinforcement** is something that increases the chance that a behavior will occur again. **Positive reinforcement** is a positive stimulus that increases the probability that the desired behavioral response will occur. For dogs, positive reinforcement can be a treat, a toy, and/or praise given immediately after a correct response. **Negative reinforcement** is the cessation of an aversive stimulus in order to induce the desired behavioral response. **Negative punishment** is the withholding, when an incorrect response is given, of something the animal desires (Owens 1999). For dogs, a negative punishment is the withholding of a treat when it does not respond correctly when asked. This is much like suspending a teenager's telephone privileges. Negative punishment should *not* be interpreted as physical punishment for an incorrect response. Modern training programs most often use a combination of positive reinforcement and negative punishment techniques, such as giving and withholding rewards. The technique of **shaping** can be used to guide a dog into a proper response, especially when that task is somewhat complex. With this method, the animal is reinforced for successive approximations until the end response is achieved. A general response is first accepted as an approximation and is rewarded, the next reinforcement is given immediately after a better approximation, and so on until the dog does the sought-after response.

Dog trainers do not have to be professionals; dog owners also can—and should—be adept at training their dogs (Thixton & Abbott 1999). Even puppies can accomplish learning, although oftentimes not to the extent or perfection of adults. Conversely, older dogs may have difficulty learning particular tasks (Adams et al. 2000). Consistency is important; if the trainer is vague with commands or expectations, then the dog's performance will be poor and inconsistent. Commands that are simple and easy for the dog to detect are best (Owens 1999; Thixton & Abbott 1999), and it is easiest to start with basic learning tasks (such as *sit, stay, down,* and *come*) before advancing to others. Researchers have found that the sooner reinforcement follows a response, the better learning progresses (Klein & Mowrer 1989). Beginning trainers often reinforce a behavioral response too late and learn that consistent training is not as easy as it may appear. Reinforcement should be within a second to be most effective (McConnell 1996). Many dogs benefit from knowing when a training session is finished; thus a word like "free" can be used to release the dog from further duty. Ending a training session can be rewarding in itself, so the session should end after a set of correct responses, even if this means returning to some basic commands, such as *sit* and *stay.* Keep these principles in mind as you begin this introduction to dog training using positive reinforcement and shaping techniques.

MATERIALS

Stopwatches, data sheets, dog on its leash for each group of students, dog treats, dog toys, plastic bags, and newspaper.

PROCEDURE

After your instructor has had a professional trainer assess the dogs for safety around students and has assigned an owner and dog to your group, spend the first few minutes introducing yourself to the dog. Let the dog approach you as you hold your hand to the dog below its chin for a sniff while avoiding a direct stare (Owens 1999). In addition, turning sideways and perhaps crouching while the dog comes to you will prevent you from looming over it (Pets, Part of the Family 2000). Greeting a dog in this way ensures that you do not send an aggressive social signal. Talk to the owner to see what commands and signals are already used with the dog so that you can be consistent. Ask the owner to give these commands while you record the results. Take the dog for a walk and give it a chance to relieve itself, bringing along a plastic bag or newspaper to clean up after it. BE SURE to clean up any messes. Failure to do so can lead to parasite spread and general unpleasantness on campus. After this settling-down period, find a low-traffic area on campus to begin training. Either an indoor or an outdoor site may be used; however, because it is critical that distractions be minimized so that the dog can focus on the training session, an indoor site is preferable.

Start by determining what will motivate your dog to follow your new command by testing several alternative rewards (such as praise, a toy, or a treat) using a command it already knows. Use a stopwatch to measure latency—the time in seconds it takes to perform the behavior after the command is given. Once you identify the most effective reward, you may begin to condition the new behavior. Begin by training your dog to roll over, or train it to shake hands if it already knows how to roll over. A "trick" is preferable to an obedience exercise that the owner may already have taught in a particular way.

To train the dog to roll over, you should use positive reinforcement during shaping. This is very effective if administered promptly after each successive approximation of the final response. Remember that a prompt reinforcement is one administered within a second of the correct behavior. If possible, the instructor may demonstrate the steps needed to shape a dog to roll over. For example, if the dog lies down, it should be reinforced right away. When told to roll over again, the owner may get the dog lie down and bring its nose toward its tail with a treat as a lure; this new approximation of a "roll over" should be immediately reinforced. Continue to reinforce small increments of movement of the head toward the tail so that progress is encouraged and rewarded. Once the dog has its head pointing toward its tail, rub the chest to cause the top foreleg to rise. Reinforce this key step very well. Once the dog can get this close to rolling over reliably, the lure can be passed across the chest to the other side of the body, and the dog's head and body should follow through. Never force the dog, and take frequent breaks to reduce stress and enhance learning. The training sessions should be enjoyable for both you and the dog. Record the latency during the training process. During training, latency cannot be measured from the time the command is given to the time the response is completed, because the behavior is shaped in steps that change as training progresses. Rather, latency during training should be defined as the time in seconds that it takes for the dog to begin to perform the behavioral sequence. After acquisition of the behavior, record latency from the time the command is given to the time the response is completed. Measure how many training trials it takes for the dog to learn to do the new behavior. Several short training sessions are preferable to one long training session, so take frequent breaks to walk the dog.

HYPOTHESES AND PREDICTIONS

One hypothesis could be that age affects ease of acquisition of a new skill. This hypothesis would predict that young dogs acquire new skills more quickly than old dogs. The class will have to see what age range is available and to form a specific prediction based on the ages of the dogs that are present. Another hypothesis is that prior learning facilitates current learning. This hypothesis predicts that dogs that already know many commands will learn a new behavioral skill faster than dogs that know only a few commands. As you consider your experimental design to test these hypotheses, keep in mind potential interactions of variables. One potential problem is that age and prior knowledge of behavioral skills may interact. How would you distinguish between these two hypotheses? One solution could be to find dogs of the same age that know different numbers of skills, or, conversely, to find dogs of different ages that know the same number of skills.

DATA RECORDING AND ANALYSES

After the owner gives a few "known" commands to the dog, record the results. Also record the dog's age. You will share these results with the class later. With these pilot data in mind, state the biological hypothesis you are testing (such as "Age affects ability to learn a new trick"). You should

Table 13.1 **Sample data sheet for training and practicing a new behavioral skill.**

Training		Practice After Acquisition	
Trial Number	Latency (Seconds)	Trial Number	Latency (Seconds)
1		1	
2		2	
3		3	
4		4	

decide what data are necessary to examine the predictions that arise from this hypothesis. One prediction might be that older dogs have longer learning latencies than younger dogs. Another prediction is that older dogs require more trials than younger dogs to learn a new behavioral skill. A data sheet similar to Table 13.1 may be useful to organize how you will study the process of training a dog to do a new trick, such as rolling over or shaking hands. You will probably need more training trials than the four given in the example, so make sure you leave enough space. Remember to take frequent breaks during training!

If you decide to measure latency, consider that latency is usually expected to decrease or shorten as proficiency of performance of the new skill increases (Taylor & Lupker 2001). However, some researchers have shown that latencies may decrease without a comparable increase in accuracy (Zakay & Tuvia 1998; Bentall et al. 1999). Other researchers have found that latency can increase or decrease depending on the type and difficulty of the task (Wixted et al. 1997). Because of the potential variability in latency scores, you should use some caution when interpreting latency.

Record the number of training trials used to train your group's dog and the latency time for each trial. Then calculate an average latency time to report to the class. After the dog is consistently performing its new behavior, carry out a number of additional trials, record the latency time for each, and calculate an average. Consistent performance of a behavior should be operationally defined. Some authors have chosen the value of 80% as a criterion for a consistent response (Owens 1999). You may want to consider the use of your own operational definition as a class. Graph latency as a function of trial number for your group's dog. Do you expect latency to be longer or shorter for older dogs? After a brisk walk and play break, will the dog repeat its new behavior when commanded? Will the dog perform its new behavior for any member of the training group?

Now pool the data from the entire class by making a chart like that shown in Table 13.2. Check with your instructor to see whether statistical analyses are recommended. Analysis of the data to test your biological hypothesis requires the examination of a statistical null hypothesis and its alternative. In the case of latency, the null hypothesis is that old and young

Table 13.2 **Sample data sheet for collating the class data.**

Dog	Age of Dog	Number of Training Trials	Average Latency During Training (Seconds)	Average Latency After Acquisition of New Behavior (Seconds)
1				
2				
3				
4				

dogs have the same average latency. The alternative to the null hypothesis is that old and young dogs have different average latencies. This is a two-tailed test. Another alternative is that old dogs have longer latencies than young dogs (a one-tailed test). To test the prediction that latencies during training for dogs 1 year old and younger differ from those for dogs older than 1 year, use a Mann–Whitney U test.

QUESTIONS FOR DISCUSSION

Did latency increase or decrease during training? Why? Did latency differ for different rewards?

Did amount of previous training experienced by each dog influence the success of subsequent training? How would you test this hypothesis?

Design a new hypothesis and experiment to extend the results of your current experiment.

ACKNOWLEDGMENTS

We thank the many students and dog owners who have participated in our labs. We are also grateful to Patricia McConnell, Charles Deutch, Ken Yasukawa, and Bonnie Ploger for their valuable comments on various versions of the manuscript.

LITERATURE CITED

Adams, B., Chan, A., Callahan, H., Siwak, C., Tapp, D., Ikeda-Douglas, C., Atkinson, P., Head, E., Cotman C. W. & Milgram, N. W. 2000. Use of a delayed non-matching to position task to model age-dependent cognitive decline in the dog. *Behavioural Brain Research*, **108**, 47–56.

Bentall, R. P., Jones, R. M. & Dickins, D. W. 1999. Errors and response latencies as a function of nodal distance in 5-member equivalence classes. *Psychological Record*, **49**, 93–115.

Craig, J. V. 1981. *Domestic Animal Behavior: Causes and Implications of Animal Care and Management.* Englewood Cliffs, NJ: Prentice-Hall.

Fox, M. W. & Bekoff, M. 1975. The behaviour of dogs. In: *The Behaviour of Domestic Animals.* 3rd ed. (ed. E. S. E. Hafez). London: Bailliere Tindall, pp. 370–409.

Fraser, A. F., ed. 1990. Publication overview of Applied Animal Behaviour Science, volumes 1–25, 1974–1990. *Applied Animal Behaviour Science,* **25,** 191–345.

Grier, J. W. & Burk, T. 1992. *Biology of Animal Behavior.* St. Louis, MO: Mosby Year Book.

Kilgour, R. & Dalton, C. 1984. *Livestock Behaviour: A Practical Guide.* Boulder, CO: Western Press.

Klein, S. B. & Mowrer, B. R., eds. 1989. *Contemporary Learning Theories: Instrumental Conditioning Theory and the Impact of Biological Constraints on Learning.* London: Lawrence Erlbaum.

Markowitz, H. 1982. *Behavioral Enrichment in the Zoo.* New York: Van Nostrand.

McConnell, P. B. 1996. *Beginning Family Dog Training.* Black Earth, WI: Dog's Best Friend.

Monaghan, P. 1984. Applied ethology. *Animal Behaviour,* **32,** 908–915.

Owens, P. 1999. *The Dog Whisperer.* Holbrook, MA: Adams Media.

Pets, Part of the Family. 2000. *Petspeak.* New York: St. Martin's Press.

Pryor, K. 1984. *Don't Shoot the Dog!* New York: Simon & Schuster.

Taggart, M. 1986. *Sheepdog Training: An All-breed Approach.* Loveland, CO: Alpine Publications.

Taylor, T. E. & Lupker, S. J. 2001. Sequential effects in naming: a time-criterion account. *Journal of Experimental Psychology: Learning, Memory, & Cognition,* **27,** 117–138.

Thixton, S. & Abbott, S. 1999. Training your dog. learnFREE, Inc. http://www.trainingyourdog.com

Wixted, J. T., Ghadisha, H. & Vera, R. 1997. Recall latency following pure- and mixed-strength lists: a direct test of the relative strength model of free recall. *Journal of Experimental Psychology: Learning, Memory, & Cognition,* **23,** 523–538.

Zakay, D. & Tuvia, R. 1998. Choice latency times as determinants of post-decisional confidence. *Acta Psychologica,* **98,** 103–115.

CHAPTER 14

Paternal Care and Its Effect on Maternal Behavior and Pup Survival and Development in Prairie Voles (Microtus ochrogaster)

BETTY MCGUIRE
Department of Biological Sciences, Smith College,
Northampton, MA 01063, USA

INTRODUCTION

In most mammals **parental care** is performed exclusively by females (Clutton-Brock 1991). Whereas female reproductive success is limited by the time and energy invested in offspring, male reproductive success is typically maximized by mating with multiple females rather than by mating with one female and helping her care for the young. This pattern among mammals has been explained by several characteristics of the taxon, including (1) internal fertilization, which increases the likelihood that males will be disassociated from their young and may reduce **certainty of paternity**, and (2) internal gestation and lactation, which necessitate a major role for the female in parental care and restrict the ability of a male to aid offspring early in development (Trivers 1972; Williams 1975). Nevertheless, paternal care has been reported among carnivores, primates, and rodents (Kleiman & Malcolm 1981), and this raises the possibility that in some situations, males maximize their reproductive success through provision of parental care. In such situations, paternal care may substantially increase the survival of offspring and thereby offset the costs of missed matings (Gubernick et al. 1993).

Prairie voles (*Microtus ochrogaster*) are small rodents that inhabit grasslands and agricultural fields of the midwestern United States. Live trapping and radiotelemetric monitoring of field populations indicate that individuals live in communal groups typically comprised of a male–female pair, their **philopatric** offspring, and a few unrelated adults (Getz et al. 1993; Getz & Carter 1996). Communal groups are most common in late autumn and winter, whereas male–female pairs and single female units predominate in spring and summer. The precise mating dynamics among members of communal groups are unknown; however, members of male–female pairs show traits associated with behavioral **monogamy** (Carter & Getz 1993; Getz & Carter 1996).

Paternal care is often associated with a monogamous mating system and with nest sharing by males and females (Clutton-Brock 1991; Dewsbury 1985); thus field data indicate the potential for paternal care in prairie voles. Direct observations of parent–young interactions are impossible under field conditions because prairie voles nest in underground burrows. Observations under seminatural conditions in the laboratory reveal that male prairie voles exhibit high levels of paternal care (McGuire & Novak 1984; Oliveras & Novak 1986; Thomas & Birney 1979). Males and females spend similar amounts of time in the nest with young; they coordinate their visits to and from the nest such that pups are rarely left unattended; and males brood, groom, and retrieve pups at levels similar to those of females.

The purpose of this laboratory exercise is to determine whether paternal care evolved in prairie voles because such care increases male reproductive success through enhanced survival and development of offspring. Effects on offspring could result either directly from paternal care or indirectly from enhanced maternal condition as a result of reduced maternal workload. Thus you will also examine whether presence of the male influences level of maternal care.

MATERIALS

The following are the minimum materials required for conducting the lab exercise.

This exercise requires 2 breeding pairs of prairie voles and 2 seminatural environments (hereafter often called pens). One bale of hay and a 2.2-cubic-foot bag of peat moss will be sufficient for 2 pens (additional hay and peat moss will be needed for maintenance of animals in cages before and after they have been observed in the pens). Each pen should have 2 water bottles and food (a mix of rodent chow, rabbit chow, coarse cracked corn, sunflower seeds, and whole oats). Each observer will need a timer, stopwatch, pencil, and clipboard. Mechanics' creepers (1 for under each pen) substantially enhance comfort during data collection.

Voles can be marked with human hair dye or a commercial fur dye (in this exercise, only the male member of the breeding pair should be marked). The commercial fur dye requires Nyanzol dye crystals, 95% ethanol, and hydrogen peroxide (over the counter). To prepare the dye, you will need nonlatex gloves, a 10-milliliter graduated cylinder, a small (25- 50-milliliter) beaker, a teaspoon for measuring dye crystals, and a device to heat the mixture. Dye is applied with a Q-tip or small paintbrush.

In addition to the above materials that are directly related to the exercise, several supplies are needed for animal housing and care before and after the exercise. Voles should be housed in plastic cages (26 × 51 × 30 centimeters) with stainless steel lids. Peat moss and pine shavings should be used for bedding and hay for cover. Each cage will need a water bottle and food (use the mix described in the previous paragraph). It is necessary to wear gloves when handling voles; small containers can be used to transfer animals when changing cages or placing animals into pens.

PROCEDURE

First Lab Meeting

During this lab you will set up the pens, individually mark the male vole(s) with dye, and introduce the voles to the pens. The particular steps in each of these procedures are detailed in this section. You will also choose focal animals to observe. Outside of class (that is, not during the lab period), you will conduct informal observations of the voles and develop behavioral categories.

Setting up a Seminatural Environment

To set up a pen, place a clean garbage bag in a large garbage can and add 8 scoops of peat moss (1 scoop = 7.5 cups) and 8 cups of water, and mix. The peat should be moist but not too wet. Then empty the moist peat into the pen and spread it evenly across the surface of the Plexiglas to form a substrate 3 centimeters thick. Construct runways by clearing the peat down to the Plexiglas; runways should be about 5 centimeters wide. Although the voles will modify these runways, it is essential to construct initial runways to permit immediate viewing of voles from below the pens. It works best to clear the peat around the sides of the pen and then make some connecting runways. Add food to one corner of the pen that has been cleared of peat. Hang water bottles in two other corners (do not hang a bottle over the food, because dripping water will spoil the food) and make sure that sipper tubes are low enough for adults *and weanlings* to reach (sipper tubes should be about 2 centimeters from the Plexiglas floor). Finally, add a layer of hay cover 10 centimeters thick.

Dyeing Males and Introducing Pairs or Females to Seminatural Environments

To mix the human hair dye for marking the male vole, follow the directions on the box. To mix the commercial fur dye for marking the male vole, first put on nonlatex gloves and add 1/6 teaspoon of Nyanzol dye crystals to 3 milliliters of 95% ethanol; heat this mixture until the dye crystals dissolve. Second, remove the mixture from the heat source and add 4 milliliters of OTC hydrogen peroxide; if the mixture fizzes, then too much hydrogen peroxide has been added, and you should start over because this mixture will not result in a permanent dye mark. Third, allow the mixture to cool for 1 minute, and then apply dye to the belly of the male prairie vole with a Q-tip or small paintbrush. APPLICATION OF DYE SHOULD BE PERFORMED BY THE INSTRUCTOR OR BY SOMEONE EXPERIENCED IN HANDLING SMALL MAMMALS. Do not apply dye to the female at all, because she is lactating and dye might irritate her nipples. When dyeing the male, avoid his mouth, eyes, nose, and genitals. Fourth, allow the male to "recover" for 5–10 minutes after the dye has been applied, and then check to ensure that the dye mark is black and obvious. Finally, place a breeding pair into one pen and a female from another breeding pair into another pen. Ideally, you will be placing animals in pens on about day 20 of a 22-day gestation period; this schedule gives the animals sufficient time before the pups are born to adjust to their new environment, modify runways, and build a nest.

Choosing Focal Animals to Observe

Two possibilities exist with respect to your choice of focal animals. First, you may choose to observe the adult female of the pair and the single adult female; these observations permit comparison of the maternal behavior of females housed with their mate and that of females housed without their mate. Such observations address the question of whether presence of the male results in reduced maternal care and workload. Alternatively, you may choose to observe the adult male and adult female of the pair; these observations will document level of parental care by males and permit comparison of the parental behavior of males and females. Choose one of these two options.

Conducting Informal Observations and Developing Behavioral Categories (to be completed outside of class time)

You will conduct informal observations of the voles from below the pens in order to become familiar with their behavior. Such informal observation (no data sheets permitted!) should last at least 30 minutes for each of the

Table 14.1 **Continuous time measures of the behavior of adult prairie voles.**

Treatment (pair or single female): _____ Number of pups:_____ Age of pups: _____
Focal animal (male or female): _____ Date:_____ Time: _____

Behavior	Check Mark For Each Occurrence	Total Frequency
Nurse		
Contact Pup(s)		
Contact Mate		
Groom Pup		
Groom Mate		
Groom Self		
Build Nest		
Retrieve Pup		
Approach Pup		
Withdraw From Pup		
Approach Mate		
Withdraw From Mate		
Eat or Drink		
Build Runways		
Inactive Alone		

Duration (in seconds) for time spent in natal nest by adult male or female: _____
Developmental milestones for pups (indicate yes or no):
Fur: _____ Eyes open: _____ Out of nest: _____
Eat solid food: _____ Attach to nipple: _____

two pens and should be done at least 1 day after the voles are introduced to the pens. Of course, patterns of parental behavior will not be displayed by the adults until after the young are born; however, nonsocial behavior (such as Build Runways) and interactions between mates (such as Groom Mate) will be displayed before and after pups are born. List the patterns of behavior that you observe, and then spend a few more minutes observing the voles and adding any additional categories to your list. Check your list with the sample data sheet (Table 14.1) and modify your list as necessary.

Once you have settled on the behavioral categories that you wish to measure, define each category. Such definitions should be unambiguous and easily understood by another student conducting observations. For example, the category Approach Pup might be defined as "an adult approaches a pup in the runway at a distance of less than 15 centimeters." You may wish to spend more time observing the voles and modifying your categories and definitions.

In addition to defining categories of adult male or female behavior, also define some developmental milestones for the pups. Because pups are not individually marked, such milestones can be described in terms of when they first occur (e.g., the age at which the first pup of a litter is observed out of the nest) or when they last occur (e.g., the oldest age at which a pup from the litter was observed attached to a nipple). Check your list of milestones against the data sheet (Table 14.1) and modify it as necessary. Be sure that your definitions are unambiguous. For example, if you choose to record when fur appears, will you score when the first pup has fur or when all pups have fur? Must fur cover the entire body or simply be visible on any part of the body? You may design your own data sheet or use the sample data sheet (Table 14.1).

Bring your list and definitions of behavioral categories and developmental milestones to your next lab meeting. Also bring the data sheet you intend to use during formal data collection.

Second Lab Meeting

In this lab you will discuss your behavioral categories and developmental milestones (each student will probably have a somewhat different list, and that is fine). As a class you will discuss the design of your data sheet and how to score the behavior of each of your focal animals during 30-minute observation periods. You will conduct formal observations of your focal animals outside of class time (on each of 3 days, you will observe each of your two focal animals for a 30-minute period). Remember that some students will score the behavior of the adult female of the pair and of the adult female without her mate, and others will score the adult male and adult female of the pair. Here are some general suggestions regarding the mechanics of data collection: (1) Use a timer to monitor the length of the observation period. (2) Use a stopwatch to measure the absolute duration for the spatial location "In the Natal Nest" for the adult male or female (the natal nest is the nest containing the pups). (3) Measure absolute frequency for all other patterns of behavior (e.g., Groom Pup, Build Runways). The total duration for "In the Natal Nest" should be expressed as minutes per 30-minute observation period. Total frequencies for all other patterns of behavior should be expressed as total occurrences per 30-minute observation period. In addition to scoring the behavior of your focal animals, note the number of pups present relative to the litter size at birth, as well as any developmental milestones reached by pups during your observation.

Third Lab Meeting

In this lab, you will discuss your experiences collecting data and any questions or problems that may have arisen. You will also discuss statistical issues associated with repeated measures and litter effects.

Outside of class, you will continue formal data collection (on each of 3 days, you will observe each of your two focal animals for a 30-minute period).

Prairie voles provide care to their young for 3 weeks after birth. Here is a brief summary of what you will do during each of the 3 weeks.

Week 1

Conduct preliminary observations, and develop and define behavioral categories and developmental milestones.

Week 2

Practice collecting data (score the behavior of each of your focal animals for one 30-minute period) and then begin formal data collection. With respect to the latter, select three days (such as days 8, 10, and 12) and score the behavior of each of your two focal animals for one 30-minute period on each day.

Week 3

Continue formal data collection by again selecting three days (such as days 14, 16, and 18) and scoring the behavior of each of your two focal animals for one 30-minute period on each day.

This 3-week schedule will enable you to monitor how behavior changes with age of the young. In the case of the pair, a new litter may arrive when the pups are 22 or 23 days old.

Removal of Animals from the Pens

When data collection on a particular family group has been completed, remove the animals from the pens and clean the pens. First, carefully remove the hay cover and discard it. (Make sure that animals are not hiding in the hay!) Second, catch each animal and place the family together in a cage with appropriate food, bedding, and a water bottle. Pups are weaned at about 20 days of age and can be housed separately from their parents at this time; if the adult female is pregnant, it is best to house the older offspring in a separate cage to prevent overcrowding. Remove and discard all peat from the pen, and clean the Plexiglas surface with soap and water. Clean the water bottles.

HYPOTHESES AND PREDICTIONS

This exercise will test the hypothesis that paternal care evolved in prairie voles because care by males increases male reproductive success through increased survival and enhanced development of offspring. Such effects on offspring may result directly from paternal care or indirectly from enhanced maternal condition as a result of reduced maternal workload and more time spent foraging. Accordingly, you might make the following predictions: (1) Survival of pups will be greater in litters reared with fathers than in litters reared without fathers. (2) Pups reared with fathers will develop more rapidly than pups reared without fathers. And (3) females housed with their mates spend less time in maternal behavior and more time out of the nest foraging than do females housed without their mates.

Are there other hypotheses that could be tested with the present exercise? If so, develop predictions for each. What hypotheses might be tested by comparing male and female parental behavior? What predictions would you make?

DATA RECORDING AND ANALYSES

A sample data sheet (Table 14.1) is provided for your use, or you may design your own. Formal statistical analyses are usually not part of this exercise because of the small number of animals for which observations are typically completed. The long-term nature of the observations (each female or pair is maintained in a pen and is available for observation for approximately 20 days) and the limited number of pens available typically result in complete observations for a single pair and female; a minimum of 6 pairs and 6 females is usually required for formal statistical analyses of such data. Nevertheless, some statistical issues are raised by the experimental design and should be discussed. For example, many statistical tests assume that individual data points are independent of one another. Are repeated measurements of the same individual—for example, time spent in the nest by a particular female on days 1, 5, 10, 15, and 20—independent? What is the sample size in this example? Is it number of observations (5) or number of subjects (1)? Similarly, would body weight measurements from 3 pups within the same litter be independent? Is the sample size the number of pups (3) or the number of litters (1)? How might you address these two issues of independence—repeated measurements of individuals and "litter effects"—in your experimental design, data collection, and statistical analyses?

Although formal statistical analyses are not appropriate here, you may still wish to graph some of your data. One possibility is to make a bar graph comparing the mean number of minutes (from your 6 separate observations across weeks 2 and 3) that females with and without their mate spent in the natal nest (one bar represents data from the female of the pair, and the other bar represents data from the single female). If, on the other hand, you chose to score the behavior of the adult male and female of the pair, then one bar would represent the male data and the other bar the female data. Include error bars in your graphs. Alternatively, you might make a line graph depicting the number of minutes spent in the natal nest by each of your focal animals in relation to age of the young; in this case, minutes would be on the *y*-axis and days would be along the *x*-axis. Because you will probably have data on pup survival and development for only two litters (one litter of pups reared with fathers and one litter of pups reared without fathers), these data are best presented in the text or in a table.

QUESTIONS FOR DISCUSSION

Females housed with their mates typically mate during postpartum estrus and thus are pregnant while caring for young. In contrast, females separated from their mates before giving birth have no opportunity to mate during postpartum estrus and thus are not pregnant while caring for their litter. Why is the current experimental design problematic? How might you change the design to address the problem(s)?

Seminatural environments, such as the one used in this laboratory exercise, are designed to mimic, to some extent, the natural environments in which animals are found and thereby to promote the display of species-typical behavior. Did providing space and hay cover promote the display of species-typical behavior in prairie voles? How do your findings compare to what is known about free-living prairie voles or to findings derived from studies of prairie voles under more restricted laboratory conditions? What changes could be made to the present seminatural environment to mimic field conditions more closely? Were there costs (such as occasional poor visibility) associated with observing the animals in the pens? Overall, did the benefits of using such a system outweigh the costs?

You examined the effects of male presence on the development and survival of young to weaning. What other characteristics of offspring might be measured to assess the effects of male parental care? Which characteristics would be most useful and why? How might you assess the effects on the mother of help from her mate?

What conditions of the physical and social environment might promote parental care by free-living male prairie voles?

How do your findings compare with what is known about the effects of paternal care in other rodents? (See the "Additional Suggested References" section.)

LITERATURE CITED

Carter, C. S. & Getz, L. L. 1993. Monogamy and the prairie vole. *Scientific American*, **268**, 100–106.

Clutton-Brock, T. H. 1991. *The Evolution of Parental Care.* Princeton, NJ: Princeton University Press.

Dewsbury, D. A. 1985. Paternal behavior in rodents. *American Zoologist*, **25**, 841–852.

Getz, L. L. & Carter, C. S. 1996. Prairie-vole partnerships. *American Scientist*, **84**, 56–62.

Getz, L. L., McGuire, B., Pizzuto, T., Hofmann, J. E. & Frase, B. 1993. Social organization of the prairie vole (*Microtus ochrogaster*). *Journal of Mammalogy*, **74**, 44–58.

Gubernick, D. J., Wright, S. L. & Brown, R. E. 1993. The significance of father's presence for offspring survival in the monogamous California mouse, *Peromyscus californicus*. *Animal Behaviour*, **46**, 539–546.

Kleiman, D. G. & Malcolm, J. R. 1981. The evolution of male parental investment. In: *Parental Care in Mammals* (ed. D. J. Gubernick & P. H. Klopfer), pp. 347–387. New York: Plenum Press.

McGuire, B. & Novak, M. 1984. A comparison of maternal behavior in the meadow vole (*Microtus pennsylvanicus*), prairie vole (*M. ochrogaster*) and pine vole (*M. pinetorum*). *Animal Behaviour*, **32**, 1132–1141.

Oliveras, D. & Novak, M. 1986. A comparison of paternal behavior in the meadow vole *Microtus pennsylvanicus*, the pine vole *M. pinetorum* and the prairie vole *M. ochrogaster*. *Animal Behaviour*, **34**, 519–526.

Thomas, J. A. & Birney, E. C. 1979. Parental care and mating system of the prairie vole, *Microtus ochrogaster*. *Behavioral Ecology and Sociobiology*, **5**, 171–186.

Trivers, R. L. 1972. Parental investment and sexual selection. In: *Sexual Selection and the Descent of Man*, 1871–1971 (ed. B. Campbell), pp. 136–179. Chicago: Aldine.

Williams, G. C. 1975. *Sex and Evolution*. Princeton, NJ: Princeton University Press.

ADDITIONAL SUGGESTED READING

Bart, J. & Tornes, A. 1989. Importance of monogamous male birds in determining reproductive success. *Behavioral Ecology and Sociobiology*, **24**, 109–116.

Dudley, D. 1974. Contributions of paternal care to the growth and development of the young in *Peromyscus californicus*. *Behavioral Biology*, **11**, 155–166.

Elwood, R. W. & Broom, D. M. 1978. The influence of litter size and parental behavior on the development of Mongolian gerbil pups. *Animal Behaviour*, **26**, 438–454.

Hartung, T. G. & Dewsbury, D. A. 1979. Paternal behavior in six species of muroid rodents. *Behavioral and Neural Biology*, **26**, 466–478.

McGuire, B. 1997. Influence of father and pregnancy on maternal care in red-backed voles. *Journal of Mammalogy*, **78**, 839–849.

Shilton, C. M. & Brooks, R. J. 1989. Paternal care in captive collared lemmings (*Dicrostonyx richardsoni*) and its effect on development of offspring. *Canadian Journal of Zoology*, **67**, 2740–2745.

Wang, Z. & Novak, M. A. 1992. Influence of the social environment on parental behavior and pup development of meadow voles (*Microtus pennsylvanicus*) and prairie voles (*M. ochrogaster*). *Journal of Comparative Psychology*, **106**, 163–171.

Wuensch, K. L. & Cooper, A. J. 1981. Preweaning paternal presence and later aggressiveness in male *Mus musculus*. *Behavioral and Neural Biology*, **32**, 510–515.

CHAPTER 15

The Effect of Prenatal Visual Stimulation on the Imprinting Responses of Domestic Chicks: An Examination of Sensitive Periods During Development

WENDY L. HILL
Department of Psychology, Lafayette College, Easton, PA 18042, USA

INTRODUCTION

The purpose of this laboratory experiment is to investigate the relationship between perceptual development and social attachment by exploring the effects of premature visual experience on the imprinting responses of domestic chicks (*Gallus gallus domesticus*).

Background

Early in life many animals form a social attachment to specific objects, which thereafter become important in eliciting particular behaviors. This phenomenon is called **imprinting**. Imprinting is prominent in **precocial** birds (the term *precocial* means that their chicks are relatively independent at hatch, covered with down, eyes open, and soon able to feed themselves). As an example, both ducklings and goslings, soon after hatching, imprint on their mothers. This imprinting results in the young following their mother away from the nest and maintaining close proximity to her. In a classic study, Lorenz (1937) showed that goslings could be induced to follow an object other than their mother if they were exposed to it during a demonstrable **critical period**. A critical period is a time, usually early in development, during which some event (such as an experience or the presence of a hormone) has a long-lasting effect. Lorenz determined that when goslings saw a moving object during approximately the first 24 hours following hatch, they developed an imprinting response to that object regardless of what it was; indeed, for some birds the primary moving object observed was Lorenz himself, and as a result, they formed a social attachment to the experimenter.

Lorenz's findings led researchers to attempt to pinpoint the exact timing of the critical period. This was usually accomplished by exposing birds to an object at various ages and subsequently measuring their preference for the object (Hess 1973). When ducklings and chicks are exposed to a moving object for a limited period, the proportion that develop a preference for the object initially increases and then typically declines with age (see Fig. 15.1). Some investigators suggested that the conditions underlying the occurrence of the critical period for imprinting are the direct result of a highly invariant sequence of maturational events (such as age calculated from the beginning of embryonic development or number of hours since hatch). Others proposed

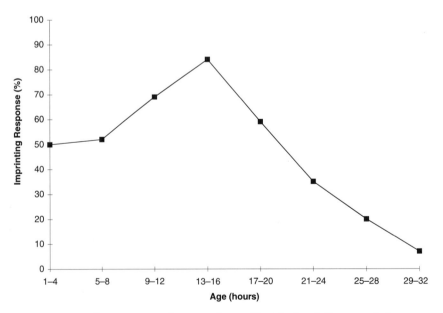

Figure 15.1 **The percentage of mallard ducklings giving an imprinting response during four test conditions as a function of age. (Adapted from Hess 1959)**

that the timing of the critical period resulted from changes brought about through the interaction between the developing organism and its sensory environment. In support of the latter interpretation, Moltz & Stettner (1961) found that ducklings 48 hours old, an age that was thought to be past the critical period for imprinting, could form an attachment to a moving object if they had been deprived of the ability to perceive visual forms since hatching; thus the age at which an imprinting model would be effective in eliciting following could be extended under certain circumstances. As a consequence of this and other research, some prefer to use the term **sensitive period**, which implies that things develop more easily at some ages than at others, rather than the term *critical period*, which suggests an all-or-nothing effect.

Imprinting is acquired readily during a sensitive period because the nervous system is "primed" to be altered in a narrowly defined way during this time (Bolhuis 1991). Bateson (1979) suggests that part of this sensitivity is the result of changes in the efficiency of the visual system. For example, visual experience with patterned light has a general facilitating effect on the development of visually guided behavior and may serve to strengthen neural connections in areas that mediate imprinting (Bolhuis 1991). The neural and sensory systems of precocial chicks, having developed rapidly during incubation, are quite mature at hatch. The onset of sensory functioning in precocial chicks proceeds according to a pattern common to all vertebrates: the tactile system develops first, followed by the vestibular, chemical, and auditory systems and then the visual system. Given this apparent regularity, it is interesting that premature stimulation of the visual system, a later-developing sensory system, can alter the development of the auditory system, an earlier-developing sensory system. Ordinarily, avian embryos would not experience patterned visual stimulation until after hatching, but by employing a simple but effective procedure, Lickliter (1990) was able to expose bobwhite quail (*Colinus virginianus*) to patterned visual stimulation prior to hatching. In his experiment, Lickliter took advantage of the fact that near the end of incubation, avian embryos penetrate the **air space** (the region between the shell membrane and the hard surface of the egg—see Fig. 15.2) with their bills, inflate their lungs for the first time, and start to breathe. Lickliter took quail in this phase of incubation and removed the shell and the inner shell membrane in the area surrounding the air space so that the chicks' heads were visible, while the rest of their bodies were still inside the egg. The chicks were then exposed to patterned visual stimulation, a 15-watt light pulsing at 3 cycles per second, until they hatched. Lickliter (1990) demonstrated not only that the visual system was more developed for the quail chicks who had received prenatal visual stimulation but also that the maturation of the earlier-developing auditory system was accelerated in these birds.

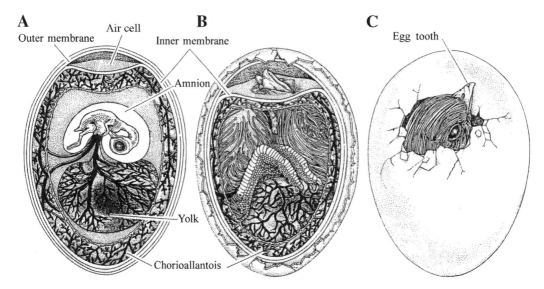

Figure 15.2 **The incubation period for domestic chicks lasts 21 days. A. 5-day-old embryo. The chorioallantois protrudes from the embryo and covers the inner shell membrane with a net of capillaries that supply the embryo with oxygen and remove carbon dioxide. B. A 19-day-old embryo. Note the tucked position of the chick. At this age the embryo penetrates into the air sac with its beak and inflates its lungs for the first time. C. A chick that has broken through an eggshell with its egg tooth. (From Rahn et al. 1979)**

Given that Moltz & Stettner (1961) found that the sensitive period for imprinting could be extended by depriving chicks of patterned visual stimulation, we will examine whether the period can be shortened by giving premature visual experience. Using the procedure described in Lickliter (1990), domestic chicks will be given prenatal visual stimulation and, following hatching, will be exposed to a mechanized toy. The strength of their subsequent imprinting response to the toy will then be assessed. The imprinting responses of these chicks will be compared to those of chicks that did not receive early visual stimulation.

MATERIALS

Each group of two or three students will need two 19-day-old chick embryos; small forceps; a dental drill bit or a small, sharp nail; cotton swabs; and a stopwatch. In addition, the lab requires a strobe light, tabletop incubator, imprinting model, imprinting chamber, and plastic leg bands for newly hatched chicks.

PROCEDURE

The laboratory experiment will take place over three sessions: one session to conduct the egg-opening procedure and expose the embryos to prenatal visual stimulation, one session to expose the chicks initially to the imprinting model, and a final session to test the chicks' imprinting responses. These sessions will occur within a 4-day period.

In preparation for the experiment, fertile eggs from domestic hens have been placed in two tabletop incubators. Chick embryos need 21 days to develop: you will be given eggs containing 19-day-old embryos. At this age, the chicks will have begun their arduous task of hatching out of the egg. First, they break through both the inner and outer shell membranes. Once into the air chamber, chicks inflate their lungs for the first time and start to breathe. The chicks then start to chip away at the eggshell using the **egg tooth**, a hard calcification on the tip of the beak (Fig. 15.2). Chicks first break through the shell by creating a small hole on the blunt side of the egg. When this small hole occurs, the egg is said to be **pipped**. It is best to use pipped eggs in the following procedure, although you can use chicks that have yet to crack through the eggshell, as long as they have broken through the inner membrane of the egg and have started to breathe.

Take one egg and remove the top portion of the shell, using the forceps to break off small pieces around the hole created by the pipping. If the egg is not pipped, then make a small hole yourself in the blunt side of the egg, using a dental drill bit or a small, sharp nail. Carefully put pressure on the dental drill bit or nail, and slowly twist it back and forth so that you puncture through the shell. Work slowly and carefully with your forceps as you break off pieces of egg shell. You will begin to see the head of the chick and the chick's right eye. Avian embryos maintain an invariant postural orientation during development (Freeman & Vince 1974), so only the right eye of the chick will be visible (Fig. 15.2). Your chick may also vocalize, which is normal (Hess 1972). As you pull the shell membrane from around the chick's head, there may be a small amount of bleeding. Use the cotton swabs to wipe up any blood.

Once the chick's head has been exposed, place the egg back into the incubator. After everyone has placed their eggs in the incubator, a strobe light will be positioned 4 centimeters above the incubator. The strobe light has a 15-watt bulb set to pulse at 3 cycles per second. The chicks in this group will receive prenatal visual stimulation by being exposed to the patterned light for the remaining hours prior to hatch. Both Gottlieb (1988) and Heaton & Galleher (1981) have shown that the egg-opening procedure does not effect the bird's survivability, so you should enjoy good hatching success. The class will repeat the egg-opening procedure with another set of 19-day-old embryos. These eggs will be placed in a different incubator and will not receive patterned visual stimulation. These chicks will serve as the control group. After hatch, numbered leg bands will be placed on all of the chicks, and they will be housed together in brood units in continuous darkness; we expect the darkness to prevent the chicks from imprinting on one another.

Within 24 hours after hatch, each lab team will train two chicks, one from each group, to the imprinting model, a brightly colored mechanized toy with motorized legs. When turned on, the toy will ambulate around the test chamber. You will be blind to group assignment and so will be unaware of which of your chicks was exposed prenatally to patterned visual stimulation and which was not. Place the model in the imprinting chamber within the grid numbered 1. Make sure the model is on the outer edge of the track inside the chamber with its side touching the wall. This will help the model stay in a circular path as it moves around the chamber. Once the model is correctly positioned, place the first chick you are training next to the model and begin timing. This training session will last 1 hour; this much time is needed for a strong imprinting response (Bateson et al. 1973). During the session, record qualitative data about the chick's behavior. You might also want to practice recording the number of grids between the model and the chick, using the part of the model's body closest to the chick as your reference point. In addition, see how well you can count the number of distress vocalizations produced by the chick. Distress vocalizations are easily identified: they are sharp, intense calls produced by chicks. After the session is finished, return the chick to its brood unit and test a second chick.

One day following the imprinting exposure session, each team of students will come in to test the strength of their subjects' imprinting responses. Just as in the training session, each chick will be exposed individually to the imprinting model; however, the test session will only last 20 minutes. During this session, you will measure the distance between the imprinting model and the chick by counting the number of grids between the model and the chick at 30-second intervals. Use the part of the model's body closest to the chick as your reference point for this measure. One experimenter should keep track of the distance data. Another experimenter will count the number of distress vocalizations produced by the chicks. Every minute, the number of distress vocalizations produced by the subject should be written on the data sheet. The third experimenter will hold the stopwatch to keep track of time and quietly inform the others when each 30-second interval and each 1-minute interval are over. Your two dependent measures will help you to determine the strength of the imprinting response among the subjects. The stronger the imprinting, the closer the chick will stay to the model. Similarly, when young chicks are in the presence of imprinted objects, they do not emit many distress vocalizations.

DATA RECORDING AND ANALYSES

Record your data using the data sheets that follow. For data analysis, the class data will be pooled. Each chick will contribute one data point for the distance measure and one data point for the distress vocalizations to the pooled class data set. From the pooled data, calculate the mean and standard deviation of each dependent variable for each group. Summarize your data by producing two bar graphs, one for each dependent variable. On the bar graph present the means for the control group and the visual stimulation group. The y error bars should indicate standard deviations about the means.

The experimental hypothesis for this experiment is that the sensitive period for developing an imprinting response will be shortened by giving prenatal visual stimulation to domestic chick embryos. If this hypothesis is true, then chicks from the experimental group will be past the sensitive period for developing imprinting and will stay farther away from the imprinting model and produce more distress calls than chicks from the control group. The null hypothesis is that prenatal visual stimulation has no effect on the sensitive period for developing an imprinting response. If the null hypothesis is true, then there will be no difference between the imprinting responses of the chicks in the experimental group and those of the chicks in the control group. Using the pooled data set, determine whether the control chicks and the chicks exposed to light differ by calculating an independent t test for each dependent variable. If your sample size is small, then instead of calculating t tests, use a Mann–Whitney U test for each of your dependent variables.

QUESTIONS FOR DISCUSSION

Why is there a sensitive period for imprinting? What is the adaptive significance of this behavior?

Given the results found by the class, what type of follow-up experiment would you conduct? What information would you seek to gain via this future experiment?

Researchers have found that some strains of domestic fowl do not imprint as well as other strains. For example, strains that have been selected for egg laying, which do not become "broody" and can no longer hatch chicks themselves, show weak imprinting responses. What does this tell you about artificial selection regimes that focus on one particular trait and about the complexity of genetic influences over behavior?

If your instructor assigns it, read Bolhuis (1991) before this discussion. According to the information presented in Bolhuis (1991), what neural regions influence the development of imprinting? How might the prenatal visual stimulation affect these regions?

LITERATURE CITED

Bateson, P. P. G. 1979. How do sensitive periods arise and what are they for? *Animal Behavior*, **27**, 470–486.

Bateson, P. P. G., Rose, S. P. R. & Horn, G. 1973. Imprinting: lasting effects of uracil incorporation into chick brain. *Science*, **181**, 576–578.

Bolhuis, J. J. 1991. Mechanisms of avian imprinting: a review. *Biological Reviews*, **66**, 303–345.

Freeman, B. M. & Vince, M. A. 1974. *Development of the Avian Embryo*. London: Chapman and Hall.

Gottlieb, G. 1988. Development of species identification in ducklings: XV. Individual auditory recognition. *Developmental Psychobiology*, **21**, 509–522.

Heaton, M. B. & Galleher, E. 1981. Prenatal auditory discrimination in the bobwhite quail. *Behavioral and Neural Biology*, **31**, 242–246.

Hess, E. H. 1959. Imprinting. *Science*, **130**, 133–141.

Hess, E. H. 1972. "Imprinting" in a natural laboratory. *Scientific American*, **227**, 24–31.

Hess, E. H. 1973. *Imprinting: Early Experience and the Developmental Psychobiology of Attachment*. New York: Van Nostrand.

Lickliter, R. 1990. Premature visual stimulation accelerates intersensory functioning in bobwhite quail neonates. *Developmental Psychobiology*, **23**, 15–27.

Lorenz, K. 1937. The companion in the bird's world. *Auk*, **54**, 245–273.

Moltz, H. & Stettner, L. J. 1961. The influence of pattern-light deprivation on the critical period for imprinting. *Journal of Comparative and Physiological Psychology*, **54**, 279–283.

Rahn, H., Amos, A. & Paganelli, C. V. 1979. How bird eggs breath. *Scientific American*, **240**, 46–55.

Prenatal Visual Stimulation and Imprinting Data Sheet

Dependent Variable: Distance Between Chick and Imprinting Model

Experimenters' names: _____

Subject number: _____

At the end of each 30-second interval, record on the following table the number of grids between the imprinting model and the chick.

0:30		1:00	
1:30		2:00	
2:30		3:00	
3:30		4:00	
4:30		5:00	
5:30		6:00	
6:30		7:00	
7:30		8:00	
8:30		9:00	
9:30		10:00	
10:30		11:00	
11:30		12:00	
12:30		13:00	
13:30		14:00	
14:30		15:00	
15:30		16:00	
16:30		17:00	
17:30		18:00	
18:30		19:00	
19:30		20:00	

Total number of grids across the 40 samples: _____

Average (total/40): _____ (This average will contribute to the pooled class data set.)

Prenatal Visual Stimulation and Imprinting Data Sheet

Dependent Variable: Distress Vocalizations

Experimenters' names: _____

Subject number: _____

Count the frequency of distress vocalizations each minute. You may want to keep track of these vocalizations by making a tick mark each time one is made.

Minute	Number of Distress Vocalizations
1	
2	
3	
4	
5	
6	
7	
8	
9	
10	
11	
12	
13	
14	
15	
16	
17	
18	
19	
20	

Total number of distress calls across the 20 minutes:_____

Average (total/20): _____ (This average will contribute to the pooled class data set.)

CHAPTER 16

Development of Thermoregulation in Altricial Rodents

GAIL R. MICHENER AND T. DIC CHARGE
*Department of Biological Sciences, University of Lethbridge, Lethbridge, Alberta,
Canada T1K 3M4*

INTRODUCTION

Animals are broadly classified into two groups, **poikilotherms** and **homeotherms**, on the basis of the extent to which they can regulate body temperature (T_b) independently of ambient temperature (T_a). Virtually all invertebrates, fish, amphibians, and reptiles are poikilotherms, whereas birds and mammals are homeotherms. Poikilotherms are characterized by a variable body temperature that fluctuates with daily and seasonal variations in environmental temperature and by a very limited ability to generate metabolic heat for thermoregulatory purposes. Homeotherms are characterized by a high and constant core body temperature regardless of environmental temperature; in the case of placental mammals, including humans, core T_b is 37–38°C. Independence of T_b from T_a in homeotherms is achieved through **physiological thermoregulation**, particularly the ability to generate sufficient internal metabolic heat to offset the rate of heat loss from the body surface. Organisms that use internally generated metabolic heat to thermoregulate are called **endotherms**, whereas those that depend primarily on external sources of heat are called **ectotherms**. Both endotherms and ectotherms use **behavioral thermoregulation** to adjust rates of heat gain and heat loss from the body surface. Behavioral thermoregulation includes modifications of body posture to alter the amount of exposed surface area, social cooperation such as huddling, and movement into preferred microenvironments to take advantage of heat exchange by convection, conduction, and radiation.

All adult mammals are endothermic homeotherms, but infants of some species have limited abilities to achieve a constant T_b through physiological thermoregulation. Mammalian infants can be broadly categorized at birth as either **precocial** or **altricial**. Precocial infants (such as fawns and lambs) are typically born as relatively large singletons or twins with a covering of fur and a well-developed nervous system; they can see and hear at birth and they are sufficiently mobile to follow the mother within hours of birth. In contrast, altricial infants (such as kittens and rat pups) are relatively small at birth and are members of a litter that is confined to a nest for the first few days or weeks of life. Altricial infants have closed eyes and ears, little or no fur, and limited mobility at birth. Precocial and altricial infants also differ in their thermoregulatory abilities at birth. Precocial infants can use physiological thermoregulation to maintain a warm internal body temperature even in a cool environment. However, altricial infants have only a limited ability to defend their internal body temperature by physiological means during the first few days of life. Instead, altricial infants depend on their mother's warmth for heat gain and on a combination of huddling with littermates and the insulation provided by the nest to minimize heat loss in the mother's absence (Blumberg & Sokoloff 1998).

The ability to thermoregulate physiologically develops gradually as altricial infants grow, become furred, and acquire the ability to generate sufficient heat by thermogenesis to offset heat loss. By briefly isolating infants from the mother, littermates, and nest material, it is possible to identify three stages in the development of physiological thermoregulation by altricial rodent infants (Hill 1976; Okon 1970). The youngest infants lose heat faster than they can generate heat, so their body temperature declines rapidly even at moderate ambient temperatures (such as 25–30°C). As they age, infants pass through a transitional stage in which they can maintain homeothermy when exposed to moderate ambient temperatures before attaining the final stage when they have adequate heat-generation and heat-retention capabilities to maintain homeothermy even at cool ambient temperatures (such as 10–15°C). Acquisition of full thermoregulatory ability usually coincides with the age at which infants are weaned and leave the nest.

Although altricial infants cannot thermoregulate by physiological means alone, they do use behavioral thermoregulation to keep the body close to the temperature that is optimal for growth and development (Blumberg & Sokoloff 1998; MacArthur & Humphries 1999). Infants are **thermotaxic**—that is, they detect and approach heat sources (such as the mother, littermates, and warm nest material) and they huddle against warm objects to gain heat and minimize heat loss (Alberts 1978; Dolman 1980; Kleitman & Satinoff 1982). Additionally, infants give distress calls to alert the mother when they are stressed. Infants of many rodent species produce **ultrasonic vocalizations** in the range of 30–50 kHz when cold-stressed (Okon 1970; De Ghett 1978; Jans & Leon 1983; Blumberg & Stolba 1996). Although these vocalizations are inaudible to most humans, they can be heard by the mother rodent and elicit her return to the nest. Maximum rate of ultrasonic calling occurs a few days after birth, and calling ceases once infants can maintain homeothermy via endothermy.

MATERIALS

Animals

This exercise uses litters of rodents of four different ages between newborn and weaning.

Equipment and Supplies per Student Group

Each group of students requires 6 glass jars with lids for use as containers to hold 1 infant per jar and a supply of paper toweling to be cut to size and layered into each container to absorb moisture and urine. If containers are to be partially immersed in water baths, obtain weighting material (such as lead shot) to place under the paper toweling. Each group of students requires a waterproof marker such as a wax pencil to label glass containers, a nontoxic marker pen to mark temporarily the skin or fur of individual infants, a pair of rubber-tipped large forceps to aid in removing infants from containers, a hand magnifying lens or dissecting microscope to view pelage quality and density, a metric ruler to measure length and width (diameter) of infants and to measure pelage length, and a stopwatch to time the righting response. A blank data sheet is needed for each assigned infant (see the "Data Recording and Analyses" section for information on designing a data sheet).

Each group of students requires two glass bowls to hold infants for the huddling experiment, materials (such as warming pads, ice packs, and a long, thin box) to construct a thermal gradient that can be warmed to 30–35°C at one end and cooled to 15–20°C at the other end, and cleaning supplies such as a spray bottle with a 10% bleach solution to clean containers between trials.

Equipment and Supplies per Classroom

Each classroom requires controlled-temperature air cabinets or water baths set at three different ambient temperatures, thermometers for insertion into one reference container per air cabinet or water bath, and temperature-recording devices such as thermistors with small-bore animal probes to measure the body temperature of infants. Each classroom requires a triple-beam balance on which to weigh infants.

Optional Equipment

An ultrasonic detector can be used to determine whether infants are producing ultrasounds.

PROCEDURE

Effect of Ambient Temperature on Body Temperature

The instructor will assign infants so that at the end of the exercise, each class of students will have data for a minimum of 5 infants for each age and ambient temperature. For example, if 4 ages of infants and 3 ambient temperatures are available, at least 60 infants will be tested per class of students. If 10 pairs of students are available, each pair will be assigned 3 infants of one age and 3 infants of another age, such that 1 infant of each age can be tested at each of 3 ambient temperatures. The instructor may vary the assignment of infants depending on the number of student groups and the availability of infants and equipment.

Prepare a container for each infant rodent that you have been assigned by placing several layers of paper toweling in the bottom of each container. If the container is to be partially immersed in a water bath, place weights under the paper toweling and check that the container is sufficiently weighted to remain upright in the water bath even when a subject infant stands upright. On each container, use a waterproof marker to identify the litter to which that infant belongs, the age of the infant, and the ambient temperature to which the infant is to be exposed.

Remove each assigned infant one at a time from the cage, place a small mark on the skin or fur to identify its litter of origin, and then immediately measure its body temperature. Gently place the temperature-recording probe on the skin in the axilla (armpit) of the infant, ensuring that the tip of the probe is completely covered by the infant's forelimb and trunk. As soon as the temperature reading has stabilized, record the infant's temperature and place the infant in its assigned container. If an ultrasonic detector is available, listen for the production of ultrasounds just before closing the lid on the container; if an ultrasonic detector is not available, determine whether the hearing range of any class members enables them to hear high-frequency calls produced by infants. Quickly continue this process until all assigned infants are placed in a container; then place each container at the appropriate ambient temperature and immediately note the time. In cooperation with other student groups, ensure that at least one reference jar containing a thermometer, but no infant, is placed at each ambient temperature. For containers that are placed in water baths, ensure that the water level remains below the level of the lid on the container.

The instructor will inform you whether you will be collecting data at 15- or 20-minute intervals. After this interval has elapsed, make brief notes on the appearance of each infant; observe the color of the skin and the posture of the body, and look for signs of shivering or other muscular activity, including locomotion and twitching of the limbs. Leaving the container in its ambient environment, remove the lid and listen for ultrasounds; then remove the infant and quickly measure its body temperature. Record the body temperature and time, and then immediately return the infant to its correct container. If the infant has fouled the paper toweling with urine or feces, remove the contaminated paper and replace it with fresh paper toweling. Record the air temperature in the reference jar. Repeat these procedures at predetermined intervals. For example, if you have been told to collect data at 15-minute intervals until 75 minutes have elapsed, you will have data for 0, 15, 30, 45, 60, and 75 minutes.

After you have obtained the final body temperature, use a stopwatch to time how quickly the infant can right itself when placed on its back; then weigh each infant and record its weight to the nearest gram. Place each infant on a ruler; then measure and record to the nearest millimeter both body length from the tip of the snout to the base of the tail and body diameter at mid-torso. Make a note of the infant's developmental condition, including the distribution of fur on the ventral and dorsal body surfaces; the appearance of extremities such as limbs, digits, and ears; and whether the eyes are sealed closed or open. To quantify pelage characteristics, drape a piece of clear plastic over the infant's back such that a hole 2 millimeters in diameter punched in the plastic is positioned between the infant's shoulder blades. With the aid of a magnifying lens or dissecting microscope, estimate the number and length of hairs in the hole.

Return each infant to its correct home cage. If the mother is still in the home cage, note whether she retrieves the infant. If the mother is not in the cage, time how long it takes for the infant to reach the nest across a standardized distance.

Thermotaxis

Use materials provided by the instructor to make a thermal gradient with a temperature differential of about 15–20°C between the opposite ends; the warm end should not exceed 35°C, and the cool end should not decline below 15°C. Ensure that other conditions (such as light intensity) are consistent along the gradient so that the only variable between the two ends is the temperature difference. Mark the midpoint between the two ends, and then place an infant at the midpoint. Record the infant's age and the time it was placed on the gradient. After 10 minutes, use a metric ruler to measure the distance in millimeters from the nose of the infant to the midpoint. Record this distance. If the infant is on the warm side of the gradient, record the distance as a positive value; if it is on the cool side, record the distance as a negative value. Return the infant to its correct nest. Wipe the surface of the gradient apparatus to remove odors, and then repeat this process with each assigned infant. The thermotaxis tests can be conducted during the intervals between measurements of body temperature of infants in the previous experiment.

Huddling

If sufficient infants are available, use several littermates to make a cluster of infants in one glass bowl, and place a single infant from the same litter in a similar container by itself. Use a marker pen to place a spot on the back of one of the infants in the cluster. Place the containers on a bench at room temperature and compare the behavior of the marked individual in the cluster with that of the solitary individual over a 10-minute interval. Then measure the T_b of the marked individual and the solitary individual.

HYPOTHESES AND PREDICTIONS

This laboratory exercise is designed to explore questions about the effects of infant age on both physiological and behavioral thermoregulatory abilities. In particular, altricial infants are predicted to have limited ability to maintain body temperature physiologically when they are young and isolated from the family group, and they are predicted to acquire this ability as they age, coincident with increasing body mass, changing body shape, and increasing pelage. Within the restrictions of their locomotory capability, infants are predicted to offset their limited physiological thermoregulatory ability through behavioral means, such as seeking sources of warmth and huddling against warm objects.

DATA RECORDING AND ANALYSES

Effect of Ambient Temperature on Infant Body Temperature

Because you are testing infants of different ages at several ambient temperatures, the data on the effect of ambient temperature on body temperature can be organized as an age-by-ambient-temperature matrix. To help you understand the format of the data, construct a matrix table showing all the age-by-temperature combinations that the class will have tested at the end of the experiment.

Prepare a data sheet on which to collect and organize the data acquired for each subject infant throughout the experiment. Use the following information as a guide to designing a data sheet, and then ask the instructor to check the data sheet before commencing the experiment. The instructor may ask you to prepare the data sheet in advance of the lab, in which case use a spreadsheet or word processor application to generate a master sheet on a computer. After the instructor approves the data-recording sheet, make one sheet for each infant you have been assigned. For each infant, you will need to record the following just once: the age, mass, linear dimensions (length and diameter), pelage characteristics, and developmental condition of the infant. Then, for each time interval starting with time 0 minutes, you will record the clock time, the infant's T_b, the T_a of the reference jar, the occurrence of ultrasonic vocalizations, and the infant's posture and movements. You may also wish to include an area for comments in order to record any unanticipated observations.

Once you have returned infants to their home cages and cleaned up the equipment, calculate the average T_a of the reference jar during the experiment. Inspect the data for each infant, and determine its T_b at thermal stability (the T_b that varies by no more than 1–2°C for the remainder of the experiment) and the time at which this T_b was attained. Then calculate the rate of cooling. Report this information, along with each infant's age, to the instructor for eventual distribution to the entire class.

The goal of this portion of the laboratory exercise is to test a biological concept about thermoregulatory capabilities by asking whether physiological processes alone are sufficient for an isolated infant to maintain a high and constant T_b regardless of its age and ambient conditions. Predictions about the effect of infant age on the ability to thermoregulate at different ambient temperatures can be formulated in two ways for statistical testing. One way is as a null hypothesis that states that age has no effect on thermoregulatory ability, in which case all infants are predicted to have a similar T_b (dependent variable) regardless of age or T_a (independent variables). If young and old infants do have the same thermoregulatory capability, then statistical testing of the null hypothesis would demonstrate that T_b does not differ significantly with age or T_a. An alternative hypothesis states that thermoregulatory capability is age-dependent, with thermoregulatory capability increasing with increasing age. If older infants do have better thermoregulatory capability, then statistical testing of the alternative hypothesis would demonstrate that older infants have a significantly higher T_b than younger infants, the difference being most extreme at cooler ambient temperatures. Whereas the null hypothesis involves a two–tailed statistical test, the alternative hypothesis involves a one–tailed statistical test because this hypothesis predicts the direction of the difference—specifically, that older infants will have a higher T_b.

Once the combined class data are returned to you, calculate the mean and standard deviation for stabilized T_b for every age-by-ambient-temperature combination. Prepare a graph that allows you to visualize the mean stabilized T_b for each age-by-ambient-temperature combination by plotting mean stabilized T_b (the dependent variable) against T_a (the independent variable). This graph will have several lines, one for each age tested.

Unless the instructor advises you to use ANOVA (analysis of variance), plan to test the data using Student's unpaired *t* test for independent samples. You can organize the data analysis in two ways, and each is useful. Hold a group discussion to decide which of the two methods of data analysis you will use. One method of analysis is to compare mean stabilized T_b between each age group at a given T_a—for example, by comparing the T_b of 0-day-old and 7-day-old infants that were exposed to 20°C. This within-temperature/across-age comparison compares how the age classes perform at each T_a, which in turn illustrates changes in thermoregulatory competence as infants age. If you select this analysis, Student's *t* test for independent samples is used to determine whether the mean stabilized T_b values between infants of two ages (say, 0- and 7-day-olds) at a given T_a (say, 20°C) are statistically similar. You can repeat this statistical comparison for each pair of age combinations at each T_a. However, inspection of the data should suggest to you that only some of the many possible test comparisons need to be performed in order to establish whether or not stabilized T_b at a given T_a is dependent on infant age. The alternative analysis is to compare mean stabilized T_b within the same age group across the different ambient temperatures—for example, by comparing the T_b of 0-day-old infants that were exposed to 20°C and 25°C. This within-age/across-temperature comparison assesses the ability of infants of different ages to maintain T_b at each T_a. If you select this analysis, Student's unpaired *t* test is used to compare the mean stabilized T_b between the two ambient temperatures (say, 20°C and 25°C) for a given infant age (say, 0 days old). You can repeat this statistical comparison for each pair of T_a combinations at each infant age. Again, inspection of the data should suggest to you that only some of the many possible test comparisons need to be performed in order to establish whether stabilized T_b at a given age is dependent on T_a.

Thermotaxis and Huddling

Report the following information from the thermotaxis and huddling experiments to the instructor for eventual distribution to the entire class: infant age and the distance moved along the thermal gradient, and infant age and T_b of the two test infants, one from a huddle and one alone. For the thermotaxis experiment, use the class data to calculate the mean and standard deviation of the distance moved by infants of each age. For the huddling experiment, use the class data to calculate the mean and standard deviation of the T_b of huddled infants and of solitary infants for each age.

The goal of this portion of the laboratory exercise is to test a biological concept about thermoregulatory capabilities by asking whether behavioral processes contribute to maintenance of T_b, especially in young infants that are unable to maintain homeothermy by physiological means. Analysis of the data to test this biological hypothesis requires examination of a statistical null hypothesis and its alternative. For the thermotaxis experiment, the null hypothesis is that distance moved on the thermal gradient is similar for all ages of infants, and for the huddling experiment, the null hypothesis is that T_b is similar for huddled and isolated infants of the same age. To determine statistically whether the distance moved by infants (dependent variable) differs significantly with age (independent variable), use Student's t test for independent samples to compare mean distance moved between each pair of age classes. To determine statistically whether T_b (dependent variable) differs significantly with group size (independent variable), use Student's t test for independent samples to compare the mean T_b of huddled and solitary infants for each age class. If thermotaxis and huddling behaviors are influenced by age, then several alternative biological explanations could be proposed and the predictions tested statistically. For example, use of a heat source for thermoregulation may decrease with increasing age, leading to the prediction that younger infants move closer to the warm end of a thermal gradient than do older infants and to the prediction that the difference in T_b between huddled and solitary infants is greater for younger than for older infants.

QUESTIONS FOR DISCUSSION

Is skin temperature in the axilla a representative measure of core body temperature? What other temperature measurements could be taken on infants?

Does the ratio of surface area (SA) to volume (V) change as infants grow? If so, in what way does this change in SA:V alter the infant's rate of heat exchange with the environment?

Do the distribution, quantity, and quality of fur on the body surface change as infants grow? If so, in what way do these changes in insulative covering alter the rate of heat exchange with the environment? Once fur has grown, can the insulative value of the pelage be altered?

Are mammalian metabolic rates and growth rates dependent on body temperature? If so, what are the costs and benefits of declines in body temperature during postnatal development?

If declines in body temperature are costly, what behavioral adaptations do infants exhibit to minimize their exposure to low ambient temperatures during postnatal growth?

What maternal behaviors contribute to providing a warm ambient temperature for infants during periods when mothers leave the nest to forage? Do males contribute thermal care to altricial infants?

What is the normal thermal environment of infants in natural circumstances? Do you expect the thermoregulatory abilities of infants to have been attuned to this normal environment through natural selection?

What are the advantages of infants' attaining physiological thermoregulatory competence by the time they are ready to leave the natal nest?

If you were given the task of "designing" a precocial infant rodent that could thermoregulate physiologically at birth, what behavioral and morphological attributes would you give it? What limitations would this design place on the mother rodent?

In some species of seals, infants are born on ice floes and so face an extreme thermal challenge immediately after birth. What additional adaptations would you expect to find in seals that you would not expect in precocial infants born in more benign environments?

Would you categorize human infants as precocial or altricial?

For which other classes of vertebrates in addition to mammals is the categorization of infants as altricial or precocial biologically meaningful?

LITERATURE CITED

Alberts, J. R. 1978. Huddling by rat pups: group behavioral mechanisms of temperature regulation and energy conservation. *Journal of Comparative and Physiological Psychology,* **92,** 231–245.

Blumberg, M. S. & Sokoloff, G. 1998. Thermoregulatory competence and behavioral expression in the young of altricial species—revisited. *Developmental Psychobiology,* **33,** 107–123.

Blumberg, M. S. & Stolba, M. A. 1996. Thermogenesis, myoclonic twitching, and ultrasonic vocalization in neonatal rats during moderate and extreme cold exposure. *Behavioral Neuroscience,* **110,** 305–314.

De Ghett, V. J. 1978. The ontogeny of ultrasound production in rodents. In: *The Development of Behavior: Comparative and Evolutionary Aspects* (ed. G. M. Burghardt & M. Bekoff). New York: Garland, pp. 343–365.

Dolman, T. M. 1980. Development of thermoregulation in Richardson's ground squirrel, *Spermophilus richardsonii. Canadian Journal of Zoology,* **58,** 890–895.

Hill, R. W. 1976. The ontogeny of homeothermy in neonatal *Peromyscus leucopus. Physiological Zoology,* **49,** 292–306.

Jans, J. E. & Leon, M. 1983. Determinants of mother-young contact in Norway rats. *Physiology and Behavior,* **30,** 919–935.

Kleitman, N. & Satinoff, E. 1982. Thermoregulatory behavior in rat pups from birth to weaning. *Physiology and Behavior,* **29,** 537–541.

MacArthur, R. A. & Humphries, M. M. 1999. Postnatal development of thermoregulation in the semiaquatic muskrat (*Ondatra zibethicus*). *Canadian Journal of Zoology,* **77,** 1521–1529.

Okon, E. O. 1970. The effect of environmental temperature on the production of ultrasounds by isolated non-handled albino mouse pups. *Journal of Zoology, London* **162**, 71–83.

ADDITIONAL SUGGESTED READING

Leon, M. 1986. Development of thermoregulation. In: *Developmental Psychobiology and Developmental Neurobiology.* Vol. 8: *Handbook of behavioral neurobiology* (ed. E. M. Blass). New York: Plenum Press, pp. 297–322.

Maxwell, C. S. & Morton, M. L. 1975. Comparative thermoregulatory capabilities of neonatal ground squirrels. *Journal of Mammalogy,* **56**, 821–828.

Newkirk, K. D., Cheung, B. L. W., Scribner, S. J. & Wynne-Edwards, K. E. 1998. Earlier thermoregulation and consequences for pup growth in the Siberian versus Djungarian dwarf hamster (*Phodopus*). *Physiology and Behavior,* **63**, 435–443.

CHAPTER 17

Aggregation and Kin Recognition in African Clawed Frogs (Xenopus laevis)

KATHRYN L. ANDERSON[1] AND BONNIE J. PLOGER[2]
[1]Princeton High School, 807 South 8th Avenue, Princeton, MN 55371, USA
[2]Department of Biology, Hamline University, 1536 Hewitt Avenue, St. Paul, MN 55104, USA

INTRODUCTION

In most cultures, people rely on recognizing their relatives when deciding whom to marry, which children to care for, and whom to help financially. Many other animals, including a wide variety of insects, fish, amphibians, birds, and mammals, also treat their kin differently from unrelated members of the same species (Holmes & Sherman 1983; Hepper 1991). Some widely studied examples of such **kin discrimination** include nestling gulls responding to their parents' calls but not to those of adults attending neighboring nests, small mammals avoiding mating with genetically similar individuals, and tadpoles preferentially associating with siblings rather than nonsiblings (Hepper 1991). To discriminate between kin and nonkin, individuals must have **kin recognition** abilities. Just because an animal is able to *recognize* kin does not mean kin will be *treated* differently (i.e., discriminated); kin discrimination should evolve only when the benefits it confers in terms of fitness exceed its costs (Waldman 1988).

There are two main hypotheses concerning the ultimate function of kin discrimination. First, kin discrimination may occur because it enables individuals to avoid harmful inbreeding and possibly even to optimize outbreeding (Shields 1982; Bateson 1983). Second, recognizing relatives may enable individuals to help them preferentially over nonrelatives. Such nepotism is obviously advantageous when directed toward an individual's own offspring; individuals who help their offspring more than unrelated young are likely to have higher biological fitness, producing a higher proportion of descendants than less nepotistic individuals. But as Hamilton (1964) argued, an individual can also gain fitness by helping a relative produce offspring. Because they share genes with their relatives, individuals that help relatives raise offspring could pass on their genes more successfully than those that do not provide such help. Thus kin discrimination may have evolved through **kin selection**, a form of natural selection that favors traits (such as parental care and helping behavior) that increase an individual's production of offspring or of other (**nondescendant**) kin.

Kin recognition may function in the schooling behavior of larval amphibians (Blaustein & O'Hara 1986). In many species of larval frogs and toads (**anuran tadpoles**), members of the same species (**conspecifics**) aggregate into groups similar to schools of fish (Wassersug et al. 1981; Waldman 1991). Such aggregations may result from attraction of individuals to conspecifics (**social aggregation**) or simply from individuals being independently attracted to a location with a preferred environmental condition (**asocial aggregation**; Bragg 1954), such as a patch of food or a desirable level of light. True social aggregations occur in a variety of anuran tadpoles (Waldman 1991). Social aggregations may help individuals avoid predators—for example, by diluting each individual's risk of being attacked (**dilution hypothesis**; Pulliam & Caraco 1984), by decreasing an individual's need to watch for predators (**many-eyes hypothesis**; Pulliam 1973), or by providing a living shield of others likely to be attacked first (**selfish-herd hypothesis**; Hamilton 1971). Tadpole feeding activities may actually increase the accessibility of prey, such that tadpoles feeding in a school may feed more efficiently than lone individuals (Katz et al. 1981; Wassersug et al. 1981). These benefits to schooling could be increased if individuals aggregate with kin, because the benefits would be conferred on relatives, favoring kin selection (Waldman 1991).

The proximate mechanisms by which vertebrates recognize kin have been best studied in larval amphibians (Waldman 1991). Much of this research has focused on determining whether kin recognition is accomplished by **familiarity** (a form of social learning, also called **association**) and/or by **phenotype matching**. Individuals who recognize kin by familiarity treat as kin any conspecifics with whom they were raised, because these individuals are familiar (Holmes & Sherman 1983). If you recognized your siblings in this way, you would treat as kin all familiar individuals with whom you were raised, even if they were not related to you. By contrast, you would treat as nonkin all unfamiliar individuals, even if they were actually your siblings.

Phenotype matching involves the comparison of individuals on the basis of traits that are associated with genetic relatedness (Holmes & Sherman 1983). Individuals who recognize kin by phenotype matching treat as kin any conspecifics whose phenotypes are similar to those of individuals with whom they were raised, such as nestmates (Holmes & Sherman 1983). If you recognized your siblings in this way, you would treat as kin any individuals whose phenotypes matched those of your familiar nestmates, even if those individuals were unfamiliar. Thus, if you were raised with siblings, you would be able to discriminate between unfamiliar siblings and unrelated conspecifics. You could do this because the unfamiliar siblings are phenotypically similar to the siblings with whom you were raised. Phenotype matching could also involve recognizing kin by learning one's own phenotype and then comparing oneself to other individuals (Holms & Sherman 1983). With this mechanism, you could identify unfamiliar siblings even if you were raised with nonkin or in social isolation. Recognizing unfamiliar kin through phenotype matching need not be genetically based; it may instead be facilitated by nongenetic maternal factors. For example, sibling eggs may acquire chemical "labels" from their mother while they are still in her body cavity or while they reside together as embryos in the jelly mass that she extrudes around her eggs (Waldman 1981). Whether based on familiarity or on phenotype matching, kin discrimination may involve chemical, visual, vibratory, or other cues (Waldman 1991), such as auditory or tactile information.

In this lab you will design your own experiments to explore kin discrimination and aggregation in African clawed frogs (*Xenopus laevis*). These anurans are native to sub-Saharan Africa where they generally inhabit stagnant, often murky water (Tinsley et al. 1996). *Xenopus* females lay their eggs individually in separate jelly capsules, rather than enclosing all the eggs in a single mass of jelly as other frogs do (Deuchar 1975). *Xenopus* tadpoles have been reported to aggregate in the wild (Wager 1965). In the laboratory, tadpoles form social aggregations in midwater as they filter-feed in a characteristic posture with their heads angled downward, often all facing the same direction (Katz et al. 1981; Wassersug et al. 1981; Wassersug 1996). By schooling, these **planktivores** (plankton eaters) may generate a common feeding current, which might enable individuals to filter a larger volume of water than they could when feeding alone (Katz et al. 1981). Direct tests of this hypothesis are inconclusive (Wassersug 1996), but in one study, weight at metamorphosis correlated positively with the rearing density of tadpoles, which suggests that higher densities may confer feeding advantages (Katz et al. 1981).

Unlike most frogs, *Xenopus* remain completely aquatic even as adults. Adults are most active at night, when they rely on chemical and vibratory cues to detect their prey (Elepfandt 1996), which include aquatic inverte- brates and dead animals (Wassersug 1996). Adult eyes, specialized for seeing in air, are probably adapted to detecting above-water predators (Elepfandt 1996), which include birds and people (Tinsley et al. 1996).

Most research on kin recognition and aggregation in amphibians has been conducted on tadpoles that feed at the bottom of temporary ponds (Waldman 1991). Only a few studies have explored social aggregation and kin discrim- ination in *Xenopus*. From these studies, we know that *Xenopus* tadpoles can orient to neighbors in complete darkness, probably by using vibratory cues sensed by their lateral line organs (Katz et al. 1981). However, when such cues are prevented, vision alone is sufficient for the formation of social aggregations of tadpoles, but not of newly metamorphosed froglets (Wassersug and Hessler 1971). Whether *Xenopus* froglets or adults will form social aggregations when additional cues are available remains unexplored. Further research is needed to determine the role of other sensory cues (such a olfactory or auditory cues) in *Xenopus* schooling at various developmental stages. The role of kinship in these social aggregations is also poorly known. Tadpoles raised with siblings reach larger sizes than those raised alone (Katz et al. 1981). Kin recognition in *Xenopus* has been examined only in unpub- lished experiments that showed tadpoles discriminated between unfamiliar siblings and unfamiliar nonsiblings (Blaustein & Waldman 1992). Because so little is known about aggregation and kin discrimination in *Xenopus*, the experiments that you conduct may be original research.

For your experiments, you have available tadpoles reared in three dif- ferent conditions (Fig. 17.1). One group of tadpoles ("A Only") all came from one brood (the A brood) and were reared with their siblings as soon as they hatched. Another group of tadpoles ("B Only") were also reared with their siblings, but they all came from a different brood (the B brood) with different parents who were unrelated to the parents of the A brood.

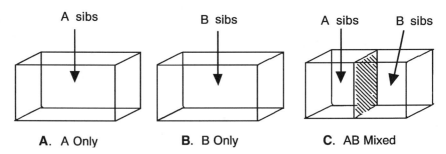

Figure 17.1 **Tanks containing treatment groups of A only, B only, and AB mixed. A mesh divider runs down the center of the AB mixed tank.**

The final group ("AB Mixed") included tadpoles from the A brood and tadpoles from the B brood. These AB Mixed tadpoles were placed together as soon as they hatched and had visual and chemical contact with each other through the water, but they were separated with a divider to facilitate identification of members of each kin group. After conducting experiments with these tadpoles, you may have time to explore further aspects of the development of aggregation and kin discrimination in newly metamorphosed froglets.

MATERIALS

The following items are needed for each group of students: a test tank (Fig. 17.2) with 2 or more tank dividers (porous, nonporous, clear, opaque); 2–4 liters of conditioned/dechlorinated water; 3 containers of tadpoles, one of which will have siblings from brood A (A Only), one of which will have siblings from brood B (B Only), and one of which will have an even mixture of siblings from broods A and B (AB Mixed) separated by a porous tank divider. Each student group will also need 4 empty tadpole containers marked "used test tadpoles" (1 container for each of the following: A Only, B Only, A tadpoles from AB Mixed, and B tadpoles from AB Mixed), one 200-milliliter beaker, a ruler and china marker, and 2 stopwatches.

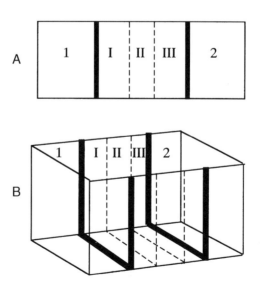

Figure 17.2 **Design of the test tank. The dashed lines indicate china marker lines, whereas the dark, bold lines indicate the placement of tank dividers. A. Top view. B. Side view.**

WHOLE-CLASS EXPERIMENT

Procedure

First, the whole class will test if tadpoles prefer one side of the tank to the other. This is done by placing a test tadpole in the center of the test tank (Fig. 17.2) with the outside regions containing tadpoles of the same kin group as the test tadpole. Therefore, the test tadpole should show no preference for one set of siblings over another. This experiment serves two purposes. First, it will be a good way for you to learn the basic procedures of the lab. Second, by doing this experiment first, we can rule out any preference on the part of the test tadpoles for a particular side of the test tank. If the tadpoles have no such preference, then any preference shown in your own experiment will be exhibited because test tadpoles prefer a particular treatment group, not just a particular side of the tank.

When setting up your experiments, you will be transferring tadpoles between various tanks. These animals are delicate, so they must be handled gently to minimize stress and prevent injury. To capture animals, gently scoop them into a 200-milliliter beaker filled with plenty of water. When placing animals in the test chamber or returning them to their home aquarium, submerge the beaker in the water and let the tadpoles swim out, rather than pouring the animals out of the beaker. Never expose tadpoles to the air or to strong water currents that could be created by sudden movements of the beaker.

Obtain one container of A Only tadpoles. Place two opaque porous tank dividers in the holders in your experimental tank to create three compartments. In the central compartment, draw two lines with the china marker to designate three areas (I, II, III) of equal width, as shown in Figure 17.2. Add dechlorinated water to the tank until it fills the tank to a height of 6–8 centimeters. After water has been added, wait 1–2 minutes. Then place A Only tadpoles in the following compartments of the experimental tank: 10 tadpoles in side 1, 10 in side 2, and 1 (the "test tadpole") in the center, handling the animals gently as we have described. Leave these tadpoles undisturbed for 5 minutes to allow them to acclimate. Minimize your movements and sounds while the animals are acclimating and while you are collecting data, because movements may cause a test tadpole to move in a direction it would not otherwise have chosen. After the acclimation period, begin your experiment by recording the time that the test tadpole spends in each of the three middle regions (regions I, II, and III) during 10 minutes. To do this, one member of your team should time how long the test tadpole is in the far-left region (region I, Fig. 17.1). Another member of your team should time how long that tadpole is in the far-right (region III, Fig. 17.1) during the 10-minute test period.

When timing, you should consider the test tadpole to have entered a region as soon as its nose crosses the boundary line into that region.

When the 10-minute test period has elapsed, place the test tadpole into an empty bowl of dechlorinated water marked "used test tadpoles—A Only." Add a new test tadpole from your stock of unused A Only tadpoles, and repeat the acclimation and testing procedure.

Hypothesis and Predictions

In the whole-class experiment, our general research question is whether tadpoles prefer one side of the tank to the other.

Data Recording and Analysis

Record your data for the whole-class experiment in a table like Table 17.1. Note that in Table 17.1 it looks like the tadpoles spent most of their time on the left side of the tank. This could be because the tadpoles really preferred the left side of the tank. Alternatively, the tadpoles may have had no preference, in which case the apparent difference was due to chance alone. For statistical analysis, these mutually exclusive possibilities must be expressed more precisely, in terms of the actual independent and dependent variables in the experiment. Thus, in your experiment, your **null hypothesis** (H_0) would be that the test tadpole does not spend different amounts of time in the two side regions (I and III in Fig. 17.1). Conversely, your **alternative hypothesis** (H_a) would be that the test tadpole spends different amounts of time in the right and left regions. This is a two-tailed test. How can you tell? Note that although tadpoles can move among three regions (I, II, and III) in the test tank, the null and alternative hypotheses compare time spent in just the two side regions (I and III). This is because a side preference will be indicated only if tadpoles spend more time in one of the two side regions.

Table 17.1 **Example of data for the whole-class experiment. These imaginary data do not reflect the expected trends of this experiment.**

Side 1 tadpole type: <u>A only</u>
Side 2 tadpole type: <u>A only</u>
Test tadpole type: <u>A only</u>

Tadpole I.D.	*Time (seconds) Spent in*			
	Region I	*Region II*	*Region III*	*Total*
1	425	15	160	600
2	300	10	290	600
3	175	20	405	600
4	250	300	50	600

You will use the Wilcoxon matched-pairs signed-ranks test (Siegel 1956) to determine whether the test tadpoles spent different amounts of time in the left and right sides of the test chamber. The Wilcoxon test is appropriate because you have two independent variables (left side and right side) that are related (paired) because the same tadpole experiences both of these treatment conditions. Pool all of the data for the whole class, and then use the Wilcoxon test to determine whether there was a side bias overall in the whole class.

SMALL-GROUP EXPERIMENT

Hypothesis and Predictions

Before doing your small-group experiment, brainstorm to create a list of 2–5 general questions about kin recognition that your group would like to explore. Then, as a group, narrow down your choices to the one that you will test.

Procedure

In your small groups, after choosing your research question, design a test for this question by making appropriate modifications to the procedure used for the whole-class experiment. For example, you could vary the tadpoles in side 1 and 2 of the test tank on the basis of number of tadpoles, kinship of tadpoles, or physical size of tadpoles. Keep in mind that in the experiment you design, the test tadpoles you use could be A or B raised with siblings only or raised with both siblings and nonsiblings. You also have the option of using transparent or opaque tank dividers or using solid or porous tank dividers on one or both sides of the tank.

Discuss how you will deal with the possibility of side bias. In the whole-class experiment, by pooling your data with the class, you tested whether there was a side bias in the whole room, which would enable you to detect, for example, whether tadpoles in *all* tanks preferred the east side. You did not, however, test whether side bias occurred within your particular experimental tank. For example, even if tadpoles used by the whole class showed no significant side bias, tadpoles in *your* tank might significantly prefer the west side, as could occur if one of your teammates consistently moved around on the east side, scaring the tadpoles. During your small-group experiment, you will need to be sure to control for side bias, and you must decide how to do so. For example, for each new test tadpole, you might switch the sides of the tank that receive each treatment. Thus, after testing your first tadpole, you would move the treatment that was on side 1 to side 2 and the treatment that was on side 2 to side 1. If wheeled carts are available, you could turn the cart around to facilitate this switch while minimizing disturbance.

If you have time, in addition to controlling for side bias during your experiments, you should conduct a separate test for side bias in your experimental tank. To do this, repeat the protocol for the whole-class experiment on four more test tadpoles. Combine these results with those you obtained for the two test tadpoles that you used earlier for the whole-class experiment, and then analyze the data for these six tadpoles as before, with the Wilcoxon signed-ranks test.

Data Recording and Analysis

Record your data in Table 17.2, or make a similar table on a computer spreadsheet. Be sure to fill in appropriate column headings to fit your experimental design. What are the independent and dependent variables in

Table 17.2 **Datasheet for the small-group experiment. Fill in this table with your own data. Fill in the type of test tadpoles (such as A Only, B Only, or A from AB Mixed) that you used. For side 1 tadpole type, fill in the type of tadpoles placed next to region I. For the side 2 tadpole type, fill in the type placed next to region III. Also fill in the identity (I.D.) of the test tadpoles you used (1, 2, etc.).**

Side 1 tadpole type: _____
Side 2 tadpole type: _____
Test tadpole type: _____

Tadpole I.D.	Time Spent in			
	Region I	Region II	Region III	Total

your experiment? Discuss how best to graph your data, and then make the graph after checking your decision with your instructor. Write a null and an alternative hypothesis for the specific independent and dependent variables that you tested. Should you do a one- or a two-tailed test? Check your answers with your instructor. Test your hypothesis with the Wilcoxon signed-ranks test to determine whether the test tadpoles spent different amounts of time in the two sides of the test chamber (regions I and III, Fig. 17.1).

FROGLET EXPERIMENT

In 6–8 weeks your tadpoles will become small fogs (froglets). If you are provided with sufficient lab time, you can repeat your small-group experimental design to see whether *Xenopus* behavior changes with metamorphosis. Alternatively, you can test a new experimental question. Because froglets are small, you can use the same test apparatus and handling methods as with tadpoles. Statistical analysis also will be the same. You should test for side bias before doing your actual experiment. Why?

QUESTIONS FOR DISCUSSION

Consider the following experimental results: Individual tadpoles associate more with familiar siblings than with unfamiliar nonsiblings but do not distinguish between siblings and nonsiblings when both are familiar. What do these results indicate about the importance of familiarity as a cue for kin recognition? Do these results indicate that genetic cues are *not* involved in kin recognition in this species?

Consider the following experimental design and results: Immediately after fertilization, each egg from the same egg mass was placed into a separate container and raised in isolation. Later, when given a choice between unrelated tadpoles and siblings, these tadpoles preferred siblings, even though the siblings had never been in contact as tadpoles. Is this conclusive evidence of phenotype matching by using genetic cues? Explain.

Using the experimental protocols of this lab, if you found that tadpoles spent more time on the side of the tank with familiar siblings than on the side of the tank with familiar nonsiblings, one obvious interpretation would be that tadpoles are *attracted* to their siblings. Another interpretation would be that tadpoles *avoid* nonsiblings. Design experiments that would enable you to distinguish between these possibilities.

If your experiments demonstrated that tadpoles are avoiding nonsiblings rather than being attracted to siblings, what would these results imply about the function of kin recognition in the species that produced these results?

Consider the following experiment: Equal numbers of tadpoles are placed in each choice region. If test tadpoles spend significantly more time on the side with the porous, clear divider than on the side with the nonporous, opaque divider, what can you conclude about the cues used for social aggregation?

In your experiments, you marked three separate regions (I, II, and III, Fig. 17.2) in the test tadpole's chamber. A simpler experimental design would have been just to draw a line in the center, separating the chamber in half. If you did this, you would not be able to compare directly the times that each tadpole spent in these regions, and you would not be able to use the Wilcoxon test as you did in your experiments. Why not? Can you think of a way to analyze data for this experimental design?

On the basis of your results, what can you conclude about the question you investigated. For example, do tadpoles aggregate or recognize kin? What cues are involved? Provide two hypotheses concerning why you got the results that you observed. What might be an adaptive value of the mechanisms that you propose? Describe an experiment you could do next to test one of these hypotheses further.

ACKNOWLEDGMENTS

We thank BJP's 2000 animal behavior class and volunteers from BJP's 1999 biology concepts 1 class at Hamline University for participating in tests of this lab. We are grateful to Bruce Waldman for his many helpful comments on an earlier draft of this manuscript. Sylvia Kerr and Kathryn Malody assisted us tremendously, helping with injecting frogs and providing information about general frog care. Funding was provided by a grant to KLA from the Lund Summer Research Program in Biology at Hamline University.

LITERATURE CITED

Bateson, P. 1983. Optimal outbreeding. In *Mate Choice* (ed. P. Bateson), pp. 257–277. Cambridge: Cambridge University Press.

Blaustein, A. R. & O'Hara, R. K. 1986. An investigation of kin recognition in red-legged frog (*Rana aurora*) tadpoles. *Journal of the Zoological Society of London (A)*, **209**, 347–353.

Blaustein, A. R. & Waldman, B. 1992. Kin recognition in anuran amphibians. *Animal Behaviour*, **44**, 207–221.

Bragg, A. N. 1954. Aggregational behavior and feeding reactions in tadpoles of the Savannah spadefoot. *Herpetologica*, **10**, 97–102.

Deuchar, E. M. 1975. *Xenopus: The South African Clawed Frog*. New York: J. Wiley.

Elepfandt, A. 1996. Sensory perception and the lateral line system in the clawed frog, *Xenopus*. In: *The Biology of Xenopus* (ed. R. C. Tinsley & H. R. Kobel), pp. 97–120. Oxford: Clarendon Press.

Hamilton, W. D. 1964. The genetical evolution of social behavior. I, II. *Journal of Theoretical Biology*, **7**, 1–52.

Hamilton, W. D. 1971. Geometry for the selfish herd. *Journal of Theoretical Biology*, **31**, 295–311.

Hepper, P. G. (ed.) 1991. *Kin Recognition*. Cambridge: Cambridge University Press.

Holmes, W. G. & Sherman, P. W. 1983. Kin recognition in animals. *American Scientist*, **71**, 46–56.

Katz, L. C., Potel, M. J. & Wassersug, R. J. 1981. Structure and mechanisms of schooling in tadpoles of the clawed frog, *Xenopus laevis*. *Animal Behaviour*, **29**, 20–33.

Pulliam, H. R. 1973. On the advantages of flocking. *Journal of Theoretical Biology*, **38**, 419–422.

Pulliam, H. R. & Caraco, T. 1984. Living in groups: Is there an optimal group size? In: *Behavioural Ecology: An Evolutionary Approach*. 2nd ed. (ed. J. R. Krebs & N. B. Davies), pp. 122–147. Sunderland, MA: Sinauer Associates.

Shields, W. M. 1982. *Philopatry, Inbreeding, and the Evolution of Sex*. Albany: State University of New York Press.

Siegel, S. 1956. *Nonparametric Statistics for the Behavioral Sciences*. New York: McGraw-Hill.

Tinsley, R. C., Loumont, C. & Kobel, H. R. 1996. Geographical distribution and ecology. In: *The Biology of Xenopus* (ed. R. C. Tinsley & H. R. Kobel), pp. 35–59. Oxford: Clarendon Press.

Wager, V. A. 1965. *The Frogs of South Africa*. Capetown, South Africa: Purnell.

Waldman, B. 1981. Sibling recognition in toad tadpoles: the role of experience. *Zeitschrift für Tierpsychologie*, **56**, 341–358.

Waldman, B. 1988. The ecology of kin recognition. *Annual Review of Ecology and Systematics*, **19**, 543–517.

Waldman, B. 1991. Kin recognition in amphibians. In: *Kin Recognition* (ed. P. G. Hepper), pp. 162–219. Cambridge: Cambridge University Press.

Wassersug, R. 1996. The biology of *Xenopus* tadpoles. In: *The Biology of Xenopus* (ed. R. C. Tinsley & H. R. Kobel), pp. 195–211. Oxford: Clarendon Press.

Wassersug, R. & Hessler, C. M. 1971. Tadpole behaviour: aggregation in larval *Xenopus laevis*. *Animal Behaviour*, **19**, 386–389.

Wassersug, R. J., Lum, A. M. & Potel, M. J. 1981. An analysis of school structure for tadpoles (Anura: Amphibia). *Behavioral Ecology and Sociobiology*, **9**, 15–22.

PART 4

Adaptation and Evolution

SECTION I

Foraging

CHAPTER 18

Diving Birds: A Field Study of Benthic and Piscivorous Foragers

JENNIFER J. TEMPLETON AND D. JAMES MOUNTJOY
Department of Biology, Knox College, Galesburg, IL 61401, USA

INTRODUCTION

Several species of water birds search for their food underwater. Most diving ducks, such as buffleheads (*Bucephala albeola*) and common eiders (*Somateria mollisima*) are **benthic** foragers that search for relatively sessile prey such as snails, amphipods, crustaceans, or aquatic vegetation located on the bottom substrate (Ehrlich et al. 1988; Ydenberg & Guillemette 1991). Other divers, such as the common loon (*Gavia immer*) and the western grebe (*Aechmophorus occidentalis*) are **piscivorous** foragers, which must pursue and capture evasive fish prey (Ehrlich et al. 1988; Ydenberg & Forbes 1988).

Diving birds have a number of physiological adaptations to help them cope with their inability to breathe under water (Butler & Woakes 1979; Hainsworth 1981), including relatively large lungs, a greater oxygen–carrying capacity in the blood, increased storage of oxygen in myoglobin, redistribution of blood flow to the heart, and a slowing of the heart rate during dives. But actively foraging divers are faced with an energetic conflict. In order to obtain food, they must enter a world in which they cannot breathe and must use energetically expensive foraging behaviors such as swimming underwater, searching for and pursuing prey, and resurfacing. How do different species of diving birds adjust their behavior to maximize their energy intake despite the physiological constraints of **asphyxia** (oxygen deprivation)?

All diving birds make repeated foraging trips below the water surface to which they must return to breathe (Ydenberg 1988). A **dive cycle**, therefore, is made up of **dive duration**, during which the bird searches for and captures food underwater, and **pause duration**, during which the bird consumes the prey on the surface and recovers from the oxygen debt of the preceding dive. In an attempt to determine how diving birds compensate behaviorally for the inability to breathe while foraging, Ron Ydenberg and his colleagues have focused on the relationship between dive time and pause time and on how this relationship differs between benthic and piscivorous foraging behavior. These foraging strategies differ not only in the types of prey encountered but also in the ease of obtaining prey items once they are discovered. For example, once a benthic forager has located a bed of mussels, snails, or aquatic vegetation, it can forage at its leisure; the aquatic invertebrates and plants cannot escape (Ydenberg & Guillemette 1991). In contrast, a piscivorous forager must take advantage of a discovered school of fish immediately if it is to capture as many fish as possible before they disappear (Ydenberg & Forbes 1988). Similarly, piscivorous foragers (such as loons) that prey on solitary fish (perch or stickleback) must chase their fish prey until it is captured. Thus, chasing prey can require extended periods underwater, with only brief breaks to "catch one's breath" at the surface. Full recovery is delayed until the prey is captured or the bird gives up (as a consequence of either satiation or lack of success).

The purpose of this field study is to compare the diving behavior of several different species of water birds. We will be asking the following questions: (1) What is the general relationship between dive time and pause time in diving birds? (2) Why might this relationship be similar between benthic and piscivorous species? (3) How do benthic and piscivorous species differ in their dive–pause relationships, and what factors might be responsible for these differences?

MATERIALS

Each group of four students will need 1–2 pairs of binoculars and/or a spotting scope, 2 stopwatches, 1 set of data sheets (provided), a clipboard and pencil, and a field guide.

PROCEDURE

Your instructor will select two species of diving birds whose behavior you will observe in the field. One will be a benthic forager, the other a piscivorous forager. Prior to departure, you will be given the opportunity to learn to identify these species. At the field site, you will work in teams of four, with team members taking turns being observers, timers, and recorders. Each team should attempt to observe diving birds in a different location from the other teams in order to reduce the possibility of sampling the same individuals. First, a flock of diving birds or a solitary individual will be located, and the species will be identified by all members of the team. Spend a few minutes simply watching the birds just to get used to using the equipment and to make sure you can recognize the different species. Then you will carry out **focal-animal sampling**, wherein single birds are selected arbitrarily for detailed observations and data collection. Of course, you should attempt to sample only actively foraging individuals.

The designated *observer* will monitor the focal bird with binoculars and will be responsible for calling out "Dive!" when the bird disappears from view and "Surface!" when the same bird reappears. The *dive timer* will be responsible for starting the stopwatch at the word "dive" and stopping it at the word "surface"; the *surface timer* will do the opposite. Both timers should report their times to the *recorder*, who is responsible for recording these durations on the datasheet provided (Table 18.1).

Observations of a single individual (for example, bird 1) will commence with the first dive it makes and will continue for a few successive dive cycles, if possible (a single dive cycle = dive duration + pause duration). Each team will try to collect data from at least three different individuals from each of the two types of foragers.

HYPOTHESES AND PREDICTIONS

The purpose of this lab is to determine whether there is a relationship between dive time and pause time in diving birds. Your hypothesis is that longer dives require longer recovery times, or, in other words, that dive time and pause time are positively correlated in both benthic and piscivorous foragers. In addition, you will be asking whether benthic and piscivorous foragers differ in their dive–pause relationships and, if so, in what way.

DATA RECORDING AND ANALYSES

The easiest way to present the data is simply to plot all the dive–pause scores collected from each individual bird as a scatter plot with dive duration (in seconds) on the x-axis and pause duration on the y-axis. You may plot the data from both types of forager on the same graph, but use different symbols for their data points. It is likely that some pause durations will occur just after birds have finished foraging and thus will involve behaviors not directly related to foraging (such as preening or sleeping). Such dive–pause scores should be omitted from the analysis. In addition, any dive–pause scores in which the dive duration is longer than some arbitrary length of time (such as 100 seconds) could be omitted, because they generally reflect the observer's missing the bird's resurfacing.

Using a statistical or graphing program on a computer, perform a separate regression analysis for each type of forager, and plot the regression lines. Record the equation for each regression line as Pause = Slope × Dive + y-Intercept. A regression analysis tests the null hypothesis that there is no relationship between dive time and pause time. The alternative hypothesis, of course, is that dive time and pause time are significantly (and positively) related. The regression analysis will calculate a coefficient of determination (r^2) and the corresponding P-value to determine whether the relationship between dive time and pause time is statistically significant. In other words, you will use r^2 to consider whether dive time provides a good "explanation" for the variation in pause time. For example, if r^2 is 0.095, the regression explains less than 10% of the total variation in pause duration. This would mean that pause time is probably not directly related to the preceding dive time, so something else must be affecting it. On the other hand, if r^2 is 0.95, the regression explains 95% of the total variation. The corresponding P-value for such a score would most certainly be less than 0.05, indicating that the relationship is significant and enabling you to reject the null hypothesis of no relationship.

QUESTIONS FOR DISCUSSION

In what way are benthic and piscivorous foragers similar in their dive–pause relationships? What might explain this similarity?

Do benthic and piscivorous foragers differ significantly in their dive–pause relationships? If so, in what way? What might explain this difference?

Do your data meet the assumption of independent data points? If not, why not? How might a lack of independence influence the outcome of the statistical analyses?

How might the presence of competitors (other individuals of the same species) influence the dive–pause relationship in benthic foragers? in piscivorous foragers?

An additional constraint on a diving bird's ability to maximize intake rate is the presence of **kleptoparasites** (see "piracy," Ehrlich et al. 1988, pp. 159–161), which attempt to steal the divers' hard-earned meals. How might group-foraging birds alter their diving behavior in the presence of a kleptoparasite? (Even if you did not witness this type of behavior, please speculate.) Would this behavior make stealing the food more difficult? Why or why not? How might the group foragers benefit from this behavior? What types of antipredator behaviors does this remind you of?

Some species of water birds that are known as "dabbling ducks," sometimes tip forward (submerging their heads) to obtain aquatic vegetation and benthic invertebrates from the bottoms of shallow ponds (Ehrlich et al. 1988, pp. 75–79). Although they are not deprived of oxygen as long as diving birds, dabblers still face an oxygen debt. Consider what the relationship between diving time and recovery time might be for dabbling ducks and how the relationship might differ between this group of foragers and the other two types of diving birds.

ACKNOWLEDGMENTS

We would like to thank students in the following classes for testing various versions of this field lab: vertebrate zoology (Bio 386) at University of Nebraska-Lincoln, 1996; ecology (Bio 215) at Bowdoin College, 1997; and behavioral ecology (Bio 379) at Franklin & Marshall College, 1998–2000. We are also grateful to Bonnie Ploger, Ken Yasukawa, and Lauren Wentz for their thoughtful remarks and constructive criticism.

LITERATURE CITED

Butler, P. J. & Woakes, A. J. 1979. Changes in heart rate and respiratory frequency during natural behaviour of ducks, with particular reference to diving. *Journal of Experimental Biology,* **79**, 283–300.

Ehrlich, P. R., Dobkin, D. S. & Wheye, D. 1988. *The Birder's Handbook: A Field Guide to the Natural History of North American Birds.* New York: Simon & Schuster/Fireside Books.

Hainsworth, F. R. 1981. *Animal Physiology.* Reading, MA: Addison-Wesley.

Ydenberg, R. C. 1988. Foraging by diving birds. *Proceedings of the XIXth International Ornithological Congress.*

Ydenberg, R. C. & Forbes, L. S. 1988. Diving and foraging in the western grebe. *Ornis Scandinavica,* **19**, 129–133.

Ydenberg, R. C. & Guillemette, M. 1991. Diving and foraging in the common eider. *Ornis Scandinavica,* **22**, 349–352.

Table 18.1 **Data sheet for recording birds' diving behavior.**

Date:_____   Time:_____   Group Members:_____

Dive Cycle	Bird Number	Species	Dive Duration	Pause Duration	Comments (e.g. Postdive Behavior, Harassment by Others)

Additional remarks: (weather conditions, other species present, etc.)

CHAPTER 19

Found A Peanut: Foraging Decisions by Squirrels

Sylvia L. Halkin
Department of Biological Sciences, Central Connecticut State University, New Britain, CT 06050-4010, USA

INTRODUCTION

Animals that minimize the energetic cost of searching for and processing their food will benefit from a higher net energy gain than animals that do not behave in this way (net energy = energy obtained from food − energetic cost of finding and processing food). However, energy is not the only relevant consideration for foraging animals. Time available for foraging is often limited. If predators pose a greater threat to animals that use some foraging methods rather than others, less energy-efficient but safer methods may be used than would be predicted if energy alone were considered. Mathematical models of the role of energy and time constraints in shaping foraging behavior were first developed by Emlen (1966) and MacArthur & Pianka (1966); the role of predation risk in shaping foraging behavior of squirrels has been modeled by Lima et al. (1985). Animals that store food for later consumption need to choose food items that are likely to survive the period of storage in good condition (Steele & Smallwood 1994). Can you think of additional factors that may lead animals to forage in ways other than would be predicted simply by considerations of max-imizing energy gain?

You will test the importance of some of these factors to local squirrels foraging on peanuts. You will offer squirrels two piles of peanuts: one pile inside intact shells (no cracks or holes) and another pile with the shells removed. In order to equalize the chances of the squirrels finding both piles, you will put them only a short distance apart, so that a squirrel foraging at either pile will also have the other pile in view. After you collect the data, you will try to explain your results in terms of models of foraging from behavioral ecology. Which of the factors discussed in the previous paragraph have you manipulated by removing the shells of some of the peanuts? What are the costs, and the benefits, of taking each kind of peanut? Do you predict that the choice of nuts will depend on what the squirrel is going to do with them?

If you make observations near sunset, try to figure out which factors are important in determining when squirrels stop foraging for the night. At what point will it be more energetically efficient to stop foraging than to continue? How may the risk of predation play a role?

MATERIALS

You will need about 1 pound of *unsalted* peanuts in the shell (roasted or unroasted) per group of 4–6 students, containers to carry peanuts that have had their shells removed (plastic 8-ounce yogurt cups work well), and 7 × 35 binoculars for each student or pair of students. These will make possible detailed observations at distances that will not disturb the squirrels and will enable you to observe squirrels that climb trees or go to other places where it is difficult for you to follow them. If you have very tame local squirrels, you may be able to get close enough to collect adequate data without binoculars. Each group of students will also need a stopwatch or wristwatch with a second hand, several pieces of white bread or about four ounces of popped unsalted and unflavored popcorn, *eight* data sheets, a clipboard, and a pencil with an eraser. Be sure to wear warm clothing, including hat and gloves and warm socks and shoes or boots, if observations will be made in a cold season and/or at a cold time of day.

PROCEDURE

In the lab, each group of 4–6 students should prepare peanuts for their two piles. To clarify terminology, in the remainder of this lab, *peanut* will refer to a shell plus the nuts that it contains, and *nut* will refer to a single peanut seed from within a shell. The class as a whole will decide whether the pairs of piles should contain the same number of nuts or should have the same total volume. If the class decides that the piles should have the same number of nuts, for each pair of nuts left inside a single peanut shell in one pile, there should be two nuts removed from their shell in the other pile. If the class decides that the piles should have the same total volume, the limiting factor on the size of the piles will probably be the number of nuts you can remove from shells in the time allotted. To eliminate a source of variation that you are not trying to test, do not use peanuts that contain only a single nut (or peanuts that contain more than two nuts) in either pile. It will be easiest to keep together the two cotyledons of nuts removed from the shell if their integument (the red "skin" holding them together) is left intact: this will require some extra care in removing the shell. It is best to obtain nuts without shells by removing the shells yourselves, rather than by separately purchasing nuts without shells. (Why do you think this is the case?) All peanuts used in the pile with shells should have completely intact shells—no holes or cracks (Why?)

Take your nuts, binoculars, bread or popcorn, clipboards, and data sheets to the place where you will be doing the experiment. If possible, groups should be separated by distances of 20–40 meters or more, to avoid disturbing one another's squirrels.

Within each group, look for actively foraging squirrels, and try to put your nut piles as close to them as possible without scaring them away in the process. There should be a distance of between one-half and one squirrel body length (including tail) between the closest edges of the two piles. If the squirrels you are studying are not accustomed to being fed by people, it will probably be helpful to place the piles within 1 meter of a tree, because squirrels are more likely to venture out to feed when there is a nearby tree to climb if danger threatens. Make sure both piles are the same distance from the tree trunk, and make each pile fairly compact (10–15 centimeters in diameter). Move to about 6–7 meters away, to a location with a clear view of the piles. Wait for 5–10 minutes; we hope that at least one squirrel will find your nut piles. If this does not happen, you may throw a few small (1- or 2-cm^2) pieces of bread or a few pieces of popcorn, to try to lure them toward the piles. Try to minimize this feeding to keep the squirrels as hungry as possible.

In each group, at any given time, one person should serve as data recorder (rotate this duty) and the other members should be watching squirrels. If an individual squirrel moves away from the piles with a nut, one group member should follow it (maintaining a distance of 6–7 meters) to discover where it goes and whether it eats or stores the nut.

HYPOTHESES AND PREDICTIONS

List at least two hypotheses about the data you will collect (in addition to the example that follows). For example, you might hypothesize that eating peanuts with shells takes more energy than eating nuts without shells and that squirrels behave as though they "know" this information (from experience or from a genetically encoded tendency to respond preferentially to nuts without shells when looking for something to eat). Write a prediction of what will happen if each of your hypotheses is correct. In the example given, you might predict that squirrels will only eat nuts without shells. In thinking about additional hypotheses to generate, note that (1) squirrels may either eat nuts or bury them for later consumption and (2) nuts may be eaten either at the nut piles or elsewhere.

DATA RECORDING AND ANALYSES

Instructions for Filling Out Data Sheets

Table 19.1 shows how to fill out data sheets. For each nut taken by a squirrel, place a check mark on your data sheet in the two columns that apply following the nut number. For nuts taken in the shell, fill out two sequential lines of the table; indicate that these two lines refer to nuts from the same shell. Note that it is possible for these two nuts to have different fates; they may be eaten at different locations, or one may be eaten and the other buried. In the last column, give information on where the nut was eaten or stored. If multiple nuts are eaten by a squirrel on a single visit to the piles or are removed from the piles on the same visit, indicate this in the right margin with brackets including all nuts taken on the same visit. When your squirrels remove nuts from the pile, record at the bottom of the sheet how many they carry at once. Start a new data sheet each time you switch data recorders.

Over the course of the lab, for 2–5 nuts from each pile, try to get data on how many seconds it takes a squirrel to eat the nut. For nuts without shells, start timing when the squirrel first puts all or part of the nut in its mouth (provided that chewing motions follow soon after first contact with the mouth), and stop timing when chewing ceases and the nut has apparently been fully consumed. To time how long it takes a squirrel to eat two

Table 19.1 **Sample data.**

Nut	In Shell	No Shell	Eaten	Stored	Where/Time to Eat/Other Information of Interest
1	√		√		5 cm from pile · same shell: 70 sec to eat both nuts · same visit to piles · same squirrel
2	√		√		5 cm from pile
3		√	√		8 cm from pile 41 sec to eat
4	√			√	Buried in lawn 8 meters from piles · same shell
5	√			√	"
6	√		√		2 meters from pile · same shell
7	√			√	Buried in lawn 10 meters from piles, with half of shell around it

nuts in a shell, start timing when a squirrel first starts to remove the shell, and stop timing when chewing stops after both nuts in the shell have been fully consumed.

Do not record data on nuts for which a disturbance such as a dog or person running by causes the squirrel to move to a new location. (Why not?) Stop collecting data when one of your piles runs out of nuts. (Why aren't data from the remaining pile, comparable to data from two piles?)

It is important that your data sheets can be easily interpreted by other people. When you have finished collecting data for the day, look over each sheet that you filled out, and be sure that any abbreviations or symbols are fully explained at the start of the sheet. Check with other group members if you are not sure about the meaning of something that one of you recorded. For peanuts with shells, check to see that you have a separate line for each nut in the shell. Exchange data sheets with a member of another group to determine whether ambiguities remain; clarify these on the data sheets before you leave.

Data Analyses

Data from the entire class will be combined for analysis. First, you will calculate the average amount of time it took a squirrel to eat the two nuts in a shell and will compare this to the average amount of time it took a squirrel to eat a single nut without a shell. Do your data indicate that it takes squirrels longer to eat nuts in shells than to eat the same number of nuts without shells? To test this statistically, you could use a Mann-Whitney *U* Test; note that your null hypothesis would be that it takes twice as long to eat a pair of nuts in a shell as to eat a single nut without a shell. Is doubling the time required to eat one nut without a shell really an accurate indication of how long it took a squirrel to acquire and eat two nuts without shells? Why or why not? Do you think that squirrels use more energy when eating nuts in shells than when eating nuts without shells? Justify your answers to these questions.

Next, the class will use information recorded in the last column on the data sheets to come up with several subcategories for "Eaten" and "Stored" nuts. For example, you may want to count nuts eaten "at" the pile (within a distance you specify) as a separate category from nuts eaten "away from" the pile. Probably a total of four or five subcategories will be most informative. (Why not have as many categories as you can think of?) You will make a summary data table, in the form shown on the next page. In this table, letters stand for different subcategories of what happened to the nuts. (Your table should have meaningful labels for each category, instead of just letters.) Should nuts within the same shell be counted as single nuts or as two nuts? Discuss this issue with the rest of the class before you fill in the table. What would you want to consider in making this decision?

Sample summary data table.

| | What Happened to Nut | | | | |
| | Eaten | | | Stored | |
	A	B	C	D	E
Nut Type					
With shell	—	—	—	—	—
Without shell	—	—	—	—	—

Decide which hypotheses you want to test with these data (your observations may have led to new hypotheses that you did not anticipate in advance). For each hypothesis that you want to test statistically, write a formal null hypothesis and an alternative hypothesis. The null is the hypothesis of no difference. In this exercise, many null hypotheses will predict no difference between what happens to nuts with and without shells, or that there will be equal numbers of nuts with and without shells in each "What Happened to Nut" category. For example, if you predicted that squirrels would only eat nuts without shells, a null hypothesis would state that the number of nuts eaten from the pile with shells will not differ from the number of nuts eaten from the pile without shells. The alternative hypothesis is that the number of nuts eaten from the pile with shells *will* differ from the number of nuts eaten from the pile without shells.

If you want to ask whether the distribution among the different subcategories differs for nuts with and without shells, you can use a Fisher exact probability test if you have access to computer software that will calculate this statistic (the SAS program is a commonly available one that will do this). Otherwise, you can do a chi-square test of independence, but in order to do this test, you may need to combine categories further. This test requires (1) expected values of at least 5 in 80% or more of your categories and (2) no categories with expected values less than 1. If only one of your subcategories has large numbers, even combined categories may not provide expected values large enough to allow you to do a chi-square test of independence.

If you want to compare the proportions of nuts with and without shells within a given category, you can do a chi-square goodness-of-fit test, provided that your sample size is at least 10 nuts in that category. If there are n nuts in the category, under the null hypothesis you would expect that $n/2$ of these have shells and that the other $n/2$ do not have shells.

QUESTIONS FOR DISCUSSION

Which factors mentioned in the Introduction seem to be important to squirrels in determining whether they choose nuts with or without shells and in determining what they do with each kind of nuts? How many of either of these kinds of nuts did a squirrel carry at the same time, and why is this information relevant? Try to design an experiment that would vary just one of the factors that seem to be important, while keeping the others constant.

If observations were made near sunset, what factors seemed to be important in causing the squirrels to stop foraging when they did?

Do squirrels need to be "aware" of the factors you have considered in order to make the "decisions" that you would predict on the basis of quantitative calculations? Does use of the term *decision* imply mental capacities that squirrels may not have?

If squirrels are not individually identified, and data from all individuals are summed together for the purpose of statistical testing, individual squirrels will be represented in the data set in proportion to the number of nuts they have eaten. If individual variations in behavior exist, and all squirrels do not eat the same number of nuts while you are watching them, the variations shown by the squirrels that eat the most nuts will be most strongly represented. How might this bias your data, and how does it limit the scope of your conclusions? Note that if individual squirrels can be identified, using the Wilcoxon matched-pairs signed-ranks test will give each individual an equal "vote" in determining the outcome of the test. For an example of how this test can be used, see the section "Changing Nominal into Ordinal Data" in Appendix C.

ACKNOWLEDGMENTS

Thanks to Peter D. Smallwood for logistical advice and insights into squirrel behavior, to David A. Spector for suggesting the "Found a Peanut" part of the title, to Bonnie Ploger for useful suggestions about statistics, and to my students at Central Connecticut State University for their data, their good ideas, and their help in testing different versions of this lab.

LITERATURE CITED

Emlen, J. M. 1966. The role of time and energy in food preference. *American Naturalist*, **100**, 611–617.

Lima, S. L., Valone, T. J. & Caraco, T. 1985. Foraging-efficiency–predation-risk trade-off in the grey squirrel. *Animal Behaviour*, **33**, 155–165.

MacArthur, R. H. & Pianka, E. R. 1966. On optimal use of a patchy environment. *American Naturalist*, **100**, 603–609.

Steele, M. A. & Smallwood, P. D. 1994. What are squirrels hiding? *Natural History*, **103**, 40–45, 90.

ADDITIONAL SUGGESTED READING

Brown, L. & Downhower, J. F. 1988. *Analyses in Behavioral Ecology: A Manual for Lab and Field.* (two related labs: Lab 7, "Food value and the foraging preferences of squirrels," pp. 28–30; Lab 5, "Prey location in squirrels," pp. 22–23). Sunderland, MA: Sinauer.

Gurnell, J. 1987. *The Natural History of Squirrels.* New York: Facts on File.

Steele, M. A. & Koprowski, J. L. 2001. *North American Tree Squirrels.* Washington, DC: Smithsonian Institution Press.

Vander Wall, S. B. 1990. *Food Hoarding in Animals.* Chicago: University of Chicago Press.

Peanut-Foraging Data Sheet

Data recorder: _____

Other Group members: _____

Date: _____ Start time: _____ End time: _____

Location: _____

The fate of each individual nut should be recorded separately, with the fates of pairs of nuts that were in the same shell recorded on sequential lines.

Nut Number	In Shell	No Shell	Eaten	Stored	Where/Time to Eat/Other Information of Interest
1					
2					
3					
4					
5					
6					
7					
8					
9					
10					
11					
12					
13					
14					
15					

Maximum number of peanuts with shells that were carried at one time: _____

Maximum number of nuts without shells that were carried at one time: _____

CHAPTER 20

Economic Decisions and Foraging Trade-offs in Chickadees

RONALD L. MUMME
Department of Biology, Allegheny College, 520 North Main Street, Meadville, PA 16335, USA

INTRODUCTION

Every moment of their lives, animals are faced with behavioral decisions. Where and in what types of habitats should they search for food? How long should they remain foraging in a particular area before moving elsewhere? What kinds of foods should they eat, and what kinds should they reject? How much time should they devote to foraging vs other activities such as breeding, resting, and avoiding predators? At what times of the day would searching for food be most profitable? These are only some of the behavioral decisions that animals must make on a regular basis. How do they make these choices? What are the ecological factors that affect their decisions?

When considering decision making in animals, you must remember that animals are almost certainly not making the sorts of conscious decisions that humans routinely make. Nonetheless, animals often behave as though they are weighing the costs and benefits of various behavioral options, and they often adjust their behavior appropriately when environmental conditions change. Although the costs and benefits of competing behavioral options ultimately must be measured in terms of effects on Darwinian fitness (lifetime reproductive success), the economics of decision making can also be investigated by using more convenient and immediate measures, such as rate of food intake or duration of exposure to potential predators, that are likely to be highly correlated with fitness. In other words, natural selection should favor animals endowed with neural machinery that enables them, when faced with competing behavioral options, to pursue a behavioral strategy that is most likely to have a positive effect on their survival or reproductive success (Krebs & Kacelnik 1991; Krebs & Davies 1993; Cuthill & Houston 1997).

In this field exercise, you will explore the economic basis of animal decision making through an experimental field study of foraging decisions in the black-capped chickadee (*Poecile atricapillus*, formerly *Parus atricapillus*). The black-capped chickadee is a small gray bird with a distinctive and unmistakable black cap and bib. It is a common permanent resident throughout the northern United States and southern Canada. During the fall and winter months, chickadees travel through woodlands and suburban areas in small flocks of 2–10 birds of both sexes and all ages. During the fall, winter, and early spring, chickadees are easily attracted to bird feeders, where sunflower seeds are one of their favorite foods. After grasping a sunflower seed in its bill, a chickadee will fly to a sturdy perch in a nearby tree and, using its feet to secure the seed to the perch, peck at the seed until it succeeds in extracting the nutritious kernel from the shell. Processing a sunflower seed and extracting the kernel may take 30 seconds or more. However, many of the seeds that chickadees take from feeders during the fall and winter are not eaten immediately but stored in small cache sites in trees for recovery later (Smith 1991).

Although bird feeders provide chickadees with a readily available source of food, visiting feeders can be dangerous. Predators such as the sharp-shinned hawk (*Accipiter striatus*), Cooper's hawk (*Accipiter cooperii*), and domestic cat have learned that bird feeders can be profitable places to search for prey, and they often attack and kill small birds at feeders (Dunn & Tessaglia 1994).

You will investigate how chickadees visiting bird feeders choose between the behavioral options of foraging on relatively profitable food items vs foraging at a relatively safe and more easily accessible site.

MATERIALS

Materials required for this field exercise include four identical feeding platforms with supporting posts and feeding dishes, whole striped sunflower seeds, and intact kernels (hearts) of striped sunflower seeds. A suitable study site—an open field or lawn adjacent to a well-defined edge of a forest or woodland and an observation post (Fig. 20.1)—is also required.

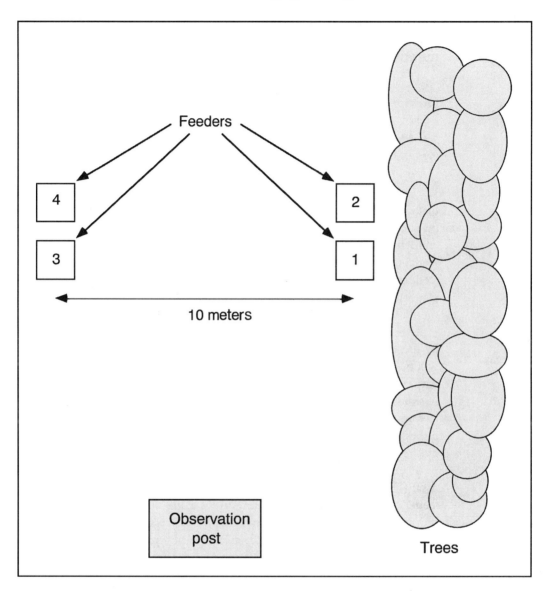

Figure 20.1 **Idealized study site showing the positions of the four feeders relative to adjacent trees and the observation post.**

PROCEDURE

Feeding platforms should be positioned at the study site as illustrated in Figure 20.1. Feeders 1 and 2 should be placed side by side at the edge of the open field or lawn about 1 meter from good cover (such as the low branches of a bushy conifer) in the neighboring tree line. Feeders 3 and 4 should be placed in the open, about 10 meters away from feeders 1 and 2 and at least

Table 20.1 **Schedule showing types of sunflower seeds to be placed in feeders during the two data collection periods. Feeder numbers correspond to those in Figure 20.1.**

	Feeder 1	Feeder 2	Feeder 3	Feeder 4
Period 1				
Days 1–4	Kernels	Whole seeds	Whole seeds	Kernels
Days 5–8	Whole seeds	Kernels	Kernels	Whole seeds
Period 2				
Days 9–16	Whole seeds	Whole seeds	Kernels	Kernels

that distance from any other trees or potential cover. Feeding platforms should be placed atop the supporting posts (1.2–1.8 meters long) and should be large enough to support a small (15- 20-centimeter) dish in which the sunflower seeds and kernels can be placed. Small drain holes should be drilled into the feeding dishes, and each dish should be securely attached to the feeding platform. In order to provide an adequate training period for the birds, feeders should be set up at least a week before data collection begins.

After the feeders have been established and are being visited regularly by the local chickadees, data collection can begin. In class a sign-up sheet will be circulated showing all the possible 1-hour time slots during which data can be collected. You should sign up to do two 1-hour data collection sessions, one during each of two data collection periods (Table 20.1). Before starting a data collection session, make sure that the four feeder dishes are equally full and stocked with the correct type of seeds, as specified by the schedule (Table 20.1).

At each session, use the data sheet (Table 20.2) to record the following data: (1) date, (2) start time, (3) type of food present at each of the four feeders, and (4) number of feeding visits made by chickadees to each of the four feeders. Chickadees almost never process or eat sunflower seeds at feeders. During the brief feeder visits, a bird typically takes a single seed or kernel in its bill and carries it to a nearby tree for eating or storage. The number of feeder visits therefore is usually a reliable indicator of the number of seeds taken.

HYPOTHESES AND PREDICTIONS

Although chickadees receive the same gross caloric and nutritive reward from sunflower kernels and whole seeds, sunflower kernels should be relatively more profitable because they require less processing time. Chickadees are also likely to benefit from choosing feeders close to cover, for two reasons. First, the risk of predation is likely to be lower for feeders located close to cover (Newman & Caraco 1987; Abrahams & Dill 1989; Todd & Cowie 1990). Second, because chickadees forage and store seed

Table 20.2 **Sample data sheet for recording feeder visits during two 1-hour observations sessions. Use hash marks (卌 means 5) to tally chickadee feeding visits to each feeder. After completing the 1-hour observation period, record the total number of visits to each feeder under the "N Total" column.**

Period 1
Name: _____
Date: _____
Start time: _____

Feeder	Seed Type	Chickadee Feeding Visits (Tally)	N Total
1	_____	_____	_____
2	_____	_____	_____
3	_____	_____	_____
4	_____	_____	_____

Period 2
Name: _____
Date: _____
Start time: _____

Feeder	Seed Type	Chickadee Feeding Visits (Tally)	N Total
1	Whole seeds	_____	_____
2	Whole seeds	_____	_____
3	Kernels	_____	_____
4	Kernels	_____	_____

primarily in trees and other woody vegetation (Smith 1991), birds choosing feeders close to trees will expend less time and energy in flight than will birds choosing feeders far from trees.

Accordingly, during period 1 of this field experiment, you will test two hypotheses: that chickadees prefer sunflower kernels over whole seeds, and that chickadees prefer feeders close to cover over feeders far from cover. During period 2 you will also examine how chickadees behave when they are forced to make *trade-offs*. Can you predict what chickadees will do when they are offered a choice between relatively profitable food (sunflower kernels) at a dangerous and inaccessible location and relatively unprofitable food (whole seeds) at a safe and accessible location?

DATA RECORDING AND ANALYSES

After you have completed your field work and filled in the blanks in your data sheet (Table 20.2), data from the entire class can be pooled and analyzed. Assuming that each visit by a chickadee to a feeder is an independent event, we can use the chi-square goodness-of-fit test to analyze food and feeder preferences during period 1. In the goodness-of-fit test, the observed number of feeder visits is compared against the number expected if the null hypothesis of no food or feeder preference is true. For example, to test the hypothesis that chickadees preferred kernels over whole seeds during period 1, you first calculate the total numbers of kernels and whole seeds taken during the period. These are your observed values. Under the null hypothesis that chickadees are indifferent to seed type, you would expect chickadees to take an equal number of both kernels and whole seeds; thus the expected value for both kernels and whole seeds is simply half the grand total of seeds of either type taken by chickadees during period 1. You can use a similar procedure to test the hypothesis that chickadees preferred feeders close to cover over feeders far from cover during period 1.

For data collected during period 2, you can use a third chi-square goodness-of-fit test to test the hypothesis that chickadees show a preference for either (distant) kernels or (nearby) whole seeds.

QUESTIONS FOR DISCUSSION

Halfway through period 1, the contents of feeders 1 and 2 were switched, and the same was done for feeders 3 and 4 (Table 20.1). Why was it important to perform this switch? Would there have been difficulties in interpreting the results if it had not been done?

Under the conditions of the field experiment, which appears to be more important to chickadees, selecting food items with low handling times or selecting feeders close to cover?

Do chickadees avoid feeding at locations far from cover because of the higher risk of predation or because of the higher energetic costs of flying to distant feeders? Are these two hypotheses mutually exclusive? Can they be tested experimentally?

During period 2, when chickadees were forced to choose between feeding on whole seeds close to cover and feeding on kernels far from cover, why might the birds feed on the (near) whole seeds at some times but on the (distant) kernels at other times? Alternatively, why would some individuals show a preference for (near) whole seeds while others show a preference for (distant) kernels? Propose two testable hypotheses that might explain this variation in food and feeder preference, and briefly describe how you would test your hypotheses.

ACKNOWLEDGMENTS

I thank Abigail Bole, Tim Vyrostek, and my Biology 103 and Biology 370 students at Allegheny College for assistance in developing and testing this field exercise. Bonnie Ploger, Ken Yasukawa, and an anonymous reviewer provided helpful comments on an earlier version of the manuscript.

LITERATURE CITED

Abrahams, M. & Dill, L. M. 1989. A determination of the energetic equivalence of the risk of predation. *Ecology*, **70**, 999–1007.

Cuthill, I. C. & Houston, A. I. 1997. Managing time and energy. In: *Behavioural Ecology: An Evolutionary Approach*. 4th ed. (ed. J. R. Krebs & N. B. Davies), pp. 97–120. Oxford: Blackwell Science.

Dunn, E. H. & Tessaglia, D. L. 1994. Predation of birds at feeders in winter. *Journal of Field Ornithology*, **65**, 8–16.

Krebs, J. R. & Kacelnik, A. 1991. Decision-making. In: *Behavioural Ecology: An Evolutionary Approach*. 3rd ed. (ed. J. R. Krebs & N. B. Davies), pp. 105–136. Oxford: Blackwell Scientific.

Krebs, J. R. & Davies, N. B. 1993. *An Introduction to Behavioural Ecology*. 3rd ed. Oxford: Blackwell Scientific.

Newman, J. A. & Caraco, T. 1987. Foraging, predation hazard, and patch use in grey squirrels. *Animal Behaviour*, **35**, 1804–1813.

Smith, S. 1991. *The Black-capped Chickadee*. Ithaca, N Y: Cornell University Press.

Todd, I. A. & Cowie, R. C. 1990. Measuring the risk of predation in an energy currency: field experiments with foraging blue tits, *Parus caeruleus*. *Animal Behaviour*, **40**, 112–117.

CHAPTER 21

Seed Selection by Foraging Birds

MONICA RAVERET RICHTER[1], JUDITH A. HALSTEAD[2], AND KIERSTIN SAVASTANO[2]

[1] *Department of Biology, Skidmore College, 815 N. Broadway, Saratoga Springs, NY 12866, USA*

[2] *Department of Chemistry, Skidmore College, 815 N. Broadway, Saratoga Springs, NY 12866, USA*

INTRODUCTION

When food is a **limiting resource**, acquisition of energy is a major determinant of whether an individual survives and reproduces. In such situations, one might expect strong selection pressure on an animal's foraging behavior. **Optimal foraging theory** attempts to predict the foraging strategies that animals exhibit in specific circumstances and relies on assumptions that include maximization of fitness by a high rate of energy intake. In many cases, factors such as predator avoidance, acquisition of an essential nutrient, or consumption of a balanced diet may more strongly influence an animal's choice among available foods than does rate of energy intake (Begon et al. 1990). However, there exist circumstances in which the energy maximization assumption might be expected to apply. For example, birds foraging in harsh winter climates have increased energy needs (Barkan 1990), and one might thus expect their food choices to be strongly influenced by the energetics of foraging. If these birds forage optimally, their choice of food should reflect energetic considerations such as maximizing net energy gain per unit time or net gain per cost expended in foraging (Pyke et al. 1977; Seeley 1986).

In this lab, you will consider optimization of energy intake, as well as other factors that might influence seed selection, and predict which seeds foraging birds will select. You will then observe the seed selection of foraging birds in the field and perform a chi-square test to determine whether the food choice of these birds is consistent with your hypothesized preference. Finally, you will consider alternative explanations for your observations and will develop the rationale and design for an additional avian food choice experiment.

Many birds will visit feeders and collect sunflower seeds. The feeders that we use in this laboratory are stocked with three visually distinct sunflower seeds: soft-hulled Feathered Friend™ black oil and Feathered Friend™ grey striped seeds and large, thick-hulled Lyric™ California white sunflower seeds. Seeds are presented simultaneously either in a three-dish feeder, each dish containing only one kind of seed, or in three identical clear-sided tube feeders, each containing one type of seed. Prior to your observations, the birds have been visiting feeders at the study site for at least a week.

Table 21.1 summarizes measurements on the three different kinds of sunflower seeds, showing the average weights for kernel, hull, and entire seed; the kernel/hull (i.e., food/waste) ratio for seeds of each seed type; the average caloric content of a seed kernel; and the percent protein and fat for each seed type. Use these values and your knowledge of foraging behavior and optimal foraging theory to make predictions about the food choice of the birds and to make explicit the rationale behind your predictions.

Table 21.1 Mass, caloric content (in gram or "small" calories), % protein, and % fat of sunflower seed components. Measurements on mass and kernel/hull ratios are reported as mean ± SD and are based on means of 15 groups, each containing 10 seeds. Caloric measurements, % protein, and % fat are reported with 95% confidence intervals.

Seed Type	Kernel (mg/seed)	Hull (mg/seed)	Entire Seed (mg)	Kernel/ Hull	cal/g of Kernel	cal/Seed	% Protein	% Fat
Black oil	28.8 ±3.8	12.2 ±1.6	41.0 ±0.5	2.374 ±0.259	5770 ±590	166	61.8 ±0.6	22.7 ±6.9
Grey striped	36.8 ±3.9	26.4 ±3.3	63.3 ±6.9	1.403 ±0.111	5510 ±700	203	53.8 ±1.4	24.6 ±4.4
California white	61.7 ±4.7	59.1 ±2.7	120.8 ±6.0	1.045 ±0.081	5220 ±610	322	52.4 ±1.6	22.11 ±3.0

MATERIALS

One to several bird-feeding sites will be established at least 2 weeks prior to your lab to ensure that birds have been feeding on the test seeds for at least a week. Each feeder site will have either one multicompartment bird feeder, with a compartment for each seed type to be used in the experiment, *or* several identical clear-sided bird feeders, each feeder dispensing a different type of seed. Each site will be stocked with up to three kinds of sunflower seeds; possibilities include Feathered Friend™ black oil seeds, Feathered Friend™ grey striped seeds, and Lyric™ California white sunflower seeds. Each student or pair of students will have a stopwatch, binoculars, clipboard, data pad or notebook for data collection, calculator for data analysis, and bird identification guide.

HYPOTHESES AND PREDICTIONS

Work in small groups and consider potential research questions about the seed choice of birds. Before you generate your hypotheses, try to open each of the different types of sunflower seeds. On the basis of your attempts, which seed type do you predict will take the most time for the birds to open? Review the data in Table 21.1. If birds maximize their caloric intake per seed ingested, which seed type do you predict that they will prefer? On the basis of the values in the table and your experience when you attempted to open the seeds, which type of seed do you predict that birds will prefer if they maximize their intake of calories per unit handling time? What other seed characteristics and morphological and behavioral attributes of birds might influence seed preference? What other optimization criteria might serve to characterize seed preference? After considering the possibilities, choose two testable hypotheses, and provide a clear statement of and rationale for each of these research hypotheses.

PROCEDURE

In the Lab

Your small group will propose and test two experimental hypotheses on avian seed choice. You will present these hypotheses, the underlying rationale, your predictions of what you expect to happen if either of your hypotheses is true, and your experimental protocol to your classmates for evaluation and suggestions. On the basis of their input, you will revise your approach as necessary.

Table 21.2 **Common feeder birds in the Northeastern United States.**

Common Name	Latin Binomial
Black-capped chickadee	*Poecile atricapillus*
Tufted titmouse	*Baeolophus bicolor*
American goldfinch	*Carduelis tristis*
House finch	*Carpodacus mexicanus*
Red-breasted nuthatch	*Sitta canadensis*
White-breasted nuthatch	*Sitta carolinensis*
Downy woodpecker	*Picoides pubescens*
Hairy woodpecker	*Picoides villosus*

Review the characteristics of the birds that visit feeders in your location at the time of year at which you are performing this experiment. Table 21.2 lists birds that we commonly find during the winter at our feeders in the northeastern United States. All of these species except the tufted titmouse are also present in the western United States and Canada. In urban settings, the house sparrow (*Passer domesticus*) will be a common feeder visitor. Pay particular attention to features that will help you to recognize these birds and to distinguish them from similar birds.

Outdoors

One or two groups of students will go to each feeding station. After you are confident that all group members can accurately identify the visiting birds, divide your group into subgroups. It is difficult to keep an accurate tally of seed choice by feeder visitors and, at the same time, to follow and record the behavior of individual birds, so each subgroup will focus on gathering a particular type of data, as follows:

(1) For each visit to a feeder, record the species of bird visiting the feeder and which type of seed it chooses. When there are not large numbers of birds visiting the feeders, you may also be able to opportunistically record observations of how individual birds handle the seeds that they have selected.

(2) Describe the behavior of birds as they select seeds and immediately after seed selection. (a) How long does it take an individual bird (record the species of bird and the type of seed eaten) to eat a seed and return to the feeder for another? Use your binoculars to follow individual birds, and record the time from when a bird takes a seed to when it returns for another. (b) Describe in detail the seed-handling behavior of the birds visiting the feeders. How do the different birds crack open each type of seed? Do they have difficulty opening any of them? Are all of the seeds eaten, or are some of them dropped or cached by the birds?

DATA RECORDING AND ANALYSIS

Recording Data

To record your data on feeder choice, devise a data collection sheet on which you record the species of each forager and the type of seed each forager selects during your sample interval, along with other data you need to test your hypothesis. An expansion of or variation on the following sample data sheet (Table 21.3) might work for a subgroup charged with recording seed selection by foragers at the bird feeders.

Discuss your data collection sheet with your instructor before you begin your observations. After you have collected your data, construct another table to summarize your results overall and by species.

One or two students in your group should focus on observing and describing in detail how different birds handle the seeds on which they forage (or, better yet, rotate your data collection assignments so that each of you has a chance to make some of these observations). Be sure to have some observers follow individual foragers and record the handling time

Table 21.3 **Sample data sheet.**

Date: _____ Group members: _____

Time and duration of sample: _____

Weather: _____

Location of feeder and description of habitat: _____

_____

Seeds Selected by Feeder Visitors:

Type of Seed Selected

Feeder Visit	Bird Species	Black Oil	Grey Striped	California White	Behavioral Observations
1	Chickadee	X			Bird perched on branch and secured seed with foot; opened seed by hammering it with beak.
2	White-breasted nuthatch		X		Bird carried seed out of view.
3	House finch		X		Bird carried seed out of view.
4	White-breasted nuthatch			X	Bird wedged seed into bark of a tree; opened seed by hammering it with beak.

Table 21.4 **Sample data sheet.**

Date: _____ Group members: _____

Time and duration of sample: _____

Weather: _____

Location of feeder and description of habitat: _____

_____

_____

Definition of handling time:

Devise a definition that makes clear which action patterns correspond to handling time. How will you treat data from situations in which a seed is selected by a bird but not successfully consumed?

Type of Seed Selected

Feeder Visit	*Bird Species*	*Black Oil*	*Grey Stripe*	*CA White*	*Handling Time (seconds)*	*Behavioral Observations*
1	Chickadee			X	62.4* *bird did not eat seed	Bird tried repeatedly and unsuccessfully to open seed, which it secured with its foot. *Bird then dropped seed, uneaten.
2	Chickadee		X		22.3	Bird perched on branch and secured seed with foot, opened seed by hammering it with beak.
3	White-breasted nuthatch			X	21.2	Bird wedged seed into bark of a tree; opened seed by hammering it with beak.

(how will you define this?) for each seed by each species of bird. A variation on the sample data collection sheet above (Table 21.4) could be used.

Discuss your observations with all members of your research group, and include detailed descriptions of the foraging behavior in your lab write-up. These behavioral observations are extremely important—they will provide you with insights for interpreting your results and suggesting future studies.

Analysis

For each of your research hypotheses, state your statistical null and alternative hypotheses regarding seed choice by foragers. Check your hypotheses with your instructor. If you wish to test your hypotheses on data from particular species, you will need to analyze the data for each species separately, using a chi-square goodness-of-fit test. This test will enable you examine preferences for two or more seed types, testing the null hypothesis that the birds of a given species show no preference among seeds. If you wish both to look at seed preferences and to compare these preferences for two or more bird species, you can do so with a chi-square test of independence.

QUESTIONS FOR DISCUSSION

Clearly state the rationale for the hypotheses that you tested, present your null and alternative hypotheses, and state which seeds you predict the birds will choose if a given hypothesis is correct.

Describe and summarize what you observed in the field. Were some parameters more difficult to measure than others, and if so, why? Present the results of your chi-square statistical test on feeder choice. Which predictions do your data support? Are there any problems with your analytical approach? Interpret your observations as they relate to your hypotheses, and discuss your interpretation.

How could you redesign your experiment to measure energy gains, handling time, and energetic costs of foraging better, and thus to test the predictions of optimal foraging theory more accurately? Can you think of other hypotheses regarding seed choice by birds? Propose a follow-up study that would enable you to address one of your research hypotheses more effectively or that would enable you to test a related idea about avian foraging behavior. Make clear the ways in which your proposed study is an extension of or improvement on the experiment on which you report here.

ACKNOWLEDGMENTS

Thanks to Jasmin Keramaty for stocking and maintaining bird feeders and for processing and weighing sunflower seeds, and to the many enthusiastic and insightful students and teaching associates (Betsy Stevens, Sue Van Hook, and Laurie Freeman) who contributed to this laboratory. Corey Freeman-Gallant has assisted students in developing follow-up projects. Wayne Richter, Bonnie Ploger, and Ken Yasukawa commented on drafts of this laboratory. Thanks to Betsy Stevens and to Wayne and Caitlin Richter for sharing their knowledge and enjoyment of birds and for encouraging the entomologist author to listen and to look upward and beyond her foraging insects. This work was supported by Skidmore College.

LITERATURE CITED

Barkan, C. P. 1990. A field test of risk-sensitive foraging in black-capped chickadees (*Parus atricapillus*). *Ecology*, **71**, 391–400.

Begon, M., Harper, J. L. & Townsend, C. R. 1990. *Ecology: Individuals, Populations and Communities.* 2nd ed. Boston: Blackwell Scientific.

Pyke, G. H., Pulliam, H. R. & Charnov, E. L. 1977. Optimal foraging: a selective review of theory and tests. *Quarterly Review of Biology*, **52**, 137–154.

Seeley, T. D. 1986. Social foraging by honeybees: how colonies allocate foragers among patches of flowers. *Behavioral Ecology and Sociobiology*, **19**, 343–354.

CHAPTER 22

Competitive Behavior of Birds at Feeders

ALISON M. MOSTROM
University of the Sciences in Philadelphia, Department of Biological Sciences,
600 South 43rd Street, Philadelphia, PA, 19104-4495, USA

INTRODUCTION

Birds congregating at either artificially or naturally abundant food resources, such as feeders, animal carcasses, or trees that produce numerous seeds, interact competitively and cooperatively (Morse 1970, 1977, 1978). The interplay of these factors can be observed as birds interact at artificial feeders. This particular set of field exercises focuses on observing **intraspecific** (involving members of the same species), **interspecific** (involving members of different species) competition, and **agonistic** (submissive and aggressive) communication among locally wintering birds as they forage at stocked feeders.

The strongest competition for limited resources arises among members of the same species. Socially dominant members of intraspecific flocks gain access to feeders (and to more protected feeders) over subordinate individuals (Baker et al. 1981; Hogstad 1988a; Caraco et al. 1989; Inman 1990; Sasvari 1992), and dominant flockmates gain access to preferred feeding microhabitats while foraging naturally (Ekman & Askenmo 1984; Ekman 1987; Hogstad 1988b). Males dominate females in the sexually dimorphic northern cardinal (*Cadinalis cardinalis*) (Nice 1927; Laskey 1944).

Females dominate males in dimorphic house finches (*Carpodacus mexicanus*) (Thompson 1960; Brown & Brown 1988; Shedd 1990; Belthoff & Gauthreaux 1991) and in dimorphic house sparrows (*Passer domesticus*) (Hegner & Wingfield 1987). In the sexually monomorphic black-capped chickadee (*Poecile atricapilla*), Carolina chickadee (*P. carolinensis*), and tufted titmouse (*P. bicolor*), males tend to dominate females during the winter (Dixon 1963; Glase 1973; Smith 1976, 1984, 1991, 1993; Brawn & Samson 1983; Mostrom 1993), but as spring progresses, black-capped females dominate their mates (Smith 1991), and Carolina chickadee males relinquish dominance over their mates so that neither member of a mated pair dominates (Mostrom 1993).

Although less intense than intraspecific competition, interspecific competition for food also adversely affects the foraging behavior of socially subordinate species (Morse 1970, 1977, 1978; Alatalo 1981; Pierce & Grubb 1981; Carrascal & Moreno 1992; Sasvari 1992; Cimprich & Grubb 1994). Judging on the basis of the **dominance relationship** (the social relationship based on the outcome of agonistic encounters) between two species, the larger usually dominates (Morse 1978; Alatalo 1981; Pierce & Grubb 1981). Additionally, bird species that shell sunflower seeds inside their beaks by manipulating the seed with their tongues (such as northern cardinals and finches), usually remain at feeders for extended periods of time, often excluding birds (such as titmice and chickadees) that usually depart the feeder with a single seed to open it elsewhere.

MATERIALS

Each student will need a pair of binoculars (7 × 35 are good for most field conditions). Each field team (3–4 students) will need its own set of data sheets (the details of which you will devise), a clipboard, a stopwatch, and a field guide to birds in your region. In order to observe the birds' behavior, your team will need access to a bird feeding station. This should be a vertical tube feeder that contains 4–6 feeding perches and feeding holes large enough for the birds to extract sunflower seeds. Suspend the feeder in a manner that discourages squirrels from using and possibly destroying it. This usually involves either suspending the feeders from a tree branch (using a long wire line) far enough from the trunk that squirrels cannot jump onto them or hanging feeders from a squirrel-proof pole or shepherd's hook. Additionally, a squirrel baffle/guard should be suspended above each feeder. The feeding arena should be established at least a few weeks in advance and stocked continually to ensure that birds will be present during your study.

HYPOTHESES AND PREDICTIONS

Intraspecific Competition of Dimorphic Species

In this field exercise, your team may wish to investigate the dominance (as determined by consistent displacement at the feeder) of one sex over another, the feeding rates of males vs females of a dimorphic species, and/or the **ritualized** communication patterns that occur during agonistic events. For example, do female house finches dominate males? Do female house finches feed uninterrupted for longer periods of time than males? What visual displays or vocalizations do females use when displacing males from a feeder or while remaining on a feeder perch despite a close approach by a male?

Interspecific Competition

In this field exercise, your team may wish to investigate the dominance of one species over another, the feeding rates of the two species, and/or the ritualized displays that occur. For example, when both house finches and tufted titmice visit feeders, does either species dominate? What effect does the presence of the dominant species have on the feeding rate or pattern of the subordinate species? What displays do the dominant species use prior to displacing individuals of the subordinate species from a feeding perch? Does the subordinate species give visual displays or vocalizations that are associated with submissive behavior or appeasement?

PROCEDURE

General Instructions

You will probably have two field sessions to conduct this exercise. During the first half of the first field session, your team and class can determine the following: (1) which species visit your feeder, (2) which species are dimorphic, (3) the displacement rate between the sexes of dimorphic species and between different species, (4) the manner of feeding by each species (remaining at the feeder to eat the seed vs departing the feeder to eat the seed elsewhere), and (5) the most common visual and vocal displays associated with agonistic behavior. From these observations, your team should choose one or two related questions to address and then design both a preliminary observational protocol and a data sheet to field-test during the second half of the first field session. The entire class may choose to address the same question or questions so that data from each team can be shared in order to maximize the number of observations. If your class does this, each team should observe birds at one feeder that no other team is observing. Further modifications to the protocol and data sheet should be made during this field test session so that the entire second field session can be devoted to data collection. All observations of birds should be made at such a distance that observers do not interfere with the birds' feeding behavior.

In order to identify the species that visit the feeder arena and to determine whether each species is dimorphic or monomorphic, use the field guides provided by your instructor. Your field team will need to watch the behavior of birds closely as one bird displaces another, forcing the latter off its feeding perch. Your team will need to define "displacement" explicitly. It is important to distinguish between a bird that is forced to leave the feeder and one that leaves the feeder because it has finished feeding. Only the former should be considered a displacement. You may simply choose to define displacement as one bird taking the exact perch of another while the second bird attempts unsuccessfully to remain on the perch and feed. Alternatively, you may note that as one bird approaches within a certain distance, a perched bird regularly departs and that if it doesn't depart on its own, it gets pushed off by the approaching bird. This may lead you to extend your definition of displacement beyond "taking the exact perch of another bird" to include "the departure of a bird as another approaches within a defined distance."

For species such as chickadees, titmice, and nuthatches, you may choose to require that the bird that was displaced did not obtain a seed before leaving the feeder. For species that remain at the feeder while opening seeds (such as finches), this requirement of not feeding before being forced off will probably be too strict a definition of displacement. Your team will need to spend time observing birds visiting your feeder to determine the best definition of displacement. The entire class may want to discuss their observations and ideas to arrive at a consensus.

You will also need to define when a perched bird fails to be displaced and successfully remains on its perch. At what distance does the approaching bird veer away from the perched bird? Does the approaching bird simply fly to another unoccupied perch on the feeder, or does it leave the feeder without feeding?

During the reconnaissance phase of this exercise, your team will also want to observe the details of how birds communicate with each other as one approaches another at the feeder. Does an aggressive bird raise feathers on its head to form a crest? Does an increase in the elevation in the crest of a tufted titmouse or a northern cardinal correspond with its subsequent aggressive behavior? Does the aggressor widen the gape of its beak or jab its beak toward its opponent? Does a bird that is about to leave a perch when confronted by another squat and cower away from its opponent? Does a submissive or a dominant bird spread and quiver its wings? Do the focal birds vocalize during agonistic events? If so, what vocalizations are most frequently associated with aggressive and with submissive behavior?

Intraspecific Competition of Dimorphic Species

The following example is provided as a guideline only. If your team is interested in addressing the question of whether female house finches dominate males, you will need to observe the displacement of males by females, and *vice versa*, at the feeder. For this question, what is the biological hypothesis? What is the independent variable? What is the dependent variable? How will you measure the dependent variable? What predictions can you make about the birds' behavior if your hypothesis is correct? Check your answers with your instructor before collecting data. While making observations relevant to this question, a three-person team could also collect data on the communication that occurs during displacements and/or the feeding rates of the sexes. One way to divide the labor between teammates would be as follows: (1) One person makes focal observations of a perched bird by noting its gender, describes any visual displays or vocalizations it makes in the presence of an approaching bird, and times the duration of its feeding bout. (2) One person makes focal observations of a bird of the opposite sex as it approaches the perched bird and describes any visual displays or vocalizations directed at the perched bird. (3) One person records all data on a single data sheet as by the two other team members call out their observations.

Interspecific Competition

The following example is provided as a guideline only. If your team is interested in addressing the question of whether either house finches or tufted titmice dominate the other species at feeders, you will need to observe displacements between these two species. For this question, what is the biological hypothesis? What is the independent variable? What is the dependent variable? How will you measure the dependent variable? What predictions can you make about the birds' behavior if your hypothesis is correct? Check your answers with your instructor before collecting data. While making observations relevant to this question, a four-person team could also collect data on the feeding rate of the subordinate species in the presence and absence of the dominant species and on communication that occurs between these two species as they interact. One way to divide the labor between teammates would be as follows: (1) One person makes focal observations of the perched bird by noting the species (and possibly gender for house finches) and any visual displays or vocalizations that this perched bird makes toward an approaching bird of the appropriate species. (2) One person times the duration of the feeding bout of the perched bird or notes the number of seeds it obtains. (3) One person makes focal observations of a bird of the opposing species as it approaches the perched bird, noting any visual displays or vocalizations directed at the perched bird. (4) One person records all data on a single data sheet as the other team members call out their observations.

DATA RECORDING AND ANALYSES

Data Recording: Some Examples

If your team is interested in relating either intersexual or interspecific dominance to displays associated with aggressive and submissive behavior, then your data sheet will need to include information about the sex (or species) of both the perched bird and the approaching bird, the outcome of the attempted displacement, and specific information about visual and/or vocal displays. For an example, see Table 22.1.

If your team is interested in relating either intersexual or interspecific dominance to feeding rate, then your data sheet will need to include information about the sex (or species) of both the perched bird and the approaching bird, the outcome of the attempted displacement by the approaching bird, the length of time the perched bird fed before departing (or the number of seeds obtained), and the reason why the perched bird departed the feeder (because it was displaced or because it finished feeding). For an example, see Table 22.2.

Data Analysis

For either of the biological hypotheses proposed here, or for your specific biological hypothesis, what are the statistical null (H_0) and alternative hypotheses (H_a)? For example, for the biological hypothesis that females house finches dominate males, the statistical null hypothesis is that females and males are equally likely to be displaced from a feeding perch. One statistical alternative hypothesis is that females displace males statistically more frequently than 50% of the encounters. Because this is a predictive alternative hypothesis, the appropriate statistical test will be one-tailed.

Table 22.1 **Sample data sheet for relating dominance to displays.**

Perched Bird				Approaching Bird			
Species	Sex	Outcome of Approach: Wins = Displaces; Loses = doesn't Displace	Visual & Vocal Displays: Head Crest? Bill Jab? Cower? Wing Quiver? Vocals?	Species	Sex	Outcome of Approach: Wins = Displaces; Loses = doesn't Displace	Visual & Vocal Displays: Head Crest? Bill Jab? Wing Quiver? Vocals?

Table 22.2 **Sample data sheet for relating dominance to feeding.**

Perched Bird					Approaching Bird		
Species	Sex	Feed: Duration (Seconds)	Feed: Number of Seeds	Why Leave? (Displaced or Finished Feeding)	Species	Sex	Outcome of Approach: Wins = Displaces Perched Bird; Loses = doesn't Displace Perched Bird

If the alternative hypothesis is not directional in its prediction, then the appropriate statistical test will be two-tailed. For example, your two-tailed alternative hypothesis to an interspecific competition hypothesis should be that one of the two species you are observing will be more likely to displace the other, but you do not specify which species. Check your H_0 and H_a with your instructor before proceeding.

In your project, you may be examining intersexual and/or interspecific relationships. You may be relating a measure of dominance (not necessarily dominance at a statistically significant level) to feeding behavior, agonistic behavior, or communication patterns. Because the potential questions are varied and numerous, you will need to suggest the most appropriate statistical test or tests for the type of data you propose to collect and analyze. You may wish to consult an illustrative guide presented in a statistics book to help you determine which statistical test is most appropriate (e.g., Ambrose & Ambrose 1995, p. 50). Be sure to consult with your instructor before actually conducting the statistical test, and be ready to explain your rationale for choosing a particular test.

QUESTIONS FOR DISCUSSION

Although statistics are useful in preventing a **type I error** (the probability of rejecting a null hypothesis when you should have accepted it), is it necessary for a bird of one sex (or species) to win a **statistically significant** number of displacements to be considered **dominant** over the other sex (or species)? Discuss the reasons for your answer. Do you feel comfortable simply stating the number (and percent) of displacements won by each sex (or species) and then relating this trend to observations made concerning feeding rates or communication between individuals (or species)?

The ability to displace other birds from a feeder has obvious fitness advantages. What might be some of the costs associated with being the most dominant individual in the flock? Subordinate individuals, on the other hand, obviously pay the energetic price of being low-ranked while feeding at artificial feeders. Why do these individuals remain with their flock?

Displacement is not the only behavior associated with agonistic encounters. What are some others that you noticed? Which behaviors would you consider aggressive and which submissive? Could you place the aggressive (and/or submissive) displays on a scale ranging from least to most aggressive (submissive)? Which displays did you observe most frequently? Where did these displays lie on your scale? Why do you think these displays are most common?

Does a visual or vocal display need to be observed in conjunction with a behavior a statistically significant number of times in order for a bird to make an association between the display and subsequent behavior? Discuss the reasons for your answer. Do you feel comfortable presenting scientific data on which no statistical tests have been conducted? Discuss the reasons for your answer.

What is the selective advantage of ritualized displays? Are these displays species-specific, or did you observe similar displays in more than one species? For example, do both tufted titmice and northern cardinals use an elevation of the head crest to signal potential aggression, or do titmice elevate the crest while cardinals lower the crest in aggressive interactions (Halkin & Linville 1999)?

Does it appear that birds are attempting to communicate about their behavior "honestly," or do they seem to be communicating in order to manipulate the behavior of others to their advantage? What selective advantages arise from honest vs manipulative communication? Are the two hypotheses mutually exclusive, or might birds be communicating in both manners?

Interspecific flocks form under natural conditions. Which species that you observed came into, and/or departed, the feeding arena together? What are the costs and benefits to individuals in single-species vs multispecies flocks? How often did you observe intraspecific flocks? interspecific flocks? How cohesive is each of the two types of flocks?

ACKNOWLEDGMENTS

I would like to thank Bonnie J. Ploger and Ken Yasukawa for their very insightful comments on this manuscript. Fred Wasserman and Catherine Bentzley provided insightful comments on earlier versions of this manuscript. Sylvia L. Halkin helped me brainstorm over this exercise during the American Ornithologists' Union meeting, August 14–20, 2000, St. John's, Newfoundland, Canada.

LITERATURE CITED

Alatalo, R. V. 1981. Interspecific competition in tits *Parus* spp. and the Goldcrest *Regulus regulus*: foraging shifts in multispecies flocks. *Oikos*, **37**, 335–344.

Ambrose, H. W. III & Ambrose, K. P. 1995. *A Handbook of Biological Investigation*. 5th ed. Knoxville, TN: Hunter Textbooks.

Baker, M. C., Belcher, C. S., Duetch, L. C., Sherman, G. L. & Thompson, D. B. 1981. Foraging success in junco flocks and the effects of social hierarchy. *Animal Behaviour*, **29**, 137–142.

Belthoff, J. R. & Gauthreaux, S. A. 1991. Aggression and dominance in house finches. *Condor*, **93**, 1010–1013.

Brawn, M. J. & Samson, F. B. 1983. Winter behavior of tufted titmice. *Wilson Bulletin*, **95**, 223–232.

Brown, M. B. & Brown, C. M. 1988. Access to winter food resources by bright- versus dull-colored house finches. *Condor*, **90**, 729–731.

Caraco, T., Barkan, C., Beacham, J. L., Brisbin, L., Lima, S., Mohan, A., Newman, J. A., Webb, W. & Witham, M. L. 1989. Dominance and social foraging: a laboratory study. *Animal Behaviour*, **38**, 41–58.

Carrascal, L. M. & Moreno, E. 1992. Proximal costs and benefits of heterospecific social foraging in the great tit, *Parus major*. *Canadian Journal of Zoology*, **70**, 1947–1952.

Cimprich, D. A. & Grubb, T. C., Jr. 1994. Consequences for Carolina chickadees and tufted titmice in winter. *Ecology*, **75**, 1615–1625.

Dixon, K. L. 1963. Some aspects of social organizaiton in the Carolina chickadee. *Proceedings of the XIIth International Ornithological Congress*, 240–258.

Ekman, J. 1987. Exposure and time use in willow tit flocks: the cost of subordination. *Animal Behaviour*, **35**, 445–452.

Ekman, J. & Askenmo, C. 1984. Social rank and habitat use in willow tit groups. *Animal Behaviour*, **32**, 508–514.

Glase, J. C. 1973. Ecology of social organization in the black-capped chickadee. *Living Bird*, **12**, 235–267.

Halkin, S. L., & S. U. Linville. 1999. Northern Cardinal (*Cardinalis cardinalis*). *In* The Birds of North America, No. 440 (A. Poole and F. Gill, eds.). The Birds of North America, Inc., Philadelphia, PA.

Hegner, R. & Wingfield, J. C. 1987. Social status and circulating levels of hormones in flocks of house sparrows, *Passer domesticus*. *Ethology*, **76**, 1–14.

Hogstad, O. 1988a. Social rank and antipredator behaviour of willow tits *Parus montanus*. *Ibis*, **130**, 45–56.

Hogstad, O. 1988b. Rank-related resource access in winter flocks of willow tits *Parus montanus*. *Ornis Scandinavia*, **19**, 169–174.

Inman, A. J. 1990. Group foraging in starlings: distributions of unequal competitors. *Animal Behaviour*, **40**, 801–810.

Laskey, A. R. 1944. A study of the cardinal in Tennessee. *Wilson Bulletin*, **56(1)**, 27–44.

Morse, D. H. 1970. Ecological aspects of some mixed-species foraging flocks of birds. *Ecological Monographs*, **40**, 119–168.

Morse, D. H. 1977. Feeding behaivor and predator avoidance in heterospecific groups. *BioScience*, **27**, 332–339.

Morse, D. H. 1978. Structure and foraging patterns of flocks of tits and associated species in an English woodland during winter. *Ibis*, **120**, 298–312.

Mostrom, A. M. 1993. The social organization of Carolina chickadee (*Parus carolinensis*) flocks in the non-breeding season. Ph.D. thesis, University of Pennsylvania.

Nice, M. M. 1927. Experiences with cardinals at a feeding station in Oklahoma. *Condor*, **29**, 101–103.

Pierce, V. & Grubb, T. C. Jr. 1981. Laboratory studies of foraging in four bird species of deciduous woodland. *Auk*, **98**, 307–320.

Sasvari, L. 1992. Great tits benefit from feeding in mixed-species flocks: a field experiment. *Animal Behaviour*, **43**, 289–296.

Shedd, D. H. 1990. Aggressive interactions in wintering house finches and purple finches. *Wilson Bulletin*, **102**, 174–178.

Smith, S. M. 1976. Ecological aspects of dominance hierarchies in black-capped chickadees. *Auk*, **93**, 95–107.

Smith, S. M. 1984. Flock switching in chickadees: why be a winter floater? *American Naturalist*, **123**, 81–98.

Smith, S. M. 1991. *The Black-capped Chickadee: Behavioral Ecology and Natural History.* Ithaca, NY: Cornell University Press.

Smith, S. M. 1993. Black-capped chickadee. In *The Birds of North America*, No. 39 (ed. A. Poole, P. Stettenheim & F. Gill). Philadelphia: The Academy of Natural Sciences; Washington, DC: The American Ornithologists' Union.

Thompson, W. T. 1960. Agonistic behavior in the house finch. Part II: factors in aggressiveness and sociality. *Condor*, **62**, 378–402.

ADDITIONAL SUGGESTED READING

The most extensive current summary of information about any bird species that breeds in North America is available in that species's monograph in *The Birds of North America*. The Birds of North America, Inc., Philadelphia, PA.

SECTION II

Avoiding Predators

CHAPTER 23

Vigilance and the Group-size Effect: Observing Behavior in Humans

Joanna E. Scheib[1], Lisa E. Cody[2], Nicola S. Clayton[3], and Robert D. Montgomerie[4]

[1]Department of Psychology, University of California, Davis, CA 95616, USA
[2]Animal Behavior Graduate Group, University of California, Davis, CA 95616, USA
[3]Experimental Psychology, University of Cambridge, Downing St., Cambridge, CB2 3EB, UK
[4]Department of Biology, Queen's University, Kingston, ON K7L 3N6, Canada

INTRODUCTION

Many animals look up and scan the environment while they are eating. This scanning and alert behavior is called **vigilance** and can serve many functions (e.g., Lima 1990), the best understood one being predator detection (Caraco et al. 1980; Glück 1987; Lendrem 1983). A widely studied phenomenon is the "**group-size effect**," whereby **vigilant behavior**, operationalized as the frequency and/or duration of scans, decreases as the group size increases. Such a pattern has been observed in many nonhuman animals, including birds, mammals, and fish—species in which individuals associate with two or more conspecifics and thereby constitute a group (e.g., Bertram 1980; Caraco 1979; Godin et al. 1988; Holmes 1984; Roberts 1996; Studd et al. 1983; Sullivan 1984; but also see Treves 2000). The group-size effect has also been observed in humans (Barash 1972; Wawra 1988;

Wirtz & Wawra 1986), despite the virtual absence of predation in modern human societies. Thus the group-size effect in humans may be a relic of selection in the evolutionary environment of our ancestors.

The group-size effect has generally been attributed to the influence of predation risk on the behavior of individuals. For example, because every group member benefits when a predator is detected, individuals can reduce their own scanning rate, without decreasing the probability of predator detection, if all group members do some scanning for predators (Pulliam 1973). This is called the **many-eyes hypothesis**. An alternative, called the **dilution hypothesis**, states that an individual's chance of being caught by a predator during a given predator attack decreases as the group size increases (Hamilton 1971). Thus an individual's predation risk is dependent not on the vigilance of its foraging partners, as in the many-eyes hypothesis, but merely on their presence. As a result, an individual's scanning rate should decline as group size increases. There are at least two other factors, however, that could explain the group-size effect in the *absence* of predation risk (or at least are not relevant to predation risk). The first suggests that vigilance is related to competition for food within the group. As foraging group size increases, so does the competition for food within a foraging patch. Thus the observed decrease in vigilance with increasing group size may be a result of the need to spend more time foraging in the face of competition with other individuals (the **food competition hypothesis**; Bertram 1978). The final hypothesis suggests that the vigilance of group members may be directed toward detection of conspecifics rather than of predators. In social groups, the detection of conspecifics might be important, for example, for signaling status, avoiding dominant individuals, attracting potential mates, or guarding current mates. To the best of our knowledge, this **conspecific detection hypothesis** has never been tested, even though it may be particularly applicable to humans.

MATERIALS

You will need 1–2 of each of the following items per group of observers: pencil and eraser, clipboard, scanning data sheets (enough for each recording period), and stopwatch or wristwatch with a timer.

HYPOTHESES AND PREDICTIONS

Do exercise (i) and your choice of any *one* of exercises (ii), (iii), and (iv). Exercises (ii) through (iv) will give you an opportunity to examine the *function* of the group-size effect in humans.

Exercises

(i) *Examining the Effect of Group Size on Vigilant Behavior*

The size of a group is hypothesized to influence vigilant behavior in humans. As group size increases, an individual's scanning frequency and duration are expected to decrease. To test this, you will need to observe the behavior of focal subjects in groups of different sizes. First identify the sizes of groups you will observe, and then determine how you will operationalize scanning behavior. What do you predict will happen if the group-size effect is observed? What do you predict will happen if the effect is not observed? Further details on how to test the group-size effect are provided in the "Procedure" section.

(ii) *Testing between the Dilution and Many-Eyes Hypotheses*

The group-size effect has generally been attributed to the influence of predation risk on the behavior of individuals. The dilution and many-eyes hypotheses may help explain how this effect happens. The dilution hypothesis states that an individual's chance of being caught by a predator during a given predator attack decreases as the group size increases. Thus an individual's predation risk depends simply on the presence of its foraging partners. Alternatively, according to the many-eyes hypothesis, an individuals predation risk is dependent not on the mere presence of foraging partners but on their vigilance. If all group members do some scanning for predators, each individual is hypothesized to be able to reduce her or his own amount of scanning as the size of the group increases. One way to test between these alternative hypotheses is to measure how occupied the group members are (e.g., the amount of conversation). In this case, you would not need to vary the group size but would instead do all your observations on focal subjects in the same-size groups. What do you predict will occur if the dilution hypothesis is correct? What do you predict will occur if the many-eyes hypothesis is correct? Details on how to test these hypotheses are provided in the "Procedure" section.

(iii) *Testing between the Predation Risk and Food Competition Hypotheses*

Although the group-size effect has generally been attributed to the influence of predation risk, competition with foraging partners may also or may instead explain this effect. A general predation risk hypothesis states that as group size increases, an individual will decrease the amount and duration of her or his scanning for predators [this can be due to dilution or to many-eyes effects; see exercise (ii)]. Alternatively, the food competition hypothesis states that an individual will decrease scanning as group size increases, not because of reduced predation risk but because of the need to spend more time foraging in the face of competition with other individuals. One way to test between these alternative hypotheses is to compare the vigilance of focal subjects in groups who are sharing their food to that in groups who are not sharing. In this case, you would not need to vary the group size but would instead do all your observations on focal subjects in the same-size groups. What do you predict will occur if the predation risk hypothesis is correct? What do you predict will occur if the food competition hypothesis is correct? Further details on how to test these hypotheses are provided in the "Procedure" section.

(iv) Testing between the Predation Risk and Conspecific Detection Hypotheses

The conspecific detection hypothesis is another alternative explanation of the group-size effect that is not related to the influence of predation risk. This hypothesis states that an individual will decrease scanning behavior as group size increases, not in response to predation risk but because of the possibility of detecting conspecifics who might be passing by. One way to test between the predation risk and conspecific detection hypotheses is to compare the vigilance of focal subjects in groups who are exposed to high and low amounts of people traffic (e.g., how many people are walking within sight of the focal group or how crowded the eatery is). In this case, you would not need to vary the group size but would instead do all your observations on focal subjects in the same-size groups. What do you predict will occur if the predation risk hypothesis is correct? What do you predict will occur if the conspecific detection hypothesis is correct? Further details on how to test these hypotheses are provided in the "Procedure" section.

PROCEDURE

(i) Examining the Effect of Group Size on Vigilant Behavior

Work in pairs. For each pair of observers, one person will watch the scanning behavior of focal subjects eating in groups of 1–5 individuals; the other observer will record the data on the data sheet. To avoid possible confounds, we recommend that you focus on foraging groups that are entirely women or men and that the watcher be the same sex as your subjects (a woman observes women, and a man observes men). Select one kind of eating place in which you will observe your subjects, and focus on places where people are either sharing or not sharing food. To minimize potential confounds, make sure you record the time of day and location of your observations and keep these the same for each trial, because scanning behavior and group dynamics may differ among the various locations and with the time of day.

It will take approximately 6 hours to conduct the observations for this lab. Spend the first hour collecting pilot data so that you will be confident of how best to record your data and can assess what the problems and pitfalls might be. Collect pilot data for at least two individuals from groups of three different sizes (e.g., individuals alone, with one other person, and with two or more people). Think about what time of day it would be best for you to collect your data, bearing in mind not only your schedule but also when the probability will be highest of collecting good data. For example, you are more likely to see people eating, both alone and in groups, around common mealtimes—a somewhat obvious but nonetheless important consideration. Once you have a well-designed protocol, you can use the other 5 hours for collecting your data.

In each group that you study, you will be collecting vigilance data from only one member of the group. This individual is called the focal subject and is chosen randomly. Select your observation site carefully so that you can easily watch your focal subject and he or she is not aware of being watched. For this reason, we suggest that you sit at least 3 meters away, in a position where you can see your focal subject clearly. To test for the group-size effect, observe the scanning behavior of focal subjects (1) eating alone, (2) eating with another individual, and (3) eating in a group of either 3, 4, or 5 individuals (choose *one* of group size 3, 4, or 5). You will need to observe one focal subject for 5 minutes in at least 10 different groups per group size studied (not including pilot data).

Key variables to record for each foraging group include the number and sex of people eating at the table, the sex of the focal subject, the facial orientation of the focal subject with respect to you (head on, side on, etc.), the number of times the focal subject looked up and scanned (with head or eyes only), and the length of time spent scanning. Make sure that for each scanning bout recorded during a trial, you record the time at which the subject first looked up and the time at which the subject resumed looking down. Looking directly at another person at the same table should not be counted as scanning behavior. Use your pilot observations to decide how you will measure scanning behavior, and what criteria you will use. One consideration might be how much the looking-down gaze has to change to be recorded as the end of a scan. Another consideration will be how to record scans that are very short. If scans are less than 2 seconds in length, the recording durations can become very imprecise. How will you overcome this problem? Be consistent and use the same criteria and definition of scanning for every trial. You might also want to think about the order of testing and to ensure that you do not make all your observations of one group size during a single block of time.

Once you have determined how you will measure scanning behavior, each pair should perform an interobserver reliability test (do this while collecting pilot data). This test ensures that both observers measure and record behavior in the same way. To perform this test, both members of a pair measure and record the scanning behavior of focal individuals during the same 5-minute sample for each of the three group sizes. Members of each pair then compare results to ensure 90–95% agreement. Pairs should continue reliability tests until this level of agreement is achieved.

Plan to make use of partial sample periods only if they are at least 3 minutes long (sample periods may be shorter than 5 minutes as a result of changes in group-size, the departure of your focal subject during observation, etc.). Calculating the scanning rate (scans per minute) helps to control for variation in length of sample periods. Also, if a 5-minute sample interval ends while your focal subject is engaged in a scanning bout, you may include this bout in calculating the focal subject's scanning rate, but you should not include it in your calculation of the mean scanning bout duration for that subject.

Things to Think About

In this lab, you will need to decide how you are going to watch the subjects, when and where, and what factors you should try to control for. Please consider the privacy and welfare of your subjects and be as nonintrusive as possible, for their sakes and yours! Be prepared to deal with people who question you about why they are being observed. Your instructor may provide suggestions on how to do this.

(ii) Testing between the Dilution and Many-Eyes Hypotheses

Follow procedure (i), with the following modifications. (1) One person will observe and record the scanning behavior of the focal subject, while the second will observe and record the amount of conversation in the group. (2) Study only one group size between 2 and 5 individuals. You will still need to observe 30 focal subjects for a period of 5 minutes per subject, but this time make sure that at least 15 of the subjects belong to groups in which the members are talking, because you will need to compare the scanning behavior of subjects in groups exhibiting low vs high amounts of conversation. You can define low vs high amounts of conversation by categorizing the groups that fall below the median conversation time as the low-conversation groups and those that fall above the median as the high-conversation groups. As in procedure (i), do all your observations at the same time of day, focus on places where people are either sharing or not sharing food, and collect data on one focal subject per group. (3) Conversation time needs to be measured and then is calculated as the total time that group members spent talking in the 5-minute sample interval. Remember that looking directly at another person at the table does not count as scanning behavior.

(iii) Testing between the Predation Risk and Food Competition Hypotheses

Follow procedure (i), with the following modifications. Study only one group size between 2 and 5 individuals. You will still need to observe 30 focal subjects for a period of 5 minutes per subject, but this time make sure that 15 of the subjects belong to groups in which the members are sharing food, because you will need to compare the scanning behavior of subjects who are sharing food to that of those who are not. Be sure to do all your observations in the same eating place and where people are likely to be sharing food (e.g., pizza, fondue, or Chinese food). As in procedure (i), do all your observations at the same time of day and collect data on one focal subject per group.

(iv) Testing between the Predation Risk and Conspecific Detection Hypotheses

Follow procedure (i), with the following modifications. (1) Study only one group size between 2 and 5 individuals. You will still need to observe 30 focal subjects for a period of 5 minutes per subject, but this time make sure that 15 of the focal groups are in crowded (high-traffic) eateries and the other 15 are fairly isolated (low-traffic places where there are few other groups). As with procedure (i), do all your observations at the same time of day, focus on places where people are either sharing or not sharing food, and collect data on one focal subject per group. Also record the number of actual interactions (talking, waving, nodding, smiling) with passersby. (2) For each focal group, record the number of passersby as an index of the amount of traffic during your observation period (e.g., number of people walking within 5 meters of the focal group).

DATA RECORDING AND ANALYSES

Sample data sheets for each exercise are provided at the end of this chapter. Once you have recorded all your data, you can calculate the mean frequency of scanning per minute (rate) and the mean scanning bout duration for all focal subjects.

(i) Examining the Effect of Group Size on Vigilant Behavior

Calculate the mean scanning rate and the mean scanning bout duration for each experimental condition (e.g., for group sizes 1, 2, and 4 people). Because you are using mean scores to summarize the data, also calculate a measure of variance, such as the standard error, for each of the two mean scores in each condition. These means and measures of variance will help you compare behavior across the three conditions and will give you an idea whether your data support your experimental hypothesis (vigilance decreases with increasing group size) or the null hypothesis (vigilance does not differ across conditions). To test whether there is a statistical difference in the subjects' vigilance across the three conditions, use one-way analysis of variance (ANOVA).

(ii) Testing between the Dilution and Many-Eyes Hypotheses

Calculate the mean scanning rate and the mean scanning bout duration for each experimental condition (that is, for the high-conversation groups and for the low-conversation groups). Because you are using mean scores to summarize the data, also calculate a measure of variance, such as the standard error, for each of the two mean scores in each condition. These measures will help you compare behavior in the two conditions and will give you an idea whether your data support the many-eyes hypothesis (less vigilance is observed among the high-conversation groups than among the low-conversation groups) or the dilution hypothesis, which in this case is also the null hypothesis (vigilance does not differ between conditions). To test whether there is a statistical difference in the subjects' vigilance in the low- vs high-conversation groups, use independent-samples *t* tests.

(iii) Testing between the Predation Risk and Food Competition Hypotheses

Calculate the mean scanning rate and the mean scanning bout duration for each condition (that is, for groups who share food and for those who do not). Because you are using mean scores to summarize the data, also calculate a measure of variance, such as the standard error, for each of the two mean scores in each condition. These measures will help you compare behavior in the two conditions and will give you an idea whether your data support the food competition hypothesis (less vigilance is observed in groups who are sharing food than in those who are not) or the predation risk hypothesis, which in this case is also the null hypothesis (vigilance does not differ between conditions). Use independent-samples *t* tests to determine whether there is a statistical difference between the scanning behavior of subjects who share food and that of those who do not.

(iv) Testing between the Predation Risk and Conspecific Detection Hypotheses

Calculate the mean scanning rate, the mean scanning bout duration, and the mean number of interactions per passersby for each condition (that is, for groups in high-traffic and low-traffic areas). Because you are using mean scores to summarize the data, also calculate a measure of variance, such as the standard error, for each of the two mean scores in each condition.

These measures will help you compare behavior in the two conditions and will give you an idea whether your data support the conspecific detection hypothesis (more vigilance is observed in groups located in high-traffic areas than in groups located in low-traffic areas) or the predation risk hypothesis, which in this case is also the null hypothesis (vigilance does not differ between conditions). Use independent-samples *t* tests to test whether there is a statistical difference between the subjects' vigilance in low- and high-traffic areas.

For All Exercises

Provide graphical representation(s) (such as a table or figure) of your results. Please remember to label the axes and to include standard error bars in your figure. Note that you will need to do a separate test for each of your dependent measures (scanning rate and bout duration). Please see your instructor if you need help analyzing your data.

QUESTIONS FOR DISCUSSION

If you do not find an effect of group size on vigilant behavior in humans, what implication does this hold for proceeding with exercises (ii), (iii), and (iv)?

Are the dilution and many-eyes hypotheses necessarily mutually exclusive? Justify your answer.

Why did you use only one group size when testing between (1) the dilution and many-eyes hypotheses, (2) the predation risk and food competition hypotheses, and (3) the predation risk and conspecific detection hypotheses?

What do we assume when we use the amount of conversation as a method of testing between the dilution and many-eyes hypotheses?

What do we assume when we use sharing food vs not sharing food as a way to test between the predation risk and food competition hypotheses?

What do we assume when we use high-traffic vs low-traffic areas as a way to test between the predation risk and conspecific detection hypotheses?

Why is it important to examine the behavior of individuals in single-sex groups? Why should the observer be the same sex as the focal subject?

Aside from the hypotheses used in the exercises, what alternative explanation(s) could account for the group-size effect in humans? Might these hypotheses be applicable to other animals?

What further experiments would you conduct to better understand the group-size effect in humans?

Interpret your results in an evolutionary framework.

Do you think results obtained in these exercises, with modern humans, reflect the behavior of humans throughout their evolution?

Which of the hypotheses and results do you think are unique to humans?
Which of the hypotheses and results would you expect to find in nonhuman animals?

ACKNOWLEDGMENTS

This lab was originally designed by Montgomerie for his behavioural
ecology and sociobiology course at Queen's University in the late 1980s
and was modified and expanded by Scheib, Cody, and Clayton for their
animal behavior courses at UC Davis in the late 1990s. All authors contributed to the preparation of this manuscript. We thank D. Owings and
L. Miyasoto for their assistance with this lab during the UCD winter
animal behavior course. We thank Matthew Campbell, Liza Moscovice,
Bonnie Ploger, and Ken Yasukawa for helpful comments on an earlier draft
of this paper.

LITERATURE CITED

Barash, D. P. 1972. Human ethology: the snack-bar security syndrome. *Psycho-logical Reports*, **31**, 577–578.

Bertram, B. C. R. 1978. Living in groups: predators and prey. In: *Behavioural Ecology* (ed. J. R. Krebs & N. B. Davies), pp. 64–96. Oxford: Blackwell Scientific.

Bertram, B. C. R. 1980. Vigilance and group size in ostriches. *Animal Behaviour*, **28**, 278–286.

Caraco, T. 1979. Time budgeting and group size: a test of theory. *Ecology*, **60**, 618–627.

Caraco, T., Martindale, S. & Pulliam, H. R. 1980. Avian time budgets and distance to cover. *Auk*, **97**, 872–875.

Glück, E. 1987. An experimental study of feeding, vigilance, and predator avoidance in a single bird. *Oecologia*, **71**, 268–272.

Godin, J.-G., Classon, L. J. & Abrahams, M. V. 1988. Group vigilance and shoal size in a small characin fish. *Behaviour*, **104**, 29–40.

Hamilton, W. D. 1971. Geometry for the selfish herd. *Journal of Theoretical Biology*, **31**, 295–311.

Holmes, W. G. 1984. Predation risk and foraging behavior of the hoary marmot in Alaska. *Behavioral Ecology and Sociobiology*, **15**, 293–301.

Lendrem, D. W. 1983. Predation risk and vigilance in the blue tit (*Parus caeruleus*). *Behavioral Ecology and Sociobiology*, **14**, 9–13.

Lima, S. L. 1990. The influence of models on the interpretation of vigilance. In: *Interpretation and Explanation in the Study of Animal Behavior*, Vol. 2 (ed. M. Bekoff & D. Jamieson), pp. 246–267. Boulder: Westview Press.

Pulliam, H. R. 1973. On the advantages of flocking. *Journal of Theoretical Biology*, **38**, 419–422.

Roberts, G. 1996. Why individual vigilance declines as group size increases. *Animal Behaviour*, **51**, 1077–1086.

Studd, M., Montgomerie, R. D. & Robertson, R. J. 1983. Group size and predator surveillance in foraging house sparrows (*Passer domesticus*). *Canadian Journal of Zoology*, **61**, 226–232.

Sullivan, K. A. 1984. The advantages of social foraging in downy woodpeckers. *Animal Behaviour*, **32**, 16–22.

Treves, A. 2000. Theory and method in studies of vigilance and aggregation. *Animal Behaviour*, **60**, 711–722.

Wawra, M. 1988. Vigilance patterns in humans. *Behaviour*, **107**, 61–71.

Wirtz, P. & Wawra, M. 1986. Vigilance and group size in *Homo sapiens*. *Ethology*, **71**, 283-286.

ADDITIONAL SUGGESTED READING

Bertram, B. C. R. 1978. Living in groups: predators and prey. In: *Behavioural Ecology* (ed. J. R. Krebs & N. B. Davies), pp. 64–96. Oxford: Blackwell Scientific.

Lima, S. L. 1990. The influence of models on the interpretation of vigilance. In: *Interpretation and Explanation in the Study of Animal Behavior*, Vol. 2 (ed. M. Bekoff & D. Jamieson), pp. 246–267. Boulder: Westview Press.

Roberts, G. 1996. Why individual vigilance declines as group size increases. *Animal Behaviour*, **51**, 1077–1086.

Treves, A. 2000. Theory and method in studies of vigilance and aggregation. *Animal Behaviour*, **60**, 711–722.

Wirtz, P. & Wawra, M. 1986. Vigilance and group size in *Homo sapiens*. *Ethology*, **71**, 283–286.

Sample Data Sheet, Exercise (i)

Sample data sheet for (i), vigilance and group size				
Copy 10 sheets for each group size, Use 1 sheet per focal subject.				
Behavioral codes:				
Subject's facial orientation relative to observer = (H) head on, (S) side on				
Use military time for scanning durations (e.g., 8:30:45 a.m. = 08:30:45, 5:25:23 p.m. = 17:25:23).				
Length of focal sample will either be 5 minutes or somewhere between 3 and 5 minutes.				
Group 1: One individual				
Subject:	Sex:	Orientation:	Focal length:	
	Day:	Location:	Appr. dist. from subject (m):	
Scans	*Start*	*Stop*		
1				
2				
3				
4				
5				
6				
7				
8				
9				
10				

Sample Data Sheet, Exercise (i)

Group 2: Individuals		f = focal, o = other (Make observations only for the focal.)			
Subject:	Sex(f, o):	Orientation:	Focal length:		
	Day:	Location:	Appr. dist. from subject (m):		
Scans	*Start*	*Stop*			
1					
2					
3					
4					
5					
6					
7					
8					
9					
10					

Sample Data Sheet, Exercise (i)

Group 3: Individuals		f = focal, o = other (Make observations only for the focal.)			
Subject:	Sex(f, others):	Orientation:	Focal length:		
	Day:	Location:	Appr. dist. from subject (m):		
Scans	*Start*	*Stop*			
1					
2					
3					
4					
5					
6					
7					
8					
9					
10					

Sample Data Sheet, Exercise (ii)

Sample data sheet for (ii), testing dilution vs many-eyes hypotheses				
Measuring scanning				
Copy enough sheets for 30 subjects.				
Behavioral codes:				
Subject's facial orientation relative to observer = (H) head on, (S) side on				
Use military time for scanning durations (e.g., 8:30:45 a.m. = 08:30:45, 5:25:23 p.m. = 17:25:23).				
Length of focal sample will either be 5 minutes or somewhere between 3 and 5 minutes.				
	f = focal, o = other			
Subject:	Sex (f, o):	Orientation:	Focal length:	
	Day:	Location:	Appr. dist. from subject (m):	
Scans	*Start*	*Stop*		
1				
2				
3				
4				
5				
6				
7				
8				
9				
10				
11				
12				
13				
14				
15				
Subject:	Sex (f, o):	Orientation:	Focal length:	
	Day:	Location:	Appr. dist. from subject (m):	
Scans	*Start*	*Stop*		
1				
2				
3				
4				
5				
6				
7				
8				
9				
10				
11				
12				
13				
14				
15				

Sample Data Sheet, Exercise (ii)

Sample data sheet for (ii), testing dilution vs many-eyes hypotheses					
Measuring Conversation					
These data are collected simultaneously with scanning behavior, so Subject 1 here					
is the same as Subject 1 on the scanning data sheet.					
Copy enough sheets for 30 subjects.					
Subject:				f = focal, o = other	
Conversation Bout	*Start*	*Stop*	*Duration (s)*	*f, o?*	
1					
2					
3					
4					
5					
6					
7					
8					
9					
10					
11					
12					
13					
14					
15					
	Total duration (s) =				
Subject:				f = focal, o other	
Conversation Bout	*Start*	*Stop*	*Duration (s)*	*f, o?*	
1					
2					
3					
4					
5					
6					
7					
8					
9					
10					
11					
12					
13					
14					
15					
	Total duration (s) =				

Sample Data Sheet, Exercise (iii)

Sample data sheet for (iii), testing predation risk vs food competition hypotheses				
Copy enough sheets for 30 subjects.				
Behavioral codes:				
Subject's facial orientation relative to observer = (H) head on, (S) side on				
Use military time for scanning durations (e.g., 8:30:45 a.m. = 08:30:45, 5:25:23 p.m. = 17:25:23).				
Length of focal sample will either be 5 minutes or somewhere between 3 and 5 minutes.				
Group size = individuals		f = focal, o = other		
Subject:	Sex (f, o):	Orientation:	Sharing food (y/n):	Focal length:
	Day:	Location:	Appr. dist. from subject (m):	
Scans	Start	Stop		
1				
2				
3				
4				
5				
6				
7				
8				
9				
10				
11				
12				
13				
14				
15				
Subject:	Sex (f, o):	Orientation:	Sharing food (y/n):	Focal length:
	Day:	Location:	Appr. dist. from subject (m):	
Scans	Start	Stop		
1				
2				
3				
4				
5				
6				
7				
8				
9				
10				
11				
12				
13				
14				
15				

Sample Data Sheet, Exercise (iv)

Sample data sheet for (iv), testing predation risk vs conspecific detection hypotheses			
Measuring scanning			
Copy enough sheets for 30 subjects.			
Behavioral codes:			
Subject's facial orientation relative to observer = (H) head on, (S) side on			
Interactions with passersby may be scored as (T = talked, S = smiled,			
N = nodded toward, W = waved at, P = physical contact)			
Use military time for scanning durations (e.g., 8:30:45 a.m. = 08:30:45, 5:25:23 p.m. = 17:25:23).			
Length of focal sample will either be 5 minutes or somewhere between 3 and 5 minutes.			
	f = focal, o = other		
Subject:	Sex (f, o):	Orientation:	Focal length:
	Day:	Location:	Appr. dist. from subject (m):
Scans	*Start*	*Stop*	*Interactions*
1			
2			
3			
4			
5			
6			
7			
8			
9			
10			
11			
12			
13			
14			
15			
Subject:	Sex (f, o):	Orientation:	Focal length:
	Day:	Location:	Appr. dist. from subject (m):
Scans	*Start*	*Stop*	*Interactions*
1			
2			
3			
4			
5			
6			
7			
8			
9			
10			
11			
12			
13			
14			
15			

Sample Data Sheet, Exercise (iv)

Sample data sheet for (iv), testing predation risk vs conspecific detection hypotheses			
Measuring traffic			
These data are collected simultaneously with scanning behavior, so Subject 1 here			
is the same as Subject 1 on the scanning data sheet.			
Copy enough sheets for 30 subjects.			
Subject:			
Interval (Every 30 s)	Number of People Walking within 5 m of Group		
1			
2			
3			
4			
5			
6			
7			
8			
9			
10			
	Average number of people =		
Subject:			
Interval (Every 30 s)	Number of People Walking within 5 m of Group		
1			
2			
3			
4			
5			
6			
7			
8			
9			
10			
	Average number of people =		

CHAPTER 24

The Function of "Chat" Calls in Northern Mockingbirds (Mimus polyglottos): Vocal Defense of Nestlings

Cheryl A. Logan
Departments of Psychology and Biology, University of North Carolina at Greensboro, P. O. Box 26170, Greensboro NC 27402-6170, USA

INTRODUCTION

While they are in the nest and dependent on adults for food, the nestlings of many species of **altricial passerine** birds produce begging calls. In many species, begging calls may enhance adult feeding rates, and they are therefore considered adaptive for both adults and young. However, although they aid in the development of young, begging calls may also increase **predation** on nestlings by enabling predators to locate the nests and prey on the young (e.g., Redondo & Arias De Reyna 1988). Therefore, even though they are clearly adaptive, if they increase predation rates on nestlings, begging calls also involve important evolutionary costs.

Adult northern mockingbirds (*Mimus polyglottos*) commonly produce a short call termed "the chat," which is uttered in a number of different contexts. During the breeding season (March through June), the chat calls of adult males increase after eggs hatch and while dependent young remain in the nest. Females produce the call too, and in both sexes the call is common when eggs or young are vulnerable to predation (Breitwisch 1988). In urban environments, over 50% of mockingbird young are taken by predators, and nestling mockingbirds produce a loud begging call that could be used by predators to locate active nests. Adult mockingbirds could increase their reproductive success if they used vocal signals to silence the calls of young when predators appear in the territory. In doing so, they might decrease the likelihood that predators could locate their nests using begging calls of the young, thereby decreasing the risk of predation without sacrificing the adaptive stimulating effect of calls on adult feeding.

Mockingbirds are a **monogamous, multiple-brooded** species of songbird. Breeding pairs inhabit all-purpose territories approximately 1 to 4 hectares in size, which provide the important resources for both adults and their young. In the southeastern United States mockingbirds do not migrate, and the breeding season may be very long. During this time, breeding pairs will attempt to produce as many as six to eight broods of young in a season, although adults rarely rear more than three successful broods per season. For each brood, a clutch of approximately three eggs is incubated by the female for about 12 days. After hatching, young mockingbirds remain in their nests for about 12 more days, being fed by both parents. The **overlap of successive broods**, in which adults begin work on a new brood while the young of the older brood are still dependent on their parents, is common (Derrickson & Breitwisch 1992; Logan et al. 1990; Zaias & Breitwisch 1989), although fledglings from each brood leave the territory as the young of the next brood hatch.

From about 5 days of age, each mockingbird nestling produces a loud begging call that probably stimulates feeding by the parents. The production of chat calls by adults increases greatly at this time (young mockingbirds themselves will chat after they leave their natal nest). During both the incubation and the nestling periods, young are preyed upon by several predators, including snakes, hawks, crows, squirrels, and domestic and feral cats, and chat production also increases when a potential predator, such as a cat, enters the territory. After leaving the nest, young mockingbirds (now termed **fledglings**) remain in the territory for approximately 3 weeks. However, the risk of predation drops greatly about 1 week after fledging (Zaias & Breitwisch 1989). Before this time, adult mockingbirds will aggressively attack even very large predators as they approach an active nest, and the male is usually the more aggressive nest defender (Breitwisch 1988). "Hew" calls are commonly produced during an attack and when a predator is very near the nest. The chat calls, however, are rarely produced near the nest. Instead, adults chat when the potential predator is still some distance from the nest. This difference is consistent with the hypothesis that the adult chat call functions to prevent the predator from locating the nest by silencing the begging calls of the young. This experiment will examine the occurrence of adult chat calls and of nestling begging calls in the presence vs the absence of a predator stimulus. Comparison of these conditions should determine whether the begging calls of mockingbird nestlings decrease in the presence of adult chats. If this is the case, the comparison provides evidence that the adult mockingbirds' chat calls function to silence the begging calls of their young when adults detect the presence of predators.

MATERIALS

Observers will count by ear the chat calls produced by adults and the begging calls produced by young. The observers will need stopwatches to time a baseline period and a period of stimulus (predator) presentation, clipboards and pencils to record responses, and binoculars for identification and detailed observation. One student will serve as a stimulus predator in the test phase of the study. Mockingbirds readily treat humans as predators, and employing a student as predator avoids the risk of actual predation on active nests.

PROCEDURE

Field Context

Field experimentation requires careful selection and monitoring of the field context in which an experiment is conducted. For example, your experiment on mockingbirds cannot be successful unless the mockingbirds you test are mated and have nestlings of the appropriate age. Your experiment should be conducted in the territories of mated pairs that have active nests with at least two nestlings approximately 7 to12 days of age. As long as the young are in the nest and have been observed to produce begging calls, they are of the appropriate age.

The number of young in the nest may vary from one to four. This experiment uses nests with only two or three young. This will minimize variability that results from combining tests on pairs with only one nestling (where feeding occurs less often, fewer calls are emitted, and defense against predation may be less intense) and those with four nestlings (where large numbers of begging calls occur and nest defense may be much more intense).

The best times to conduct the tests are from 800 to 1000 hours or from 1700 to 1800 hours, when mockingbirds feed their young often. Because begging usually occurs in the context of feeding, it is important to maximize the likelihood that a feeding will occur during the baseline period. This will enable you to measure the difference in begging in the absence of predators vs when a predator is nearby and to relate changes in begging to adult chats.

Mockingbirds are very **sexually monomorphic**, and you will not be able to tell the male from the female in the mated pair. This is not a problem because both adults will chat. But you must note whether one or both adults call during the baseline and the test periods. Be sure you know where the nest containing the young of your test pair is located. Adults usually do not respond at all to predators that are located either outside of their territorial boundaries or at a great distance from an active nest.

Design and Method

You will test adult mated pairs once in the presence and once in the absence of a stimulus predator. Therefore, your design is a repeated-measures design in which behavior is observed before and after the introduction of a crucial stimulus. In this case, the stimulus is the predator. You will be working in groups of two to four individuals. One of you will serve as predator; the others will time the tests and record the behaviors of adults and nestlings.

Baseline

Prior to the beginning of the experiment, position yourselves near, but outside, the territory boundary with a good view of the nest and in an open area that enables you to see most of the territory. Observers should always remain outside the edge of the territory and as far from the nest as possible, because observers too may be considered predators by the birds. If observers stand too close to the nest, they may mimic the effect of the predator stimulus and confound the differences between conditions. Although individual pairs differ in their tolerance, a distance of 13 to 15 meters from the nest should produce no reaction from the adults. Using the stopwatch, begin 6 minutes of baseline observation to assess each dependent measure under the condition in which no predator is present. Pay closest attention to the number of chat calls produced by the adults and to the number of begging calls produced by the young. Details on how to measure and record the dependent variables are described in the "Data Records and Analyses" section.

Predator Test

At the end of the 6-minute baseline, restart the stopwatch and have the "predator" walk directly toward the nest, stopping at a distance of 5 meters from the base of the bush containing the nest. The predator should remain at this location for the remainder of the 6-minute test period, while observers assess the dependent variables. At the end of the test, the predator and all observers should quickly leave the territory. The appearance of a natural predator during your test will disrupt the test and make your results uninterpretable. If this occurs, abort the test and resume again half an hour after the predator leaves or when feeding has resumed.

HYPOTHESES AND PREDICTIONS

Try to develop hypotheses that follow from the foregoing considerations. For example, if the chat calls function in nest defense, they should increase during the predator-present period. However, this does not indicate that the calls function specifically to silence young. If the calls function to silence the nestlings, the number of chats should increase during the predator-present period, and there should be a correlated drop in the number of begging calls. If instead the calls serve some other antipredator function, such as harassing the predator, chat calls could rise with no related drop in the occurrence of begging calls.

Your experimental hypothesis that chats function to silence young in defense against predation (the silencing hypothesis) predicts that, relative to baseline, chats should increase during the predator-present condition and begging calls should decrease during the predator-present condition.

DATA RECORDING AND ANALYSES

Ethogram of Behaviors to be Measured

During both baseline and test, you will record the following behaviors in the adults and the young on a data sheet divided into twenty-four 15-second time periods. The codes to use in recording each response are indicated in bold type.

Adult chats   **C**—explosive broad-band calls of approximately 120 milliseconds in duration. Record the number of chats occurring in each 15-second period throughout the entire observation period. For example, if 4 occur in 15 seconds, record "C × 4"; if no chats occur in the next 15-second period, leave that space blank, etc.

Adult hews   **H**—longer, lower-frequency calls approximately 500 milliseconds in duration; they are often made when the adults attack a predator. If they occur, they may mean that your predator is too close to the nest. Record the number of hews occurring in each 15-second period. For example, if 4 occur in 15 seconds, record "H × 4"; if no hews occur in the next period, leave that space blank, etc.

Begging calls   **Bg**—high-pitched calls made by young in the nest. Their rate increases as adults approach the nest with food.

Parental feeding   **F**—a parent perches at nest and delivers food (usually fruit or insects) to the young. Begging calls will occur before and during feeding, stopping abruptly when the food is actually received.

Devise a data sheet with two 6-minute time lines (one for baseline and one for test), and divide each into twenty-four 15-second periods. Record the number of calls of all kinds and all feedings that occur in each 15-second period. Dividing your data sheet into 15-second time periods will make it easier to keep track by ear of the number of calls produced. Chats commonly occur at a rate of about 1 call every 1 to 2 seconds. This would give you a maximum call rate of about 8 to 15 chats per 15-second period, though in most cases fewer will occur.

Complete two analyses on each of your two main dependent variables, adult chats and nestling begging calls. For each, calculate the total frequency of occurrence for the baseline and the test periods. Subtract the test value from the baseline value for each adult pair or nest. Then pool your data with the whole class so that you can compare the number of calls produced by all pairs during baseline with the number of calls produced during the predator test. Because you are unlikely to have a large number of breeding pairs from which to select your test subjects randomly, each of your analyses should employ nonparametric tests. Use the Wilcoxon matched-pairs signed-rank test (one-tailed) to assess the likelihood that the differences in the number of chats during the baseline period vs the period with the predator present would have occurred by chance. Complete the same analysis for the occurrence of begging calls in the young. But remember that the young are likely to beg only if the parents feed. If chats serve a silencing function, then the occurrence of chats should increase significantly, and the occurrence of begging calls should decrease significantly, during the predator test.

Because hews naturally occur during direct predator attack, they appear to be directed at predators. For this reason, they are unlikely to silence the young. However, analysis of the total number of hews produced by the adults during the test will provide a subsidiary test of your hypothesis. To show that your predator was effective, use the Wilcoxon matched-pairs signed-rank test (one-tailed) to compare the total number of adult hews that occurred with the predator present to the number that occurred during baseline. If your predator was convincing, more hews should occur with a predator present than during the baseline period.

Small sample sizes and large individual differences between pairs may make it difficult to detect reliable group differences with the foregoing analyses. To examine the data in a way that is less sensitive to individual differences in the number of chats or begs produced, also analyze the percentage of pairs in which the following patterns occurred. (1) Adults did not chat during baseline but did produce chats with the predator present. (2) Adults chatted during baseline but did not chat with the predator present.

(3) Adults chatted during both conditions. (4) Adults chatted during neither condition. The critical comparisons are those for patterns (1) and (2). Use the McNemar test (one-tailed) to determine whether the percentages of these 2 outcomes would have occurred by chance. Then, using the same comparison of patterns, calculate the McNemar test (one-tailed) to compare the conditions under which begging calls were produced by young in the nests. With chats, a significantly greater percentage in the first grouping would constitute evidence in favor of the silencing hypothesis. With begging calls, a significantly greater percentage in the second grouping would constitute evidence in favor of the silencing hypothesis.

QUESTIONS FOR DISCUSSION

Interestingly, mockingbirds rarely chat when a crow approaches the nest, but they nearly always chat when a cat approaches. Discuss the impact of predator type on the silencing function. If your data indicate that chats function to silence young in tests using a human predator, what would be the effect of a cat or a crow on the function of the call? How would the call's function be influenced by the sensory modalities used by predators to locate nests? How do predator type and the strategy a predator uses to detect nests affect your interpretation?

Your hypothesis assumes that adults chat in response to predators, but not in response to the young's begging calls. Which comparisons in your study could you use to clarify the cause-and-effect relationship between chats, begging, and predators? How might you use playback of recorded begging calls to confirm that adults chat not to the begging call itself, but only to the presence of a predator that appears at a time when the young might beg?

What do you predict would be the effect of a predator's distance from the nest on the parents' antipredator strategy? Silencing is a low-risk strategy for the adults. How would decreased distance affect the utility of the call and the costs and benefits of more costly defense? And how might a distance variable be affected by different sensory modalities used by different types of predators?

What do you predict will be the effect of increased nestling/fledgling age on the use of the chat call? Why?

Remembering that mockingbirds are a multiple-brood species, what would be the effect on chats of increases or decreases in the number of young in the nest? of the time during the season (beginning or end) in which the brood is produced? Relate these factors to the relative value of one breeding attempt over others in the same season.

Mockingbird begging calls seem to be louder that those of other species. If this is the case, why would selection increase the amplitude of calls, given that increased amplitude is more likely to attract predators?

Mockingbirds use both chats and hew calls in the context of predation, and hews seem to occur when a predator is closer to the nest. For a given predator, at what distance should adults switch from using from chats to using hews, and how might this be affected by the type of predator?

LITERATURE CITED

Breitwisch, R. 1988. Sex differences in defence of eggs and nestlings by northern mockingbirds (*Mimus polyglottos*). *Animal Behaviour*, **36**, 62–72.

Derrickson, K. C. & Breitwisch, R. 1992. Northern mockingbird (*Mimus polyglottos*), In: *Birds of North America* (ed. A. Poole, P. Stettenheim & F. Gill), pp. 1–25. Washington, DC/Philadelphia: The American Ornithologists' Union/The Academy of Natural Sciences of Philadelphia.

Logan, C. A., Hyatt, L. E. & Gregorcyk, L. 1990. Song playback initiates nest building during clutch overlap in mockingbirds (*Mimus polyglottos*). *Animal Behaviour*, **39**, 943–953.

Redondo, T. & Arias De Reyna, L. 1988. Locatability of begging calls in nestling altricial birds. *Animal Behaviour*, **36**, 653–661.

Zaias, J. & Breitwisch, R. 1989. Intra-pair cooperation, fledgling care, and renesting by northern mockingbirds (*Mimus polyglottos*). *Ethology*, **80**, 94–110.

ADDITIONAL SUGGESTED READING

Bengtsson, H. & Ryden, O. 1983. Parental feeding rate in relation to begging behaviour in asynchronously hatched broods of the great tit *Parus major*. *Behavioral Ecology and Sociobiology*, **12**, 243–251.

Breitwisch, R., Merritt, P. & Whitesides, G. 1986. Parental investment by the northern mockingbird: male and female roles in feeding nestlings. *Auk*, **103**, 152–159.

Carlin, C. A. 1992. The use of chat calls by northern mockingbirds (*Mimus polyglottos*) for anti-predator defense. Unpublished master's thesis, University of North Carolina at Greensboro.

Mondloch, C. J. 1995. Chick hunger and begging affect parental allocation of feedings in pigeons. *Animal Behaviour*, **49**, 601–613.

CHAPTER 25

Diving and Skating in Whirligig Beetles: Alternative Antipredator Responses

ALASTAIR INMAN AND ANNE HOUTMAN[1]
Department of Biology, Knox College, Galesburg, IL 61401, USA
[1]*Department of Biology, Soka University, Aliso Viejo, CA 92656, USA*

INTRODUCTION

During the summer and fall, whirligig beetles, *Dineutus americanus* (*Coleoptera: Gyrinidae*), aggregate in groups known as rafts on the surface of ponds, lakes, and quiet streams. When disturbed by a threatening stimulus, whirligigs attempt to escape capture in one of two ways. Sometimes they skate and spin, apparently randomly, on the surface of the water. At other times they dive underwater and seek cover on the bottom, where they may remain for a minute or longer. Thus the beetles employ alternative tactics: different behaviors that have a common function—in this case, escaping from a predator.

Alternative tactics have been documented in many species, in a wide range of behaviors including mating (Arak 1988; Gross 1985; Houtman & Falls 1994), foraging (Heinrich 1979), aggressive behavior (Adams & Caldwell 1990; Hamilton 1979), and parental care (Gross 1982; Knapton & Falls 1983). The existence of alternative tactics within a single population caused evolutionary biologists to wonder why and how different behaviors that have a single function would evolve. Alternative tactics may be maintained in a population through a number of different mechanisms.

The tactics may be alternative strategies that are inherited. A **strategy** is a genetically distinct set of behavioral rules (Alcock 1998). Thus an individual uses the same strategy throughout its life. For example, a scale-eating cichlid fish has two forms, one that eats scales off the right side of its victims and one that eats scales off the left side. Individual fish inherit—and keep for life—the strategy and mouthparts twisted in the appropriate direction; they cannot switch between the alternative strategies by feeding sometimes on the left and sometimes on the right. Ordinarily, alternative strategies should not persist in a population, because if one strategy bestows even slightly higher fitness, it should increase in frequency relative to the other until only the better strategy remains. For alternative strategies to persist, their fitness must be equal on average. This will occur if the alternative strategies are frequency-dependent—that is, if the payoff to any one strategy depends on the frequency with which that strategy is adopted in the population. In the cichlids, the payoffs to the two strategies *are* frequency-dependent: as the proportion of left-feeding fish in the group increases, the payoff to left-feeding decreases. Because each strategy has a higher payoff when it is rare in the population, the two strategies are maintained by frequency-dependent selection. The group is predicted to evolve to a mixture of left-feeders and right-feeders in which the payoffs to each strategy are equal.

Alternative tactics are often part of a single conditional strategy, in which individuals inherit an ability to adopt either tactic depending on the environmental conditions they experience. For example, horned beetle males develop either into "majors," which are large and possess a large horn (used in competing with other males), or into "minors," which are smaller and have proportionately smaller horns (Eberhard 1980). The minors can't compete with the majors but achieve some success in finding mates by emerging earlier in the spring and dispersing further than the majors. Whether a male becomes a major or a minor depends on the nutrients he receives while a larvae. Thus, in this conditional strategy, adopting the tactic of being a minor is the way a small male can make the best of a bad situation. If tactics are alternatives of a single conditional strategy, then individuals may be able to switch between them. For example, male natterjack toads may either call to attract females for mating ("caller" tactic) or wait silently near a calling male and attempt to copulate with females attracted to the neighbor's calls ("satellite" tactic) (Arak 1988). Smaller males have weaker calls that are less attractive to females, and such males tend to use a satellite tactic. However, satellite males will switch to calling if there is a low density of calling males or if neighboring males have weaker calls than their own. Tactics of a conditional strategy differ from true alternative strategies in that differential reproductive success is associated with each tactic (remember that alternative strategies have equal fitness), and individuals will switch to the better tactic when environmental and/or social conditions allow it.

In this lab, you will study the alternative antipredation tactics of whirligigs. Whirligigs are easy to capture and mark, and their antipredation behavior can be easily observed by placing the beetles in a small aquarium and startling them.

MATERIALS

For this lab, you will need whirligig beetles that have been collected from a pond or stream using aquatic nets and stored in a large stock tank. You will also need aquarium nets; fingernail polish for marking beetles for individual identification; and, for each group, 2 stopwatches, 6 blank cards, and a 10-gallon aquarium half filled with pond water. The sides of the tanks should be covered with paper so that whirligig beetles placed in the tank will not be disturbed by activity outside the tank.

HYPOTHESES AND PREDICTIONS

First, the whole class will run an experiment in which you will ask whether the propensity to dive vs skate is density-dependent. Each group will observe individual beetles in 10-gallon aquaria in three conditions: alone, in a group of 5 beetles, and in a group of 20 beetles.

PROCEDURE

Half fill your aquarium with pond water, and tape paper to the sides to make sure that movement around the tank will not disturb the whirligig beetles.

From the stock tank, choose one beetle that will be the focal beetle for your first set of experiments. Gently hold the beetle on a moistened paper towel, and paint a small stripe of fingernail polish on its back. Keep the beetle on the moistened paper towel until the fingernail polish has dried, and then place it in the test aquarium. The polish will enable you to identify your focal beetle, even when it is in a group of other beetles.

You will test the same beetle twice in each of three conditions: alone, with a group of 4 other beetles, and with a group of 19 other beetles. The six trials should be run in a randomly determined order. To randomize the order of your trials, write "1" on two of your cards, "5" on two of your cards and "20" on the remaining two. Turn the cards face down and shuffle them.

Table 25.1 **Sample data sheet.**

Trial	Group Size	Time Spent Underwater	Proportion of Time Spent Underwater	Transformed Proportion of Time Spent Underwater[a]
1	20	50	0.42	0.70
2	1	36	0.30	0.58
3	1	120	1.00	1.57
4	5	30	0.25	0.52
5	20	105	0.87	1.21
6	5	0	0.00	0.00

[a]The proportion of time spent underwater is transformed using an arcsine squareroot transformation to normalize the data.

Draw the cards out one at a time, and record the resulting order on your data sheet. Your data sheet should have the column headings shown in Table 25.1, which provides an example of data collected for one focal beetle.

Set up for your first test by transferring the correct number of beetles (0, 4, or 19) from the stock tank to your test tank, using a small aquarium net. After placing all the beetles for your first trial in your aquarium, leave the beetles undisturbed for 10 minutes to acclimate. During the acclimation period, choose one student who will be your observer. Make sure that she or he is familiar with the operation of the stopwatch. Choose another student who will be the timer. The timer's job will be to time the 2-minute trial.

Make sure that both stopwatches have been reset to zero. The trial should be started by the timer starting her stopwatch as she tells the observer to begin. When told to start, the observer should begin the trial by quickly peering over the top of the aquarium, keeping his or her face directly over the center of the tank, approximately 30 centimeters above the surface of the water. Upon initiating the trial, the observer should find the focal beetle and begin to monitor its behavior. The observer should start her or his stopwatch whenever the focal beetle dives below the surface of the water and should stop the stopwatch whenever the focal beetle returns to the surface. The timer should continue to monitor the total trial time and indicate to the observer when 2 minutes have elapsed. At that time, the observer should stop her or his stopwatch and record the total time spent diving on the data sheet. At the end of the trial, return all the "companion" beetles (that is, all except the focal beetle) to the stock tank.

Using the same focal beetle, set up for the next trial by adding the correct number of new "companion" beetles to the test tank from the stock tank, allowing another 10-minute acclimation period, and running another trial. Continue in this way until you have run all six trials on one focal beetle. Use this same procedure to test as many focal beetles as your instructor has assigned, testing each focal beetle six times. If you are testing additional focal beetles, you will need to add to your data sheet a column for beetle I.D. and to include six rows (one per trial) for each beetle.

DATA RECORDING AND ANALYSES

For each of your trials, calculate the proportion of time that the beetle spent diving (by dividing the time spent diving by 120 seconds, the total trial time). By having the class combine the data from all groups, you will be in a position to analyze your data statistically. Statistical tests enable us to determine whether observed differences in the proportions of time spent diving in groups of the three different sizes (1, 5, and 20) are larger than would be expected by chance alone. Our statistical null hypothesis (H_0) is that individuals will show no change in their propensity to dive vs skate under different group sizes, whereas our alternative hypothesis (H_a) is that group size does affect the propensity to dive vs skate. This could involve beetles becoming more inclined to dive as group size increases or beetles becoming more inclined to skate as group size increases. "Parametric" statistics are powerful ways of testing many different hypotheses, but they assume that the data being tested are normally distributed (that is, that they approximate a "bell-shaped" curve). Proportional data (such as those that you collected) tend not to be normally distributed, but this can be corrected by applying an arcsine square root transformation (Sokal & Rohlf 1994). Each of the observed proportions of time spent diving should be transformed by first taking its square root and then determining the arcsine of the result. For example, if a beetle spent half its time diving, the arcsine square root transformation would be calculated as arcsine($\sqrt{0.5}$), which is arcsine(0.7071), or 0.785 radians (45 degrees). Determine the arcsine square root transformation of each of your observed proportions, and then enter your data into the class data sheet.

We wish to ask whether the mean proportion of time spent diving differs across the three group sizes. A common statistical test used to ask whether there is a difference between the means of two or more groups is the analysis of variance (ANOVA). In this case, because each beetle has been tested more than once, you will perform a two-way, "mixed-model" ANOVA with group size and individual as the factors. This analysis will yield two P-values: one that tells us whether the two group sizes differ in the proportion of time spent diving, and one that tells us whether the individual beetles differ from each other in their propensity to dive or skate. In both cases, a P-value less than or equal to 0.05 indicates that the observed difference in proportion of time spent diving is unlikely to have resulted from chance alone, so we can reject the null hypothesis. A P-value greater than 0.05 indicates that there is a better than 1 in 20 chance that the observed difference in the proportion of time spent diving resulted from chance alone, so we cannot reject the null hypothesis. Remember that the two parts to the ANOVA and the two P-values correspond to different questions. The "group size" P-value enables us to test our initial hypothesis, that the propensity to skate or dive is affected by group size. The second P-value tells us whether individual beetles differ from each other in their propensity to skate or dive. Are some beetles more dive-prone and others more skate-prone?

SMALL-GROUP EXPERIMENTS

The whole-class experiment asked whether the propensity to dive vs skate was affected by the size of the group that the focal beetle swam with. What else might affect a beetle's "decision" about which antipredation tactic to adopt? Note that whirligigs have two sets of eyes: one looking up and the other looking down. Might these be adaptations to predators? Come up with your own hypothesis about what affects the antipredation response of whirligigs, and design an experiment to test your hypothesis. Decide what the biological and statistical null and alternative hypotheses are for your experiment, as well as how you will collect and analyze your data. You may want to design your own data sheet to help organize the data you are recording. Be sure to discuss your ideas and plans for your experiment with your instructor before beginning.

QUESTIONS FOR DISCUSSION

Why do you think whirligigs evolved more than one way to respond to the threat of a predator? What selective forces might maintain these alternative tactics? Do you think there is a difference in the costs and benefits of the diving tactic vs the skating tactic? How might differences in the costs and benefits of the two tactics explain some of the results of your experiments? How would you explain individual differences between beetles in their propensity to dive vs skate? What kind of experiment could you run to test your hypothesis?

ACKNOWLEDGMENTS

Thanks to the many students who tested this exercise, and to Bonnie Ploger, Ken Yasukawa, and an anonymous reviewer for helpful comments on an earlier version of this chapter.

LITERATURE CITED

Adams E. S., & Caldwell R. L. 1990. Deceptive communication in asymmetric fights of the stomatopod crustacean, *Gonodactylus bredini*. *Animal Behavior*, **39**, 706–716.

Arak A. 1988. Callers and satellites in the natterjack toad: evolutionarily stable decision rules. *Animal Behavior*, **36**, 416–432.

Eberhard, W. G. 1980. Horned beetles. *Scientific American*, **242**, 166–182.

Gross, M. R. 1982. Sneakers, satellites and parentals: polymorphic mating strategies in North American sunfishes. *Zeitschrift für Tierpsychology*, **60**, 1–26.

Gross, M. R. 1985. Disruptive selection for alternative life histories in salmon. *Nature*, **313**, 47–48.

Hamilton, W. D. 1979. Wingless and fighting males in fig wasps and other insects. In: *Sexual Selection and Reproductive Competition in Insects* (ed. M. S. Blum & N. A. Blum), pp. 167–220. London: Academic Press.

Heinrich, B. 1979. Foraging strategies of caterpillars: leaf damage and possible predator avoidance strategies. Oecologia (Berlin) **42**, 325–337.

Houtman, A. M. & Falls, J. B. 1994. Negative assortative mating in the white-throated sparrow, *Zonotrichia albicollis*: the role of mate choice and intra-sexual competition. *Animal Behavior*, **48**, 377–383.

Knapton, R. W. & Falls, J. B. 1983. Differences in parental contribution among pair types in the polymorphic white-throated sparrow. *Canadian Journal of Zoology*, **61**, 1288–1292.

Sokal, R. & Rohlf, F. 1994. *Biometry*. 3rd ed. New York: Freeman.

ADDITIONAL SUGGESTED READING

Alcock J. 1998. *Animal Behavior*. 6th ed. Sunderland, MA: Sinauer; pp. 289–338, 556–599.

Maynard Smith, J. 1982. *Evolution and the Theory of Games*. Cambridge: Cambridge University Press.

CHAPTER 26

The Response of Tree Squirrels to Conspecific and Heterospecific Alarm Calls

ANNE HOUTMAN
Department of Biology, Soka University, Aliso Viejo, CA 92656 USA

INTRODUCTION

Animals communicate with other members of their species, or **conspecifics**, for many reasons—for example, to attract mates and to defend territories (Krebs & Davies 1993). **Alarm calls** are an important form of communication that enable animals to respond quickly to the presence of predators. If most animals within a group or area respond to another's alarm call by immediately moving to cover, any individual that does not respond will be the most likely to be taken by a predator. There should therefore be strong selective pressure for animals to respond quickly and appropriately to conspecific alarm calls. Different species within the same environment may be threatened by the same predators and thus may also benefit from listening for and responding to the alarm calls of other species (**heterospecifics**) (see Marler et al. 1992 for some examples).

In this lab, you will test whether tree squirrels respond more strongly (or even exclusively) to the alarm calls of their own species, or whether they also attend to the alarm calls of other species within their environment. The class will first work together to (1) develop hypotheses about the response of squirrels to conspecific vs heterospecific alarm calls, (2) design an experiment to test the hypotheses, (3) write up a protocol, (4) make data sheets, and (5) run the experiment. Data analysis is done with the class's combined data, and the class decides as a group how best to interpret and present these data. Small groups of students then develop their own hypotheses and design experiments to test them, using the same or a similar system.

MATERIALS

For this lab, the class must have access to an abundant local population of tree squirrels, or another sciurid. Each group will need loop tapes of alarm calls of at least two species, a loop tape of control noise, a portable tape player with speaker, a stopwatch, a clipboard to hold data sheets and protocol, data sheets, the protocol, and a meter stick or other measuring device.

HYPOTHESES AND PREDICTIONS

Whole-Class Experiment

The class should identify **hypotheses** about the response of tree squirrels to alarm calls of their own and other species. One hypothesis might be that squirrels respond more strongly to alarm calls of their own species than to alarm calls of other species, because natural selection has presumably favored squirrels' communicating clearly with others of their species. Alternatively, squirrels might respond just as strongly to the alarm calls of other species within their environment, such as bluejays and chipmunks, because the squirrels are likely to be threatened by the same predators, such as hawks and house cats.

Small-Group Experiment

After the whole-class experiment has been completed, brainstorm with two or three of your colleagues to produce several hypotheses that you could test using the methods you learned during the whole-class experiment. For example, do squirrels respond more strongly to the alarm call of another mammal (such as a chipmunk) than to an avian alarm call (such as that of a bluejay)? Does seeing a potential predator as well as hearing it change a squirrel's response?

Narrow down your choices on the basis of your interest, the accessibility of necessary subjects and materials, and the likelihood of successfully collecting data to test your hypothesis. Confirm the appropriateness of your hypothesis with your instructor.

PROCEDURE

Whole-Class Experiment

Operationalize the hypothesis that the class developed. In other words, identify a testable **prediction** of the hypothesis. The prediction should be stated as a relationship between a **dependent variable** and **an independent variable**. The independent variable is the factor that you will manipulate. The dependent variable is the factor that you predict will be affected by the independent variable and is the factor you will monitor. An example of a testable prediction is that squirrels will give the strongest behavioral response (dependent variable) to an alarm call of their own species (independent variable), a weaker response to a bluejay alarm call, and the least response to white noise (a **control condition**).

Design an experiment to test the prediction, and develop a **protocol** to run it, including a list of materials needed. A protocol spells out exactly what should be done for each step of the experiment. If you were to hand a copy of your protocol to friends who are not in the class, they should be able to run the experiment with no other input from you. Consider the following: (1) How many squirrels should you test? (2) How often should you test the same squirrel? (3) How far away from the squirrel should you be when you begin the trial? (4) How long should the trial be? (5) How will you ensure that you don't bias your observations toward your predicted results?

Be sure to include a control condition, which will help you determine whether squirrels are simply responding to your presence or any loud noise. One possibility would be to have control trials in which a tape of white noise (broad-band noise, like that between radio stations) is played. Decide exactly how you will measure the strength of the squirrel's response—there are several ways to do this. How will you define and assess the squirrel's behavior? Produce a data sheet for use during the experiment. The data sheet, which should be clear and self-explanatory, should include space for the date, the time, the names of the observers, and all relevant data.

Run the experiment! It would probably be wise to have different people responsible for different aspects of running the experiment. For example, one person could observe the squirrels and fill out the data sheet, another could be responsible for using the stopwatch, and another could run the sound equipment. But don't have too many people around, or the squirrels will be too nervous to act naturally. Groups of three or four people are ideal; each of the groups should collect data at different locations so that the squirrels aren't tested more than once. If there is not a set of playback equipment for each group, you should collect data at different times so that you can share equipment.

Small-Group Experiment

Using the hypothesis that your group developed, go through the steps that the class went through to develop the whole-class experiment. This includes operationalizing your hypothesis, designing an experiment to test your prediction, and producing a protocol and data sheet. Check your plans with your instructor, and then run the experiment.

DATA ANALYSIS AND PRESENTATION

Whole-Class Experiment

Review your data sheets to ensure that they are correct and that you will be able to read them when the class meets to analyze and interpret the experimental data. You may be asked to enter your data into a spreadsheet. If so, double-check the data you input to be sure they are accurate. Combine the entire class's data into one spreadsheet and again confirm that the data are correctly input.

Be clear on what statistical null hypotheses (H_0) and alternative hypotheses (H_a) you are testing. For example, your H_a could be that the strength of squirrel response differs to conspecific vs heterospecific alarm calls vs white noise (that is, response to conspecific ≠ response to heterospecific ≠ response to white noise). Your H_0 could be that squirrels do not differ in their response to these noises (response to conspecific = response to heterospecific = response to white noise).

Choose an appropriate statistic to test your H_a and H_0, and confirm with your instructor that this is the correct test to run. Complete the statistical analysis, double-check your results, and review your results with the instructor. For example, you might choose to run a Kruskal–Wallis test to compare the responses of squirrels to conspecific calls vs heterospecific calls vs white noise.

You now need to decide how to present your data—whether in a graph or a table, and in what type of graph or table. Think about what format would make it easiest for someone unfamiliar with your experiment to understand your results. For example, you might want to create a bar graph showing the average response of squirrels to each playback condition, with error bars showing how dispersed your data are. Confirm with your instructor that this is the best format in which to present your results. Often, this step will give you further ideas for data analysis, for issues to address in the "Discussion" section of your paper, and even for future experiments.

Small-Group Experiment

Review your data sheets and/or spreadsheet for accuracy and clarity before meeting with your professor to analyze your data. State your statistical null hypothesis (H_0) and alternative hypotheses (H_a; you should ideally have done this before running the experiment), and choose the appropriate statistical test of the hypothesis. Analyze your data after confirming your choice of statistical test with the instructor. Double-check your results, and review your results with the instructor. Finally, decide what format you will use to present your data, and confirm your decision with the instructor.

QUESTIONS FOR DISCUSSION

State your conclusions regarding squirrels' responses to (1) control conditions vs a conspecific alarm call, (2) a conspecific alarm call vs a heterospecific alarm call, and (3) a heterospecific alarm call vs control conditions.

In what social or environmental circumstances might you expect animals to respond to calls of other species? Interpret your results in light of this.

How might environmental factors influence the strength of an animal's response to an alarm call? How would you test this?

Under what conditions might you expect animals to ignore alarm calls? Can you devise an experiment that replicates these conditions?

Krebs & Davies (1993) discuss these issues.

ACKNOWLEDGMENTS

Thanks to the many students who tested this exercise, and to Alastair Inman, Bonnie Ploger, Ken Yasukawa, and an anonymous reviewer for helpful comments on the manuscript.

LITERATURE CITED

Krebs, J. R. & Davies, N. B. 1993. *An Introduction to Behavioural Ecology.* 3rd ed. Oxford: Blackwell Scientific Publications. (See especially Chapter 13 on the design of signals.)

Marler, P., Evans, C. S. & Hauser, M. D. 1992. Animal signals: motivational, referential, or both. In: *Non-verbal Vocal Communication: Comparative and Developmental Approaches* (ed. H. Papousek, U. Jurgens & M. Papousek). Cambridge: Cambridge University Press, pp. 66–86.

ADDITIONAL SUGGESTED READING

Burke da Silva, K., Kramer, D. L. & Weary, D. M. 1994. Context-specific alarm calls of the eastern chipmunk, *Tamias striatus. Canadian Journal of Zoology,* **72**, 1087–1092.

Ficken, M. S. 1990. Acoustic characteristics of alarm calls associated with predator risk in chickadees. *Animal Behavior,* **39**, 400–401.

Marler, P. 1955. Characteristics of some animal calls. *Nature,* **176**, 6–8. (This classic study of alarm calls compares several European songbirds.)

Weary, D. M. & Kramer, D. L. 1995. Response of eastern chipmunks to conspecific alarm calls. *Animal Behavior,* **49**, 81–93.

SECTION III

Agonistic Behavior

CHAPTER 27

Competition for Breeding Resources by Burying Beetles

MICHELLE PELLISSIER SCOTT
Department of Zoology, University of New Hampshire, Durham, NH 03824, USA

INTRODUCTION

Carcasses such as that of a bird or mouse are an unpredictable but extremely valuable resource; rich in protein, they are hotly contested. Conflicts for this or any limited resource are usually settled through the assessment of various asymmetries between competitors. One such asymmetry is the **resource-holding potential**, or differences in fighting ability (Maynard Smith & Parker 1976).

Burying beetles (*Nicrophorus* spp.) are some of the many animals that dispose of small carrion. This resource is required as a food source for their young. To control access to a carcass, they must outcompete many competitors, including other burying beetles, carrion-breeding flies, and scavengers. As a consequence burying beetles have evolved several behavioral and physiological adaptations to safeguard the carrion and exploit it quickly.

If more than one beetle of either sex discovers the carcass, males will fight with other males, and females will fight with females, for possession. The winning male and female become mates and cooperate to bury and prepare the carcass. Both parents help to remove the fur or feathers and to form this carcass into a ball in the underground brood chamber. The female lays eggs in the soil nearby about 30 hours after discovery of the carcass. Both parents often remain and are present to help to feed the young larvae after eggs hatch on the fifth or sixth day. Usually the brood consumes the whole carcass, bones and all. Therefore, the size of the carcass determines how many young a pair will be able to rear. This makes larger carcasses more valuable than smaller ones.

Even when the carcass is underground, it can be discovered by a burying beetle intruder. Both parents defend the young and the nest to help prevent the intruder from taking over the carcass, driving off the same-sex resident, and killing the residents' young. Thus the most important advantage of **biparental care** is to increase the probability that the brood will survive, and **intraspecific competition** has driven its evolution.

First you will determine whether the relative size of competitors predicts the outcome of competition for a carcass. Then you will investigate whether beetles are able to evaluate carcass size and whether, when given a choice, larger males and females win and bury the larger carcass. You will discover whether adaptations of burying beetles are finely tuned enough that they select a larger carcass, even though they would probably not be faced with such a choice in nature.

MATERIALS

For each replicate, you will need a large container (such as a glass aquarium measuring 75×30 centimeters) with a wire mesh lid. The container should be filled with about 10 centimeters of topsoil. You will need calipers to measure the beetles, adult burying beetles, and a selection of small carcasses of rodents or birds.

PROCEDURE

First divide the beetles by sex (see Figure 27.1) into three size categories. **Pronotal** width in millimeters (see Figure 27.2) is a good measure of "size" because it correlates well with dried weight of the beetle and with other measures such as **elytra** length. Carcasses should also be divided into weight categories (grams). Tie a piece of dental floss about 30 centimeters long to the hind leg of the carcass so that it can be more easily located after it is buried.

Experiment 1

To determine whether the larger of two burying beetles wins in a competition for a carcass, put two males and two females in the large container with a carcass. They are good fliers, so be sure to put a lid on

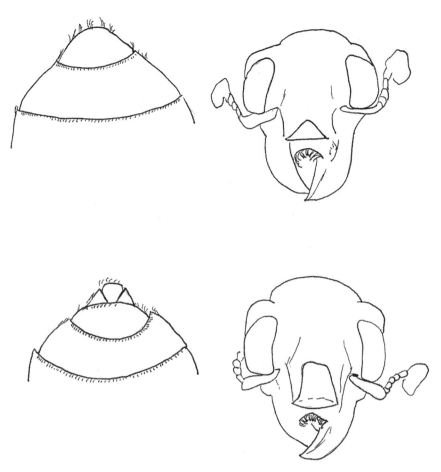

Figure 27.1 **Male and female can be differentiated by two features, which can easily be seen with a hand lens. Females are shown above and males below. On the left is a ventral view of the last abdominal segments. Females appear rounded, and males have an extra sternite that is tulip-shaped. On the right is the front of the head. The clypeus, which is just above the mandibles, between and below the eyes, is orange in some species (such as *N. orbicollis*) and black in others (such as *N. tomentosus*). In females it is short and bell-shaped. In males it is tall and trapezoidal.**

the container. The pronotal width of the two males should differ by about 0.5–0.75 millimeters, and that of the two females should also differ by 0.5–0.75 millimeters. If there is any doubt as to which is the smaller or larger beetle, use a pair of fine scissors to mark them by taking a small clip out of the pronotum or the posterior edge of the elytra. Then, 48–72 hours later, carefully follow the dental floss to the buried carcass and quickly grab all beetles that are in the brood chamber with the carcass. Note the

Figure 27.2 **Pronotal width is indicated by the arrows.**

sex and size of each. Find the other beetles; remove them from the dirt, and repeat the experiment with four different beetles. Examine each beetle for any injuries both before and after each experiment.

Experiment 2

To determine whether beetles discriminate between large and small carcasses, place three pairs of beetles and two carcasses in the container. The pronotal width of each set of three males and three females should differ by 0.5–0.75 mm so that you have a small, a medium-size, and a large pair. The carcasses should differ by 15–20 grams. Follow the same procedure as described for experiment 1 to determine the winner of each carcass 48–72 hours later.

HYPOTHESES AND PREDICTIONS

Contests for limited resources, such as a carcass, are usually resolved because of some asymmetry; relative fighting ability, often predicted by size, has been suggested to be important in burying beetles. However, there are other asymmetries that might predict winning. Suggest one or two additional hypotheses. These are *biological* hypotheses that you might test. If your hypothesis is that sizes predict fighting ability, what do you predict the outcome of the competition will be? Will you need to keep data for males and females separate or can you combine them?

Given that larger carcasses are more valuable than smaller ones, which beetles do you predict will have won and buried which carcasses in experiment 2 if larger beetles do win in competition with smaller beetles? If this is the case, what does it suggest about the beetles' ability to evaluate the resources?

DATA RECORDING AND ANALYSES

For experiment 1, data for the whole class can be recorded with simple tick marks on a table of the observed number of times the large and the small beetle of each sex were with the carcass and, accordingly, won. Analysis of the data requires us to examine two hypotheses: the null hypothesis and the alternative hypothesis (alternative to the null hypothesis). (*Note*: These are different from the biological hypotheses for contest resolution.) To analyze the data, compare the number of large to the number of small beetles that "won" (i.e., were in the burial chamber). Do separate analyses for males and females. What is the null hypothesis that you need for statistical purposes? You want to compare the observed values (number of large and small beetles that won) with the expected values generated from the null hypothesis (random outcome of competition). Use a chi-square goodness-of-fit test. Check your expected values with your instructor before finishing your analysis.

For experiment 2, make a data table for the whole class that has the following headings and columns:

	Large Males	Medium Males	Small Males
Large Carcass			
Small Carcass			

Data can be recorded with tick marks in the appropriate cells to count the number in each category. Should you do separate analyses for males and females? Why? Decide how to graph the data. What are the alternative and null hypotheses? Observed outcomes can be compared with expected number of times the large, the medium-size, and the small beetle won if the outcomes were random. Decide what statistical analysis to use, and discuss your plans with your instructor.

QUESTIONS FOR DISCUSSION

What other sorts of asymmetries could be important in animal contests? Do your results rule out these possibilities? Why do you think size is so often a good predictor of fighting ability? You have used pronotal width as a measure of size. Do you think that this correlates well with other possible measures of "size"? Would you expect that in nature, males and/or females that are able to get a carcass and breed would be larger than the average of the population? Do you think that large males would end up **assortatively mated** with large females? If so, why?

Discuss the results you got for experiment 2. If the observed results were statistically different from random, how do you think the beetles measured the size of the carcass? You can get some ideas on this question if you are able to watch the beetles during the first hour after they discover the carcass. (Most species become active and discover the carcass just when it gets dark, and you can watch them under a red light bulb.)

LITERATURE CITED

Maynard Smith, J. & Parker, G. A. 1976. The logic of asymmetrical contests. *Animal Behaviour*, **24**, 159–175.

ADDITIONAL SUGGESTED READING

Kozol, A. J., Scott, M. P. & Traniello, J. F. A. 1988. The American Burying Beetle: studies on the natural history of an endangered species. *Psyche*, **95**, 167–176.

Otronen, M. 1988. The effect of body size on the outcome of fights in burying beetles (*Nicrophorus*). *Annales Zoologici Fennici*, **25**, 191–201.

Otronen, M. 1990. The effects of prior experience on the outcome of fights in the burying beetle, *Nicrophorus humator*. *Animal Behaviour*, **40**, 980–982.

Scott, M. P. 1990. Brood guarding and the evolution of male parental care in burying beetles. *Behavioral Ecology and Sociobiology*, **26**, 31–19.

Scott, M. P. & Traniello, J. F. A. 1990. Behavioural and ecological correlates of male and female parental care and reproductive success in the burying beetle, *Nicrophorus orbicollis*. *Animal Behaviour*, **39**, 274–283.

Trumbo, S. T. 1993. Brood discrimination, nest mate discrimination, and determinants of social behavior in facultatively quasisocial beetles (*Nicrophorus* spp). *Behavioral Ecology*, **4**, 332–339.

Trumbo, S. T. & Fernandez, A. G. 1995. Regulation of brood size by male parents and cues employed to assess resource size by burying beetles. *Ethology, Ecology and Evolution*, **7**, 313–322.

CHAPTER 28

Learning to be Winners and Losers: Agonistic Behavior in Crayfish

ELIZABETH M. JAKOB[1] AND CHAD D. HOEFLER[2]
[1]*Department of Psychology, University of Massachusetts, Amherst, MA 01003, USA*
[2]*Department of Entomology and Program in Organismic and Evolutionary Biology, University of Massachusetts, Amherst, MA 01003, USA*

INTRODUCTION

Many animals engage in aggressive, or **agonistic**, interactions with conspecifics. Animals may fight over prey, territories, mates, or any other limited resource. The outcome of agonistic contests may depend on many factors. For example, animals may differ in their relative fighting ability, or each competitor may value the contested resource differently. Here we focus on the role of experience in determining contest outcome. Animals that have recently won aggressive interactions may often be more likely to win against new opponents that have not had recent winning experience; that is, winners keep winning. This phenomenon has been found in a variety of taxa, including juncos, chickens, paradise fish, red deer, and spiders (reviewed in Jackson 1991). The proximate reasons for this phenomenon are still under debate.

Crustaceans have been widely used in agonistic behavioral investigations and are excellent candidates for a test of this hypothesis. Many species are relatively large and easy to observe and measure, have well-described aggressive behaviors, are hardy, and readily interact with conspecifics. Studies of agonistic behavior in crustaceans have examined species-specific ritualized displays (e.g., Huber & Kravitz 1995), inter- and intrasexual agonistic interactions (e.g., Cushing & Reese 1998), and how fighting behavior is influenced by predation risk and time of day (e.g., Spanier et al. 1998). Similarly, crabs have been used in studies considering the energetic costs of fighting (e.g., Smith & Taylor 1993), the influence of relative and absolute body size on strategic decision making during agonistic encounters (e.g., Smith et al. 1994), and the behaviors associated with the initiation and resolution of aggressive disputes (e.g., Glass & Huntingford 1988).

Most investigations of agonistic behavior in crustaceans have focused on crayfish, which are especially easy to maintain in the laboratory. They thrive in fresh water (so controlling salinity is not a concern), can be kept at room temperature, have a wide diet, and can be collected or purchased in large numbers. The outcome and nature of crayfish fights is influenced by a multitude of factors. These include differences in reproductive status, for maternal females have been found reliably to defeat nonmaternal females and males (see Figler et al. 1995); the relative intensity of light conditions, for aggressive interactions have been demonstrated to occur more frequently in moderate levels of light vs little or no light (see Bruski & Dunham 1987); relative body size (e.g., Pavey & Fielder 1996), for larger crayfish are more likely to win; and previous contest experience (e.g., Copp 1986; Issa et al. 1999; Goessmann et al. 2000). Contest experience may affect subsequent behavior by modifying development and activity of the nervous system (e.g., Barinaga 1996; Huber et al. 1997). Because crayfish fights are decided, in part, by the relative size of the contestants, it is possible to provide crayfish with a series of training fights that generate crayfish that have had either winning or losing experiences.

The experimental design is outlined in Figure 28.1. Medium-size crayfish are matched by size. One member of each pair is trained to be

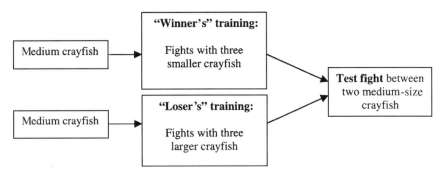

Figure 28.1 **Experimental protocol for testing the effect of training experience on fight outcome. Two crayfish matched for size are given different fighting experiences and then are fought together.**

a winner via a series of fights with small opponents, and one is trained to be a loser via a series of fights with large opponents. The two medium-size crayfish are then placed together for a test fight. If experience affects fight outcome, we should expect that the group of crayfish with "winner's" training will win significantly more often than would be expected to occur by chance.

MATERIALS

You will need crayfish in a range of sizes (at least two small, two medium-size, and two large crayfish for each group of four students). You will need three types of tanks: holding tanks (large tanks or kiddie pools for holding crayfish between class meetings), test tanks (20-gallon tanks, one for each group of students, for developing ethograms and for fighting crayfish), and isolation tanks (shoebox-size containers for isolating crayfish prior to fights in Part II). Use dechlorinated water in the tanks rather than straight tap water (water can be dechlorinated either with special chemicals or by simply leaving tap water out overnight). Your instructor will provide you with a diagram of crayfish anatomy. You will need dip nets for capturing crayfish. For measuring them, you will be provided with flexible plastic rulers, calipers, and/or a balance. Fingernail polish is excellent for giving crayfish individually identifiable marks. Finally, use pencils or pens with waterproof ink for writing down your observations (ballpoint ink smudges when wet).

HYPOTHESES AND PREDICTIONS

Many hypotheses can be tested with the data you will collect during this experiment. First, you will want to test a hypothesis about the effect of experience on winning or losing. You can also test a wide variety of other variables. For example, "winners" may not always win their training fights, and this may affect their performance in the test fight. You might also test whether there is a relationship between which crayfish initiates a fight and which wins it. In discussion with your classmates, generate a second hypothesis that can be tested with these data.

PROCEDURE

The laboratory exercise is in two parts that can be done in 1 day or over a series of labs, according to the plan of your instructor.

Part 1: Constructing an Ethogram

In this part, you will become accustomed to handling and observing crayfish, and you will develop an **ethogram** of crayfish behaviors. An ethogram is a catalog of species-typical behaviors.

You and your lab partner should begin by getting a crayfish out of the holding tank. Crayfish can give a sharp nip. Pick one up by grasping the posterior edge of the carapace, with the chelae pointed away from your body, or scoop one up with a dip net. While you have it in hand, examine the ventral surface to sex it. In males, the first swimmeret is modified as a distinctive copulatory organ. Place the crayfish in the test tank.

Familiarize yourself with the external anatomy of the crayfish by using the diagram provided by your instructor. Using the proper terms, describe the behavior of the crayfish. How does it move its chelae, antennae, and walking legs as it explores the tank? Does it generally move forward, backwards, or sideways? Does it move differently when it is startled?

Now add a second crayfish to the tank. How do they interact initially? How do they use their chelae in interactions? Look for the behaviors listed on the sample data sheet that follows (Table 28.1). Can you see other behaviors that are not listed? Look both for the more dramatic behaviors, such as tailflips and chelae strikes, and for more subtle ones, such as avoidance. Subtle behaviors are often as informative as dramatic ones. Different pairs of crayfish are likely to show different behaviors, so examine a series of them. Try pairs of crayfish of similar size as well as those of different size.

Write down a description of the behaviors you see. It is essential that your descriptions be clear and unambiguous, because in order to pool the data from the class, you must all agree. Meet with your classmates to develop a common ethogram, perhaps using the sample given in Table 28.1 as a starting point.

Consider how you will score the fights and how you will determine whether a crayfish has won a fight. Different definitions are possible, but it is very important that you agree on a standard method. One idea is to score the fight as over when one crayfish performs a *tailflip*, which previous research has established is a submissive signal (see Table 28.1 for definition). Often, however, crayfish do not perform this recognizable behavior, or an individual that has tailflipped will return to dominate its rival (e.g., Figler et al. 1995; Huntingford et al. 1995). Instead, determine in advance the duration of each trial (10 minutes works well), and score the other behaviors that you see, such as avoidance, threat, and tailtucks. You can then establish a winner and loser by adding up the number of times different behaviors were performed. Again, there are many ways to approach this problem, and part of the research process is observing animals and determining which approach seems best. Discuss your ideas with your instructor, and agree on an appropriate design with your classmates.

Table 28.1 **Sample data sheet.**

Your names:_____

I.D. number of crayfish 1: _____   I.D. number of crayfish 2: _____

Type of fight: training fight 1   training fight 2   training fight 3   test fight   (Circle one.)
Check the boxes below when a behavior occurs (you may have multiple check marks for each box).

NEUTRAL BEHAVIORS	*Crayfish 1*	*Crayfish 2*
Touching Two animals come into contact, but no aggression is observed. (*Note:* When this behavior occurs, both crayfish should receive a check mark.)		
AGGRESSIVE BEHAVIORS		
Threat A crayfish approaches the other with outspread chelae in a strike position (this often results in a retreat).		
Strike A crayfish approaches the other and strikes it with its chelae.		
Fight Two crayfish lock chelae and attempt to flip each other.		
SUBMISSIVE BEHAVIORS		
Avoidance With no discernible threatening behavior, a crayfish retreats from the opponent or gives it a wide berth.		
Tailtuck The tail is tucked under the abdomen (this does not result in rapid backward movement).		
Tailflip Rapid ventral movement of the tail results in the crayfish shooting backwards.		

FIGHT CHARACTERISTICS

Who **tailflipped** first?
What was the highest **intensity** of the fight? (Circle one.)
    no interaction   threat but no contact   contact
Who **initiated** the first contact? (Circle one.)
    crayfish 1   crayfish 2   no contact
Who **won**? (Circle one.)
    crayfish 1   crayfish 2   draw
What was the **duration**, in seconds, from the time the crayfish first interacted until the first tailflip occurred?
_____ seconds or _____ Interaction was stopped before tailflip.

One final potential problem is poor training: it is possible for a crayfish that you are trying to train as a "winner" crayfish to lose some of its training trials or for a "loser" crayfish to win some of its training trials. How best to handle this depends on the amount of time available. You can give each crayfish three training trials, regardless of their outcome, and investigate the effect of differences in training experience by teasing it out with statistical tests. Alternatively, if time allows, you can add more training trials until the crayfish reach a criterion of winning two trials in a row. It is important that your entire class (and your instructor) agree on the procedure in advance.

Part 2: Preparing the Animals

Assign the medium-size crayfish to pairs. In order to isolate the effects of experience on the outcome of fights, you will need either to control other variables as closely as possible, or to plan explicitly to include them as independent variables in your analysis. Pairing may be done by your instructor, but if you are doing the pairing, keep the following in mind: First, to control for any differences between the sexes in aggression, pair females with females and males with males. Crayfish sometimes lose appendages (in fights, during problematic molts, to predators, etc.), and it is important *not* to pair one individual that is missing an appendage with one that is fully intact. Similarly, crayfish should be paired on the basis of sexual maturity. Sexually mature individuals should be paired, and sexually immature individuals should be paired. Sexually mature males generally have enlarged chelae, and the first two pairs of swimmerets are hardened. Sexually mature females may not appear to be conspicuously different from immature females, but they are usually larger (see Huner & Barr 1991). Females should also be paired on the basis of whether they are carrying eggs, which are held beneath the body. In addition, because size is so important in determining fight outcome, pairs of test crayfish should be closely matched for size. Crayfish size can be measured in terms of weight, carapace length, or chelae length. Your instructor may suggest that you measure size in more than one way in order to test which measure best predicts the outcome of fights. (*Note*: Ideally, medium-size crayfish will be at least 10% larger than small individuals and at least 10% smaller than large individuals.) Alternatively, if time does not allow for more precise measurements, you might simply estimate the relative sizes of crayfish visually and pair those that are close. Randomly assign the members of each pair of crayfish to be trained as "winners" or "losers" by flipping a coin.

To keep track of the crayfish, mark them with identifying numbers. Dry their carapaces with a towel, and paint the number on with fingernail polish. Be careful not to get paint in the joints of the exoskeleton.

Crayfish perform best if they are isolated from conspecifics for 6–24 hours prior to the laboratory session. Otherwise, recent interactions that they have had in the holding tank may influence their behavior. (Crayfish can be measured before or after the isolation period.)

Part 3: Collecting the Data

Training Trials

Work in a group of four students, with two test tanks per group. Each group of four students is responsible for training and testing one pair of crayfish (more if time allows) that have been matched for size and sex. Two students train the "winner" in one tank by giving it three training fights with smaller opponents. At the same time, the other two students will train the "loser" in the second tank by giving it three training fights with larger opponents. It is important that the "winner" and "loser" not contact one another prior to the test fight.

Before beginning the trials, make sure that everyone in the group is confident about scoring the trials using the criteria in Table 28.1 and any additional criteria that your class developed during your preliminary observations. Behaviors may be initiated suddenly and may be quite brief, so you should be very attentive throughout the trials.

Begin the trials by placing two crayfish simultaneously into opposite ends of the tank, oriented toward one another. It is important to minimize your movement during the trials, in order not to startle or distract the crayfish. One member of the pair of observers will focus on watching the interaction and will call out the behaviors as they occur. The other member of the pair should keep track of time and take notes. Note your observations using the data sheet in Table 28.1 (your class may have agreed to modify this sheet on the basis of your preliminary observations). Put a check mark in the appropriate box whenever a behavior occurs. End the trial either when a tailflip occurs or when the maximum time that you have decided on is reached (the suggested maximum is 10 minutes).

Run a minimum of three training trials. As discussed in Part 1, "winning" crayfish may not win all of their training trials, and "losing" crayfish may not lose all of theirs. You may have decided to run only three training trials regardless of their outcome, or you may have chosen to add additional trials until a certain number of sequential successful trials is reached.

Test Trials

Here your group of four will fight the two size-matched crayfish that received different training regimes. All four of you should conduct the test trial for your pair of crayfish together. Run a single trial for each pair of crayfish.

DATA RECORDING AND ANALYSES

Use Table 28.1 as the basis for a data sheet on which you will summarize the data. You may decide to add or remove behavioral categories as you develop your ethogram in Part 1. Record your data for training trials and test trials on your data sheets.

Now consider the biological hypothesis that you are testing: What is our proposed explanation of the outcome of agonistic encounters that we are attempting to test in this study? An example of a biological hypothesis would be that prior experience affects the outcome of agonistic encounters. This hypothesis leads to several predictions. In our test, our prediction is that the experience of crayfish in three training fights will bias the outcome of a test fight. Pool your data with the rest of the class to determine whether this is the case. Write the null and alternative hypotheses for this question, and determine which statistical test to use. After checking with your instructor, do the test.

Other biological hypotheses are also possible. For example, you might expect the degree of size difference between two combatants to affect the intensity of their fight. Earlier in the laboratory period, with the help of your instructor, you should have identified a second biological hypothesis. Think about the predictions generated by this hypothesis. Write the null and alternative hypotheses for this question, and determine which statistical test to use. After checking with your instructor, do the test on the pooled data from the class.

QUESTIONS FOR DISCUSSION

Did training have an effect on the outcome of the test fights? If not, hypothesize why this might be so. For each hypothesis, propose a way to test it either by using data that you have collected already or by suggesting a new experiment.

It may seem maladaptive for the behavior of animals in fights to be influenced by previous experience: animals may withdraw from fights that they could have won. Why does this phenomenon exist in such a wide variety of species? Consider both proximate and ultimate causes.

Did your initial ethogram suffice, or did you discover new behaviors as you carried out the trials? If you were advising investigators on how to begin a behavioral project, how would you suggest that they determine how much time to invest in observing the animals before they initiate data collection?

LITERATURE CITED

Barinaga, M. 1996. Social status sculpts activity of crayfish neurons. *Science,* **271,** 290–291.

Bruski, C. A. & Dunham, D. W. 1987. The importance of vision in agonistic communication of the crayfish *Orconectes rusticus.* I: An analysis of bout dynamics. *Behaviour,* **103,** 83–107.

Copp, N. H. 1986. Dominance hierarchies in the crayfish *Procambarus clarkii* (Girard, 1852) and the question of learned individual recognition (Decapoda, Astacidea). *Crustaceana,* **51,** 9–24.

Cushing, B. S. & Reese, E. 1998. Hawk-like aggression in the Hawaiian red lobster, *Enoplometopus occidentalis. Behaviour,* **135,** 863–877.

Figler, M. H., Twum, M., Finkelstein, J. E., & Peeke, H. V. S. 1995. Maternal aggression in red swamp crayfish (*Procambarus clarkii,* Girard): the relation between reproductive status and outcome of aggressive encounters with male and female conspecifics. *Behaviour,* **132,** 107–125.

Glass, C. W. & Huntingford, F. A. 1988. Initiation and resolution of fights between swimming crabs (*Liocarcinus depurator*). *Ethology,* **77,** 237–249.

Goessmann, C., Hemelrijk, C. & Huber, R. 2000. The formation and maintenance of crayfish hierarchies: behavioral and self-structuring properties. *Behavioral Ecology and Sociobiology,* **48,** 418–428.

Huber, R. & Kravitz, E. A. 1995. A quantitative analysis of agonistic behavior in juvenile American lobsters (*Homarus americanus*). *Brain, Behavior and Evolution,* **46,** 72–83.

Huber, R., Smith, K., Delago, A., Isaksson, K., & Kravitz, E. A. 1997. Serotonin and aggressive motivation in crustaceans: altering the decision to retreat. *Proceedings of the National Academy of Science USA,* **94,** 5939–5942.

Huner, J. V. & Barr, J. E. 1991. *Red Swamp Crawfish: Biology and Exploitation.* 3rd ed. Baton Rouge: Louisiana Sea Grant College Program, Louisiana State University.

Huntingford, F. A., Taylor, A. C., Smith, I. P., & Thorp, K. E. 1995. Behavioural and physiological studies of aggression in swimming crabs. *Journal of Experimental Marine Biology and Ecology,* **193,** 21–39.

Issa, F. A., Adamson, D. J. & Edwards, D. H. 1999. Dominance hierarchy formation in juvenile crayfish *Procambarus clarkii. Journal of Experimental Biology,* **202,** 3497–3506.

Jackson, W. M. 1991. Why do winners keep winning? *Behavioral Ecology and Sociobiology,* **28,** 271–276.

Pavey, C. R. & Fielder, D. R. 1996. The influence of size differential on agonistic behaviour in the freshwater crayfish, *Cherax cuspidatus* (Decapoda: Parastacidae). *Journal of Zoology,* **238,** 445–457.

Smith, I. P. & Taylor, A. C. 1993. The energetic cost of agonistic behaviour in the velvet swimming crab, *Necora* (= *Liocarcinus*) *puber* (L.). *Animal Behaviour,* **45,** 375–391.

Smith, I. P., Huntingford, F. A., Atkinson, R. J. A., & Taylor, A. C. 1994. Strategic decisions during agonistic behaviour in the velvet swimming crab, *Necora puber* (L.). *Animal Behaviour,* **47,** 885–894.

Spanier, E., McKenzie, T. P., Cobb, J. S., & Clancy, M. 1998. Behavior of juvenile American lobsters, *Homarus americanus*, under predation risk. *Marine Biology*, **130**, 397–406.

ADDITIONAL SUGGESTED READING

Bovbjerg, R. V. 1953. Some factors affecting aggressive behavior in crayfish. *Physiological Zoology*, **29**, 127–136.

Bovbjerg, R. V. 1956. Dominance order in the crayfish *Orconectes virilis* (Hagen). *Physiological Zoology*, **26**, 173–178.

Cooke, I. R. C. 1985. Further studies of crayfish escape behaviour. II. Giant axon-mediated neural activity in the appendages. *Journal of Experimental Biology*, **118**, 367–377.

Hazlett, B., Rittschof, D. & Rubenstein, D. 1974. Behavioral biology of the crayfish *Orconectes virilis*. I. Home range. *American Midland Naturalist*, **92**, 301–319.

Jamon, M. & Clarac, F. 1995. Locomotor patterns in freely moving crayfish (*Procambaraus clarkii*). *Journal of Experimental Biology*, **198**, 683–700.

Merkle, E. L. 1969. Home range of crayfish *Orconectes juvenalis*. *American Midland Naturalist*, **81**, 228–235.

Rubenstein, D. I. & Hazlett, B. A. 1974. Examination of the agonistic behaviour of the crayfish *Orconectes virilis* by character analysis. *Behavior*, **50**, 193–216.

Soderback, B. 1991. Interspecific dominance relationship and aggressive inter-actions in the freshwater crayfishes *Astacus astacus* (L.) and *Pacifastacus leniusculus* (Dana). *Canadian Journal of Zoology*, **69**, 1321–1325.

Tautz, J. 1987. Water vibration elicits active antennal movements in the crayfish, *Orconectes limosus*. *Animal Behaviour*, **35**, 748–754.

Tierney, A. J., Godleski, M. S. & Massanari, J. R. 2000. Comparative analysis of agonistic behavior in four crayfish species. *Journal of Crustacean Biology*, **20**(1), 54–66.

SECTION IV

Courtship and Parental Care

CHAPTER 29

Costs and Benefits of Maternal Care in Earwigs

RONALD L. MUMME, JAMES O. PALMER, AND SUSAN M. RANKIN
Department of Biology, Allegheny College, 520 North Main Street, Meadville, PA 16335, USA

INTRODUCTION

Earwigs (order Dermaptera) are among the most primitive insects to exhibit parental behavior. After laying a clutch of eggs in a burrow or beneath vegetation, female earwigs place their eggs in a tight cluster and protect them from potential predators until hatching. By grooming eggs with their mouthparts, females may also retard the growth of mold and fungi. In some species the period of maternal care extends past hatching, and females may even provision their young hatchlings with food (Lamb 1976; Radl & Linsenmair 1991; Timmins 1995; Rankin et al. 1995a, 1995b, 1996). Because of their maternal care, the ease with which they can be reared in a laboratory environment, and their interesting mating behavior (Moore & Wilson 1993; Radesäter & Halldórsdóttir 1993; Tomkins & Simmons 1998; Forslund 2000), earwigs are excellent subjects for laboratory investigations in animal behavior.

Most species of earwigs are burrowing insects and are common omnivorous scavengers of woodland leaf litter and rotting logs. The most conspicuous anatomical feature of earwigs is the pair of forcepslike cerci at the tip of the abdomen. These forceps play important roles in nest defense by females, in male–male interactions, and in mating and courtship behavior

(Moore & Wilson 1993). The sexes usually can be distinguished by differences in shape of the forceps, which are typically relatively straight and close together in females but curved and pincerlike in males. In some species of long-horned earwigs (family Labiduridae), the forceps of males are asymmetrical, with the right forcep being strongly curved and having a sharp hook at its tip.

Maternal care is uncommon in insects and most other invertebrates (Clutton-Brock 1991). Why, then, did it evolve in earwigs? One plausible hypothesis is that natural selection in the past favored females that tended to remain with their clutch and guard their eggs. Such females may have reduced the loss of eggs to predators or pathogens and thus produced more surviving offspring than females that abandoned their eggs and left them to hatch without the benefit of maternal care.

Maternal care, however, is not without its potential costs. Females that guard eggs may expose themselves to increased risk of predation. In addition, because egg guarding may prevent females from feeding or otherwise interfere with the production of a new clutch of eggs, maternal care may entail a cost to the female's future fecundity. Do the benefits of maternal care outweigh these potential costs? You will explore this issue in the ring-legged earwig (*Euborellia annulipes*) by separating female earwigs from their eggs and measuring hatching success and interclutch interval (time to produce a new clutch of eggs) for guarding and nonguarding females.

MATERIALS

Materials required by each individual student or lab group: 4 female earwigs (each with a clutch of eggs), 2 male earwigs, six 240-milliliter (8-ounce) plastic containers that measure about 90 × 50 millimeters and have ventilated lids, six 2-dram shell vials that measure 45 × 15 millimeters, cotton, dry cat food, dissecting microscope, fine-hair paintbrush, plastic spoon, laboratory tape, marking pen, modeling clay, and an environmental chamber set at 28°C (helpful but not essential).

PROCEDURE

Euborellia annulipes breeds readily in the laboratory. To initiate nesting, place a mated female (or a virgin female paired with an adult male) in a container with a crumbled piece of dry cat food and with a 2-dram shell vial partly filled with water and stoppered with a firm cotton plug. The shell vial serves both as a source of water and as the nest site, and females will readily lay their eggs on the moist surface of the cotton plug. To provide a more secure nest site, the surface of the cotton plug should be about

5–10 millimeters below the lip of the vial, and the entire vial can be anchored to the side of the plastic container with a small piece of modeling clay. Females store sperm in their reproductive tract, so one successful mating is usually sufficient to fertilize several clutches (Rankin et al. 1995a). At 28°C, most mated females will produce a clutch of 20–60 white eggs within 10–12 days (Rankin et al. 1995a); females maintained at cooler room temperatures require more time.

After an adequate supply of clutches has been produced, each student (or group of students working together) should obtain four recently completed clutches. It is best if the four clutches were initiated on or about the same date. Randomly assign each clutch to one of four treatment groups (Table 29.1), and label the container with your name (or the name of your lab group) and the treatment group number. Taking care to minimize the disturbance of the females, use a dissecting scope to count the eggs in each nesting vial as accurately as possible. Record this clutch size in the Day 0 column in Table 29.1.

Table 29.1 **Sample data sheet for recording data on costs and benefits of maternal care.**

				Day					Maximum Number of Hatchlings
	0	*2*	*4*	*6*	*8*	*10*	*12*	*14*	
Treatment 1 (female present, predator absent)									
Number of eggs	—	—	—	—	—	—	—	—	
Number of hatchlings	0	—	—	—	—	—	—	—	—
Treatment 2 (female absent, predator absent)									
Number of eggs	—	—	—	—	—	—	—	—	
Number of hatchlings	0	—	—	—	—	—	—	—	—
Treatment 3 (female present, predator present)									
Number of eggs	—	—	—	—	—	—	—	—	
Number of hatchlings	0	—	—	—	—	—	—	—	—
Treatment 4 (female absent, predator present)									
Number of eggs	—	—	—	—	—	—	—	—	
Number of hatchlings	0	—	—	—	—	—	—	—	—
Removed Female T2									
Number of eggs	—	—	—	—	—	—	—	—	
Removed Female T4									
Number of eggs	—	—	—	—	—	—	—	—	

For treatments 2 and 4, use a plastic spoon or a fine paintbrush to remove the female gently from her nest vial. Take care to avoid removing or damaging any eggs in the process. Return the nest vial to its original container. Place the removed females in separate containers, each with a fresh water (nest) vial and one piece of crumbled dry cat food. Label these containers "Removed female T2" and "Removed female T4."

Because male earwigs readily eat eggs when given the chance, you will use males as potential predators. Add a single male earwig to each of the containers for treatments 3 and 4; it can be helpful if these two males have been deprived of food for 2–3 days prior to the experiment. Then place the ventilated lids on all six of your containers and house them at room temperature or, preferably, in an environmental chamber at 28°C. Over the next 2 weeks (or more) you (and/or your lab partners) should check on your earwigs every 2 days. For each of the four experimental treatments, record the number of eggs and/or hatchlings present at each visit (Table 29.1). For the two removed females, simply record the number of eggs present (if any) in subsequent visits. Add fresh water vials and cat food to all containers as needed.

HYPOTHESES AND PREDICTIONS

The data you collect will enable you to test two hypotheses about the benefits and costs of maternal care in earwigs: (1) Maternal care enhances hatching success, especially in the presence of a potential predator. (2) Maternal care increases the time required for a female to produce a new clutch of eggs.

DATA RECORDING AND ANALYSES

Once you (and your lab partners) have collected your data and have finished filling in the blanks in Table 29.1, the data from the entire class can be pooled and analyzed. The most appropriate statistical test to analyze the data on the benefits of parental care is a two-factor analysis of variance (ANOVA). The two factors that you manipulated in the experiment were (1) presence or absence of the mother, and (2) presence or absence of a potential predator (male earwig). Two-factor ANOVA will analyze the effects of both factors simultaneously and test the hypothesis that the two factors interact to a significant degree. For example, a significant interaction would indicate that the effect of maternal care on reproductive success depends on whether a potential predator is present or absent. The null hypothesis (no significant interaction) would be that the effect of maternal care on reproductive success does not depend on whether a potential predator is present or absent.

Although two-factor ANOVAs are computationally tedious, statistical software packages (and some spreadsheets) will do the heavy lifting for you.

What measure of reproductive success should be used in your analysis? One appropriate measure would be the maximum number of hatchlings counted for each clutch at any time during the monitoring period. This number probably corresponds most closely to the number of eggs that actually hatched from each clutch. Another relevant measure might be the number of hatchlings from the initial clutch that are alive when the experiment is terminated (at day 14 or later). Although it confounds hatching success and post-hatching survival, this measure is a useful indicator of the overall effect of maternal care on reproductive success.

To evaluate the cost of parental care, the number of days required for the production of a new clutch of eggs should be compared for guarding females (females in treatments 1 and 3) and nonguarding females that were removed from their initial clutches (females in treatments 2 and 4). For females that failed to produce a new clutch by the termination of the experiment (day 14 or later), you can simply record the number of days you followed them as a minimum estimate of interclutch interval. The most appropriate statistical test for comparing interclutch intervals is a *t* test for two independent samples. The null hypothesis in this analysis is that removing a female from her clutch has no effect on interclutch interval; the alternative hypothesis is that removal results in a reduced interclutch interval.

QUESTIONS FOR DISCUSSION

Do the results obtained by you and your classmates support the hypothesis that maternal care enhances reproductive success, especially in the presence of a potential predator? Do the results support the hypothesis that maternal care is costly to females in that it results in a delay in the production of a new clutch of eggs? Interpret your results in light of these hypotheses, and discuss any potential problems with your interpretations.

In this experiment you primarily evaluated the effects of maternal care on hatching success. Could maternal care have other effects on reproduction that you were unable to measure? Develop some specific hypotheses about other potential benefits of maternal care, and propose how you might test them.

In your experiment you evaluated one potential cost of maternal care, the time it takes a female to produce a new clutch of eggs. What other potential costs might be relevant? Propose some specific hypotheses about other potential costs of maternal care, and design an experiment or experiments that would enable you to test your hypotheses.

In your experiment you used the number of hatchlings to measure the effectiveness of maternal care. Why might an alternative measure of reproductive success, the *proportion* of eggs that hatched, be a more appropriate measure of the effects of maternal care?

If the results of your experiment suggest that maternal care has both costs and benefits, do the benefits outweigh the costs? How could you measure the costs and benefits in a common reproductive "currency" so that this question would be answered precisely?

ACKNOWLEDGMENTS

We thank our Biology 103 students at Allegheny College for testing this laboratory exercise. Bonnie Ploger, Paul Switzer, and Ken Yasukawa provided constructive comments on an earlier version of the manuscript. Financial support was provided by NSF grant IBN-9817340 to Susan Rankin.

LITERATURE CITED

Clutton-Brock, T. 1991. *The Evolution of Parental Care.* Princeton, NJ: Princeton University Press.

Forslund, P. 2000. Male–male competition and large size mating advantage in European earwigs, *Forficula auricularia. Animal Behaviour*, **59**, 753–762.

Lamb, R. J. 1976. Parental behavior in the Dermaptera with special reference to *Forficula auricularia* (Dermaptera: Forficulidae). *Canadian Entomologist*, **108**, 609–619.

Moore, A. J. & Wilson, P. 1993. The evolution of sexually dimorphic earwig forceps: social interactions among adults of the toothed earwig, *Vostox apicedentatus. Behavioral Ecology*, **4**, 40–48.

Radesäter, T. & Halldórsdóttir, H. 1993. Two male types of the common earwig: male–male competition and mating success. *Ethology*, **95**, 89–96.

Radl, R. C. & Linsenmair, K. E. 1991. Maternal behaviour and nest recognition in the subsocial earwig *Labidura riparia* Pallas (Dermaptera: Labiduridae). *Ethology*, **89**, 287–296.

Rankin, S. M., Palmer, J. O., Larocque, L. & Risser, A. L. 1995a. Life history characteristics of ring-legged earwig (Dermaptera: Labiduridae): emphasis on ovarian development. *Annals of the Entomological Society of America*, **88**, 887–893.

Rankin, S. M., Fox, K. M. & Stotsky, C. E. 1995b. Physiological correlates to courtship, mating, ovarian development, and maternal behavior in the ring-legged earwig. *Physiological Entomology*, **20**, 257–265.

Rankin, S. M., Storm, S. K., Pieto, D. L. & Risser, A. L. 1996. Maternal behavior and clutch manipulation in the ring-legged earwig (Dermaptera: Carcinophoridae). *Journal of Insect Behavior*, **9**, 85–103.

Tomkins, J. L. & Simmons, L. W. 1998. Female choice and manipulations of forceps size and symmetry in the earwig *Forficula auricularia* L. *Animal Behaviour*, **56**, 347–356.

Timmins, C. J. 1995. Parental behaviour and early development of Lesnei's earwig *Forficula lesnei* (Pinot). *The Entomologist*, **114**, 123–127.

CHAPTER 30

Vocal Behavior and Mating Tactics of the Spring Peeper (Pseudacris crucifer): A Field Exercise in Animal Behavior

DON C. FORESTER
Department of Biological Sciences, Towson University, Towson, MD 21252, USA

INTRODUCTION

The spring peeper, *Pseudacris* (formerly *Hyla*) *crucifer*, is one of the best-known and most common frogs of eastern North America. From the mid–Atlantic states northward, its annual breeding cycle is synonymous with the start of spring. Beginning with the snow melt of late February and the onset of the March rainy season, peepers aggregate in virtually any site that holds water. Males position themselves throughout the breeding habitat and began emitting an endless sequence of monophasic "peeps," which represent their species-specific advertisement call. The resulting breeding congress frequently includes hundreds of individuals and may be heard several kilometers away.

The behavioral ecology of *Pseudacris crucifer* is well documented (Doherty & Gerhardt 1984; Forester & Czarnowsky 1985; Forester & Lykens 1986; Forester & Harrison 1987; Forester et al. 1989; Lykens & Forester 1987; Rosen & Lemon 1974; Wilczynski et al. 1984). Spring peepers are classified as prolonged breeders (Wells 1977), which are reproductively active over a broad range of environmental temperatures. As a consequence, they represent an excellent model for a field investigation of anuran mating systems.

Male Vocal Behavior

Male peepers produce calls that differ in certain measurable properties. For example, there is variation in call duration (length in seconds), fundamental frequency (pitch), intensity (loudness in decibels), and call repetition rate (number of peeps per minute), to mention only a few parameters (for a comprehensive review of anuran vocalization, see Gerhardt 1994). Using two-speaker discrimination tests in the laboratory, Doherty & Gerhardt (1984) and Forester & Czarnowsky (1985) demonstrated that females prefer males with loud calls, long calls, mid- to low-frequency calls, and calls with a high repetition rate. Several of these properties (call repetition rate and call frequency) correlate with male body size. Because large males are older than smaller males (Lykens & Forester 1987), by selecting larger males (identified by characteristics of the mating call), females could accrue "a superior genome" for their offspring (Ryan 1980). Although this hypothesis would seem to be borne out by behavioral tests, it is falsified when one compares the mean body size of amplexed (mated) males to the mean body size of a random sample of unmated males. Forester & Harrison (1987) provided data to support an alternative hypothesis in which mating success was correlated with properties of the advertisement call that made an individual male more conspicuous to assessing females.

Satellite Behavior: An Alternative Mating Tactic

It has been hypothesized that small males produce advertisement calls that are less attractive than those of larger conspecifics (Doherty & Gerhardt 1984; Forester & Czarnowsky 1985). If this is true, then it would be adaptive for young, small males to refrain from active vocalization and to employ an alternative mating tactic during their first reproductive season, at a time when they are probably less attractive to assessing females (Forester & Lykens 1986). When a female is drawn to the advertisement call of one of the "host" males, the satellite can move quickly to intercept the female. By so doing, young males could maximize their reproductive success while channeling all available energy into growth.

MATERIALS

The following items are required: hip boots or chest waders (one pair for each student), mechanical counter (one for each group of two students), digital stopwatch (one for each group of two students), 30- or 50-meter tape (preferred) or a meter stick (one for each group of two students), plastic zip-lock bags and/or small plastic specimen jars (10 for each group of two students), flashlight with *fresh* batteries (*essential for each student*), note pad and pencil (ink tends to run if the paper gets wet), rain jacket or coat (spring evenings are often wet and cold), thermometer for recording air and water temperature, Pesola spring scale (0–10 grams) (one for each group of two students). The following items are optional: sound level meter (for recording chorus intensity), sling psychrometer (for recording relative humidity), barometer (for measuring barometric pressure), anemometer (for measuring wind speed).

PROCEDURE

Measuring Parameters of the Physical Environment

Anuran vocalizations are influenced by the physical environment. The size and strength of the breeding congress will vary from night to night and may be correlated with temperature, barometric pressure, cloud cover, humidity, and wind speed. In order to compare the activity that your team observes to that of other teams observing the same chorus on other nights, your team should make a series of basic measurements.

Before entering the chorus, record its collective sound intensity in decibels (dB) using a sound level meter. If you do not have access to a sound level meter, estimate the sound intensity using the index developed for the North American Amphibian Monitoring Program's Calling Frog Survey (http://www.mp1-pwrc.usgs.gov/amphib/Frogcall.html). The index works as follows:

0 = No frogs can be heard calling.
1 = Individual calls not overlapping.
2 = Calls are overlapping, but individuals are still distinguishable.
3 = Numerous frogs can be heard; chorus is constant and overlapping.

Also record air temperature (AT), water temperature (WT), relative humidity (RH), wind speed (WS), and barometric pressure (BP). If you do not have access to an anemometer, you may estimate the relative wind speed using the Beaufort Scale. The scale is as follows:

0 = Calm smoke rises vertically (<1 mph).
1 = Light air; rising smoke drifts (1–3 mph).
2 = Light breeze; leaves rustle; can feel wind on face (4–7 mph).
3 = Gentle breeze; leaves and twigs move around (8–12 mph).
4 = Moderate breeze; moves thin branches, raises loose paper (13–18 mph).
5 = Fresh breeze; trees sway (19–24 mph).

Note the relative degree of cloud cover (%CC) and whether the moon is contributing to the illumination of the pond. Collectively, the latter parameters may be used to estimate the relative illumination (RI) of the pond (high, medium, or low).

Practice Session

Spring peepers are small and cryptic. If you are inexperienced in locating calling frogs, you may wish to spend a few minutes honing your skills and developing a search image before you begin collecting data. Move to a remote portion of the breeding site and switch off your light. Concentrate on the call of a nearby male. Slowly move toward the sound, and when you think you are close, turn on your light and search the vegetation. On cold nights, males tend to call from the water or from vegetation near the surface of the water. On warm nights, they call from emergent vegetation (0.25–1.5 meters above the surface of the water). If your frog stops calling, turn off your light and patiently wait for him to begin calling again. If he doesn't cooperate, whistle to simulate the call of a rival male (this will usually induce him to respond, often with an agonistic trill). After a few minutes of practice, you should be ready to participate in the remainder of the exercise.

Male Vocal Behavior

It is important to remember that repeated observations of the same male represent a violation of statistical independence and will complicate analysis of your data. To ensure that no frog is accidentally observed more than once during a specific segment of the exercise, members of a group should establish mutually agreed upon landmarks to ensure separation, and groups of students should be isolated by >50 meters whenever possible.

Locate a calling male and record the characteristics of his "calling station." Is he calling from the water or from a perch exposed to the air? Remember, when the air temperature is low, males tend to call while sitting in the water; when it is warm (>12°C), they climb up on the vegetation.

Using a mechanical counter, determine how many times the male calls in a 1-minute period. Record your results in Table 30.1. Search within 25 centimeters to determine whether there is a satellite male present (see the "Satellite Behavior" section that). For any calling male with a satellite, record data in Table 30.2 in accordance with instructions in the "Satellite Behavior" section. Next measure the distance to the nearest calling male, and record this in Table 30.3 as per instructions in the "Male–Male Spacing" section. Repeat this procedure for nine additional males, being certain to avoid disrupting the observations of your classmates as you move about the site.

Figure 30.1 **Satellite association. The calling male (right) is exhibiting "high" posture. The satellite (left) is exhibiting "low" posture.**

After your data are recorded, try whistling to a calling male (for example, peep at a rate different from the one he is exhibiting). In many cases the male will emit a trilled "encounter call" designed to intimidate you. If you continue to whistle, the male should synchronize his calls with yours. Can you think of an adaptive explanation for this "antiphonal," synchronized vocal behavior? For an explanation see Rosen & Lemon 1974; Forester & Harrison 1987; Schwartz 1987.

Select a single calling male and observe him for 10 minutes. Keep a cumulative record of the amount of time he spends calling during the time interval. This will involve looking at your wristwatch and then starting and stopping the stop watch until the 10-minute time interval has elapsed.

Satellite Behavior: An Alternative to Male Vocalization

As you observe calling males, look for a noncalling conspecific male within 25 centimeters of a male that is calling. Satellite males normally employ a "low" body posture and remain motionless (Figure 30.1). If you observe a calling male with a satellite (a male that does not call for several minutes), collect both individuals. Use your spring scale to determine the mass (to the nearest 0.1 gram) of each individual. To measure a frog, gently place him in a small zip-lock baggie, attach the baggie to the clip of the scale, and allow it to hang freely (be sure to tare the scale to compensate for the weight of the bag). Record your data in Table 30.2.

The frequency of satellite behavior varies from night to night. Record the number of satellite associations that you encountered while observing

10 calling males (see the foregoing "Male Vocal Behavior" section). To determine what proportion of the total population exploits this conditional mating strategy on a given evening, pool your data with those of your classmates, and divide the number of satellites recorded by the total number of calling males observed.

Male–Male Spacing

Brenowitz et al. (1984) reported that intermale distances between calling males ranged from 40.7 to 770 centimeters (mean = 211.1 centimeters). Locate a calling male and measure the distance to its nearest calling neighbor. In some instances, the distance may be too great to measure with your meter stick. If a meter tape is available, it may be used in place of the meter stick; otherwise, you should estimate the distance. Repeat the procedure until you have intermale distances for the 10 frogs for which you recorded calling rates. Record your data in Table 30.3.

Mating Success and Male Body Size

The entire class should cooperate to search the margin of the pond for amplexed pairs of spring peepers (males and females that are engaged in the sexual embrace that precedes egg deposition). Amplexed pairs are frequently found sitting in shallow water near the shore or against the edge of tree trunks or other emergent vegetation at the water line. If five or more pairs are caught by the collective class effort, the class as a whole should collect a random sample of 20 calling males. Use a spring scale and baggie to weigh each of the calling males as well as the males and females of each amplexed pair. Record these class data in Table 30.4.

HYPOTHESES AND PREDICTIONS

The data you collect will enable you to test the following research questions: (1) Is the sound intensity of the chorus related to any environmental conditions? (2) Do call repetition rate and calling persistence vary among males? (3) Do males position themselves in the habitat in a manner that minimizes acoustic interference from conspecifics? (4) Does mass differ between satellite males (noncalling sexual parasites) and their vocalizing hosts? (5) Is mating success correlated with male body size? In order to answer these questions, you must apply appropriate statistical tests to your data set. For each of these questions, the statistical tests are relatively simple and straightforward.

DATA RECORDING AND ANALYSES

Influence of the Physical Environment on Calling Activity

Compare the environmental measurements that you made to those of other lab groups that may have visited the site on other nights. If enough data are available, calculate a Pearson's correlation coefficient (r) (Zar 1996) to determine whether there is a correlation (a predictable relationship) between the sound intensity of the chorus and one or more of the environmental parameters. Remember that the call of individual males is also affected by environmental factors.

Calling Persistence

Record the number of calls per minute (CPM) for each of 10 males in Table 30.1. Include the data collected by each of the other lab groups in the table. Calculate descriptive statistics for the data set. Then graph the mean, one standard deviation, and the range to demonstrate variation in calling rate. If data are collected on separate nights or at separate wetlands, calling rate may be influenced by a number of factors, including temperature, wind, light intensity, and so on. Use single-factor analysis of variance (ANOVA) to test the null hypothesis that there are no differences between the calling rates of males observed on different nights (or at different sites) (Zar 1996). Your alternative hypothesis is that calling rates differ between nights (or sites). If your analysis indicates a difference, you may wish to use a multiple-range test (such as the Tukey test or Duncan's test) to determine which nights (or sites) are different from one another. Each lab group calculated the amount of time a single male called during a 10-minute period. Pool these data and calculate the mean, range, and standard deviation for the data set. If the exercise was conducted on several nights, you may wish to compare calling persistence between nights using single-factor ANOVA. If so, what are your null and alternative hypotheses?

Satellite Behavior: An Alternative to Male Vocalization

Pool the class data for satellite associations in Table 30.2. Include the mass of the host and its satellite, as well as the intermale distance (IMD) between the satellite and the host. Because the males that make up each host–satellite pair are not independent, you must compare the mean mass of the two groups using a paired *t* test (Zar 1996). Determine the size relationship between satellite males and their hosts. State the null and alternative hypotheses for this test. What is the mean distance between a host and a satellite?

Male–Male Spacing

Record the distance between 10 calling males and their nearest calling neighbor (intermale distance, IMD) in Table 20.3. Pool your data with those of the other lab groups, and calculate a mean, range, and standard deviation for the pooled data set. If measurements were made on more than one night, compare IMD between nights using an independent *t* test (two nights) or single-factor ANOVA (three or more nights). If you conduct this test, what are your null and alternative hypotheses?

Mating Success and Male Body Size

Record the body mass of the male and female that make up each of the amplectic pairs in Table 20.4. If the class is successful in collecting five or more pairs, you may wish to calculate a Pearson's correlation coefficient (*r*) to determine whether there is a predictable relationship between male body mass and female body mass (Zar 1996).

Calculate a mean and a standard deviation for the mass of 20 randomly collected, calling males. Qualitatively, how do these values correspond to the same values calculated for the males recovered from amplexus? Compare the mean mass of the males taken from amplexus to the mean mass of the 20 randomly collected calling males using a two-sample independent *t* test (Zar 1996). What are your null and alternative hypotheses for this test?

QUESTIONS FOR DISCUSSION

Influence of Temperature on the Advertisement Call

What influence would decreasing temperature have on call length (seconds)? What influence would decreasing temperature have on call repetition rate (calls per minute)?

Calling Persistence

How do you explain variation in calling persistence between males? What are the costs and benefits of persistent calling?

Male–Male Spacing

How do your IMD values compare to those Brenowitz et al. (1984)? What factors might affect IMD on a given night? between nights? Is there any relationship between male spacing and calling persistence? Do structural features of the habitat influence male–male spacing?

Satellite Behavior: An Alternative Mating Tactic

How might you demonstrate that satellite behavior culminates in sexual parasitism? What is the adaptive significance of this behavior?

Mating Success and Male Body Size

Why do females exhibit a mating preference for males that are larger than the population mean? Can you envision a scenario in which selection would favor males that are significantly smaller than the population mean? Structure your answer in terms of an evolutionarily stable strategy (ESS) (Maynard Smith 1974). If satellite and calling males actually represent evolutionarily stable strategies (ESS), what relative frequencies of the two strategies should natural selection favor in a population? If calling is energetically expensive, why doesn't satellite behavior predominate? Do male spring peepers exhibit alternative mating strategies? Why are females so much larger than males in this species? How does a female assess the calls of individual males when chorus intensity is high?

ACKNOWLEDGMENTS

I wish to thank S. A. Perrill and his animal behavior classes at Butler University, as well as my own students at Towson University, for field-testing this exercise and providing feedback on it. In addition, I wish to thank the editors and an anonymous reviewer for numerous constructive comments. Their collective suggestions have resulted in significant improvements.

LITERATURE CITED

Brenowitz, E. A., Wilczynski, W. & Zakon, H. H. 1984. Acoustic communication in spring peepers: environmental and behavioral aspects. *Journal of Comparative Physiology* A, **155**, 585–592.

Doherty, J. A. & Gerhardt, H. C. 1984. Evolutionary and neurobiological implications of selection phonotaxis in the spring peeper (*Hyla crucifer*). *Animal Behaviour*, **32**, 875–881.

Forester, D. C. & Czarnowsky, R. 1985. Sexual selection in the spring peeper, *Hyla crucifer* (Amphibia. Anura): role of the advertisement call. *Behaviour*, **92**, 112–128.

Forester, D. C. & Lykens, D. V. 1986. Significance of satellite males in a population of spring peepers (*Hyla crucifer*). *Copeia*, **1986**, 719–724.

Forester, D. C. & Harrison, W. K. 1987. The significance of antiphonal vocalization by the spring peeper, *Pseudacris crucifer* (Amphibia, Anura). *Behaviour*, **103**, 1–15.

Forester, D. C., Lykens, D. V. & Harrison, W. K. 1989. The significance of persistent vocalization by the spring peeper, *Pseudacris crucifer* (Anura: Hylidae). *Behaviour*, **108**, 197–208.

Gerhardt, H. C. 1994. The evolution of vocalization in frogs and toads. *Annual Review of Ecology and Systematics*, **25**, 293–324.

Lykens, D. V. & Forester, D. C. 1987. Age structure in the spring peeper: do males advertise longevity? *Herpetologica*, **43**, 216–223.

Maynard-Smith, J. 1974. The theory of games and the evolution of animal conflict. *Journal of Theoretical Biology* **47**, 209–221.

Ryan, M. J. 1980. Female mate choice in a neotropical frog. *Science*, **209**, 523–525.

Rosen, M. & Lemon, R. E. 1974. The vocal behavior of spring peepers, *Hyla crucifer. Copeia*, **1974**, 940–950.

Schwartz, J. J. 1987. The function of call alternation in anuran amphibians: a test of three hypotheses. *Evolution*, **41**, 461–471.

Wells, K. 1977. The social behavior of anuran amphibians. *Animal Behaviour*, **25**, 666–693.

Wilczynski, W., Zakon, H. H. & Brenowitz, E. A. 1984. Acoustic communication in spring peepers. *Journal of Comparative Physiology A*, **155**, 577–584.

Zar, J. H. 1996. *Biostatistical Analysis*. 3rd ed. Upper Saddle River, NJ: Prentice-Hall.

Table 30.1 **Data sheet for recording the calls per minute (CPM) by male spring peepers. SD = standard deviation.**

Male Number	Laboratory Group					
	1	2	3	4	5	6
1						
2						
3						
4						
5						
6						
7						
8						
9						
10						
Range						
Mean						
SD						

Range (pooled data) = _____

Mean (pooled data) = _____

SD   (pooled data) = _____

Table 30.2 **Data sheet for recording the size and spatial relationship between satellite males and their calling hosts. g = grams, IMD = inter-male distance (centimeters), SD = standard deviation.**

Male Number	Host Mass (g)	Satellite Mass (g)	IMD
1			
2			
3			
4			
5			
6			
7			
Mean			
SD			

Table 30.3 **Data Sheet for recording the distance between nearest calling neighbors (inter-male distance = IMD). SD = standard deviation. All IMD measurements are in centimeters.**

Male Number	Laboratory Group					
	1	2	3	4	5	6
1						
2						
3						
4						
5						
6						
7						
8						
9						
10						
Range						
Mean						
SD						

Range (pooled data) = _____

Mean (pooled data) = _____

SD (pooled data) = _____

Table 30.4 *Data sheet for recording the body size of males and females discovered in amplexus. g = grams, SD = standard deviation.*

Pair Number	Male Mass (g)	Female Mass (g)
1		
2		
3		
4		
5		
6		
7		
8		
Mean		
SD		

CHAPTER 31

The Role of Multiple Male Characters in Mate Choice by Female Guppies (Poecilia reticulata)

DAN ALBRECHT

Department of Biology, Rocky Mountain College, 1511 Poly Drive, Billings, Montana, 59101

INTRODUCTION

Because **reproductive success** is a major component of an individual's fitness, there is great selective advantage in an animal's employing **reproductive strategies** that give it access to the "best" available breeding partner or breeding opportunity. In general, the two sexes have very different reproductive strategies, females being more discriminating in their choice of mates than males are. For example, Bateman (1948) observed that male fruit flies try to mate with almost any female, whereas females tend to interact with a number of males before they allow one to mate. These sex differences in choosiness are probably related to sex differences in the costs of making poor mating choices. Compared to males, females typically incur greater reproductive costs (egg production, gestation, incubation, lactation), and as a result, females run the risk of paying a greater cost for making poor mate choices (Trivers 1972). The costs to males are often not so great because they can inseminate multiple females in a short period (but see Gwynne 1981; Oring 1985; Rosenqvist 1990 for examples of species with **sex role reversal**). Consequently, females often choose carefully among potential mates. Given that females are the choosier sex, upon what male traits do females base their choice?

In a variety of species, females choose mates on the basis of traits such as male body size (Noonan 1983), ornamentation (Andersson 1982; Hill 1995), morphological symmetry (Thornhill 1992), or courtship vigor (Rowland 1995; Alonso 2000). In choosing mates in this manner, females may receive a variety of benefits—access to healthier, parasite-free mates, access to higher-quality breeding sites, access to mates that will provide better parental assistance, or access to males of higher genetic quality (Welch et al. 1998). However, female choice is often not based on a single male trait because most males have multiple traits and complex courtship displays (Borgia 1995; Omland 1996; Mateos & Carranza 1999). In these situations, female choice is probably based on a combination of male behavioral and morphological traits. In guppies (*Poecilia reticulata*), there is evidence that male coloration influences female choice (Endler 1983; Kodric-Brown 1985; Houde 1987; Houde & Endler 1990). However, male display behavior and male size are also thought to influence mate choice (Farr 1976, 1980; Kodric-Brown 1993, but see Houde 1988). The purpose of this lab is to investigate the relative importance of a variety of male traits in mate choice by female guppies. You will carry out a series of experiments designed to introduce you to issues related to mate choice in animals and to test hypotheses concerning the importance of different male characters in female mate choice.

Guppies are small, sexually dimorphic, live-bearing fish native to Trinidad and Tobago and to neighboring parts of South America. Because of the males' conspicuous color patterns and courtship displays, guppies are a common study organism for animal behaviorists interested in sexual selection and mate choice. This species is nonterritorial, lives in large mixed-sex groups, and has no clearly defined breeding season. Fertilization is internal, and gestations last 25–30 days, during which time females are unresponsive to males. Following parturition of multiple young (brood size is highly variable), females become receptive to males again. A courting male will position himself in front of or to the side of a female and perform a "sigmoid" display that includes bending his body into an S-shape and shuddering (Baerends et al. 1955). Displaying males are often interrupted by other males. Receptive females orient toward a displaying male and move forward. Because females breed asynchronously, the ratio of receptive females to displaying males in a population at any one time tends to be relatively low. Thus individual females have the opportunity to choose among multiple mating partners.

MATERIALS

For the following experiments, students will need access to 20 female and 40 male guppies, a 10-gallon female-choice aquarium with clear glass and one-way mirrored glass dividers, 2 stock aquaria (10 gallons or larger), Tetramin flake food, lights and timers, gravel, heaters, filters, and air pumps.

EXPERIMENTS 1–3

Each pair of students should perform only one of the following three experiments, repeating the procedure with a minimum of eight test females. General guidelines for each experimental setup are provided in this section, but students should also look at the "Data Recording and Analyses" section for details about how to record the data during each trial. Different pairs of students will perform experiments 1–3 during this first week, and experiment 4 can be performed during the second week.

Experiment 1: Female Choice for Male Color

With one-way mirror partitions positioned so that the female can see the males, but not vice versa, place a female into the central compartment of the female-choice aquarium. On either side of her, place two size-matched males that vary with respect to the amount of orange covering their bodies. The trio of fish should be allowed to acclimate for 5 minutes. After the acclimation period is over, record the position of the female in the central chamber at 10-second intervals for 10 minutes. Females can be in either of the preference zones or in the middle area (the "no preference" zone) of the compartment. In subsequent trials, males with more or less coloration should be randomly assigned to the right and left sides of the aquarium. Consult with your instructor for alternative sampling protocols.

Hypotheses and Predictions

In this experiment you will explore the general question of whether female choice is influenced by male coloration. If it is, with which male do you predict the female will spend the most time? What if female choice is not influenced by male coloration?

Experiment 2: Female Choice for Male Display

Typically, males will display to the female only when they can see the female. During this experiment, watch for and describe the male courtship display. With a one-way mirror positioned so that the female can see the male, but not vice versa, on one side of the central compartment, and with clear glass positioned on the other side, place a female into the central compartment of the female-choice aquarium. On either side of her, position two size- *and* color-matched males. The trio of fish should be allowed to acclimate for 5 minutes. After the acclimation period is over, record the position of the female in the central chamber at 10-second intervals for 10 minutes. Females can be in either of the preference zones or in the middle area of the compartment. Consult with your instructor for alternative sampling protocols.

Hypotheses and Predictions

In this experiment, you will explore the general question of whether female choice is influenced by male display behavior. If it is, with which male do you think the female will spend the most time?

Experiment 3: Female Choice for Male Size

With one-way mirror partitions positioned so that the female can see the males, but not vice versa, place a female into the central compartment of the female-choice tank aquarium. On either side of the female, place two color-matched males that vary in size. The trio of fish should be allowed to acclimate for 5 minutes. After the acclimation period is over, record the position of the female in the central chamber at 10-second intervals for 10 minutes. Females can be in either of the preference zones or in the middle area of the compartment. In subsequent trials, larger and smaller males should be randomly assigned to the right and left sides of the aquarium. Consult with your instructor for alternative sampling protocols.

Hypotheses and Predictions

In this experiment you will explore the general question of whether female mate choice is influenced by male size. If it is, with which male do you predict the female will spend the most time?

DATA RECORDING AND ANALYSES

During each 10-minute trial, the first student should quietly announce the 10-second intervals. The second student should then announce the position of the female, while the first student records the data by making tick marks on the data sheet. The student observing the female should be positioned behind a blind so as not to disturb the fish. Table 31.1 shows a sample data sheet.

Table 31.1 **Sample data sheet.**

Data and time: _____
Experiment: <u>Small male vs large male</u>   Trial: <u>1</u>
Male positions: <u>Smaller male on right side, larger male on left side</u>

Left Preference Zone (Larger Male)	"No Preference" Zone	Right Preference Zone (Smaller Male)
///// ///// ///// ///	///// ///// /////	///// ///// 10
18	///// // 32	

After running all the trials for your experiment, enter the data from each trial into Table 31.2. Next, determine the basic summary statistics (means and standard errors). Present these in a bar graph. Finally, you need to analyze the data to determine whether females showed a preference for one of the two mate options with which they were presented. In other words, did the females spend more time in one of the preference zones than in the other? But before testing for this, you must determine whether females showed any preferences at all. This can be accomplished by using a paired *t* test or a Wilcoxon matched-pairs test to compare the amount of time spent in both preference zones combined to the amount of time spent in the middle area (the "no preference" zone) of the choice aquarium. If females are spending more time in the "no preference" zone than in the preference zones, this suggests that females were not showing a preference, and no further analyses are needed. However, if females are showing significant preference, you must then compare the periods of time spent in the two preference zones to determine which mate option the test females preferred. In the case of effect of male size, the statistical null hypothesis (H_0) is that female guppies spend equal amounts of time near large and near small males. The alternative (H_a) to the null is that female guppies spend more time near a large male than near a small male (a one-tailed test). Generate the appropriate null and alternative hypotheses for your experiment and check your answer with your instructor.

EXPERIMENT 4: FEMALE CHOICE FOR MULTIPLE MALE TRAITS

You will conduct this experiment in the second week of this lab or on your own time. On the basis of experiments 1–3, you have identified how differences in male coloration, behavior, or size affect female mate choice. In this last experiment, you are asked to design experiments that examine the choice of females when they are presented with males bearing a combination of preferred and unpreferred versions of these traits.

Hypotheses and Predictions

At a broad level, the general research question of this experiment is whether one male trait consistently correlates with female preference or, alternatively, females show weaker preference when presented with males bearing a combination of preferred and unpreferred traits. When females are presented with a male possessing two or three preferred traits, do you predict an even stronger response than when she was presented with a male possessing just one preferred trait? What do you predict will happen when the males have combinations of preferred and unpreferred traits? Because females were previously presented with "isolated" male characters (variation in color alone or variation in size alone or variation in display behavior alone), do you predict that female response will be stronger when they are choosing between males that vary with respect to multiple male traits? Is there some sort of summation going on? In the first three experiments, females were presented with a limited number of male characters to which they could respond. In this last experiment, females are being exposed to multiple varying male traits.

Brainstorm to generate several different combinations of preferred and unpreferred male traits that you could use to compare female preferences. For example, if you were studying mate choice in frogs, you might compare female preference for smaller males that call frequently to their preference for larger males that call rarely. This comparison is of conflicting traits because females generally prefer larger males and males that call frequently. Generate a list of similarly conflicting comparisons that would be appropriate to test in guppies. Decide what combination of traits you will compare. What is the general research question that you will investigate?

Procedure

Design an experiment to compare the response of females exposed to multiple varying male traits. Use the same basic procedures you followed in experiments 1–3. What kinds of males and barriers will you use on each side of the female-choice aquarium? Be sure to include this information on your data sheet.

Data Recording and Analyses

Record and analyze your data following the instructions given for experiments 1–3. Be sure to test for preference vs no preference first. When you test whether females show a preference for one combination of your conflicting traits, what are your null and alternative hypotheses? Check your answer with your instructor.

QUESTIONS FOR DISCUSSION

Why is choosing a mate rather than mating at random beneficial for females?

How would you demonstrate that female choice really is beneficial?

Do females use just one character in making mate choices, or might they be combining information from multiple male traits in making their mate choices?

Is a female's location indicative of mating preference? How could you test this assumption?

Does male mate choice occur? How could you demonstrate this?

Are there other aspects of male coloration that might influence female choice? Blue spots? Iridescence? Black spots? Total amount of body surface covered by color spots? Symmetry of the color distribution on the two sides of the body? How would you test these alternative hypotheses?

Are the females showing a preference for a side of the aquarium rather than a preference for males with specific degrees of coloration? How can this be determined?

If females show no clear preference for male size, does this mean that male size has no impact on mating in guppies? How might male size influence the outcome of competition for mates between males?

LITERATURE CITED

Alonso, J. A. 2000. The breeding system of the orange-crowned manakin. *Condor*, **102**, 181–186.

Andersson, M. 1982. Female choice selects for extreme tail length in the widowbird. *Nature*, **299**, 818–820.

Baerends, G. P., Brouwer, R. & Waterbolk, H. T. 1955. Ethological studies on *Lebistes reticulatus* (Peters). I. An analysis of the male courtship pattern. *Behaviour*, **8**, 249–334.

Bateman, A. J. 1948. Intra-sexual selection in *Drosophila*. *Heredity*, **2**, 349–368.

Borgia, G. 1995. Complex male display and female choice in the spotted bowerbird: specialized functions for different bower decorations. *Animal Behaviour*, **49**, 1291–1301.

Endler, J. A. 1983. Natural and sexual selection on color patterns in poeciliid fishes. *Environmental Biology of Fishes*, **9**, 173–190.

Farr, J. A. 1976. Social facilitation of male sexual behavior, intrasexual competition, and sexual selection in the guppy. *Evolution*, **30**, 707–717.

Farr, J. A. 1980. Social behavior patterns as determinations of reproductive success in the guppy *Poecilia reticulata*: an experimental study of the effects of intermale competition, female choice, and sexual selection. *Behaviour*, **74**, 38–91.

Gywnne, D. T. 1981. Sexual differences theory: Mormon crickets show reversal in mate choice. *Science*, **213**, 779–780.

Hill, G. E. 1995. Ornamental traits as indicators of environmental health. *Bioscience*, **45**, 25–31.

Houde, A. E. 1987. Mate choice based on naturally occurring color pattern variation in a guppy population. *Evolution*, **41**, 1–10.

Houde, A. E. 1988. Effects of female choice and male–male competition on the mating success of male guppies. *Animal Behaviour*, **36**, 510–516.

Houde, A. E. & Endler, J. A. 1990. Correlated evolution of female mating preferences and male color patterns in the guppy, *Poecilia reticulata*. *Science*, **248**, 1405–1408.

Kodric-Brown, A. 1985. Female preference and sexual selection for male coloration in the guppy, *Poecilia reticulata*. *Behavioral Ecology and Sociobiology*, **17**, 199–206.

Kodric-Brown, A. 1993. Female choice of multiple male criteria in guppies: interacting effects of dominance, coloration, and courtship. *Behavioral Ecology and Sociobiology*, **32**, 415–420.

Mateos, C. & Carranza, J. 1999. Effects of male dominance and courtship display on female choice in the ring-necked pheasant. *Behavioral Ecology and Sociobiology*, **45**, 235–244.

Noonan, K. C. 1983. Female choice in the cichlid fish, *Cichlasoma nigrofasciatum*. *Animal Behaviour*, **31**, 1005–1010.

Omland. K. E. 1996. Female mallard mating preferences for multiple male ornaments. I. Natural variation. *Behavioral Ecology and Sociobiology*, **39**, 353–360.

Oring, L. W. 1985. Avian polyandry. *Current Ornithology*, **3**, 309–351.

Rosenqvist, G. 1990. Male choice and female–female competition for mates in the pipefish, *Nerophis ophidion*. *Animal Behaviour*, **39**, 1110–1116.

Rowland, W. J. 1995. Do female sticklebacks care about male courtship vigour? Manipulation of display tempo using video playback. *Behaviour*, **132**, 951–961.

Thornhill, R. A. 1992. Female preference for the pheromone of males with low fluctuating asymmetry in the Japanese scorpionfly, *Panorpa japonica*. *Behavioral Ecology*, **3**, 277–283.

Trivers, R. L. 1972. Parental investment and sexual selection. In: *Sexual Selection and the Descent of Man* (ed. B. Campbell). Chicago: Aldine.

Welch, A. M., Semlitsch, R. D. & Gerhardt, H. C. 1998. Call duration as an indicator of genetic quality in male gray tree frogs. *Science*, **280**, 1928–1930.

ADDITIONAL SUGGESTED READING

Andersson, M. 1994. *Sexual Selection*. Princeton, NJ: Princeton University Press.

Houde, A. E. 1997. *Sex, Color, and Mate Choice in Guppies*. Princeton, NJ: Princeton University Press.

Kirkpatrick, M. & Ryan, M. J. 1991. The evolution of mating preferences and the paradox of the lek. *Nature*, **350**, 33–38.

Kodric-Brown, A. & Brown, J. H. 1984. Truth in advertising: the kinds of traits favored by sexual selection. *American Naturalist*, **124**, 309–323.

TABLE 31.2 **Data sheet for female choice trials.**

Trial	Time Spent Closer to _____	Time Spent Closer to _____	Time Spent in "No Preference" Zone
1			
2			
3			
4			
5			
6			
7			
8			
Totals			

CHAPTER 32

Investigating Human Mate Choice Using the Want Ads

MARY CROWE

Department of Biology, Coastal Carolina University, P.O. Box 261954, Conway, SC 29528, USA

INTRODUCTION

Humans often forget that we are animals and that our behaviors have evolved through processes such as **natural selection**. As behavioral ecologists, we ask questions and form hypotheses about many types of animals and their behavior. One aspect of human behavior that *everyone* is interested is **mate choice**. Preferential mate choice occurs across the animal kingdom, and it means that an individual chooses to mate with individuals that possess certain characteristics (Alcock 1998). For example, peahens prefer to mate with peacocks with more "eyes" in their tails, and in many insect species larger males and females prefer to mate with each other rather than with smaller individuals. These are examples of mate choice, which is a powerful mechanism of evolution.

Males and females invest differently in offspring. Consequently, males and females differ in what they prefer in a mate and in the risks associated with a sexual encounter. From a biological perspective, human females invest considerably more time and energy in offspring than do human males.

Females carry the developing fetus for 9 months; they bear the child; and once the offspring is born, females invest even more energy during lactation (nursing). Male humans, on the other hand, make a minimal biological investment in their offspring. Granted, modern birth control has altered the costs of a sexual encounter; however, human mate choice has evolved over millions of years and, just as in other animal species, our evolutionary past influences our current behavior. As a consequence of the differential **parental investment** we have noted, human females are generally more discriminating when choosing a mate and are more careful about whom they have intercourse with (Buss 1994a).

In the late 1970s, anthropologists and psychologists began to study human mate choice from an evolutionary perspective (Buss 1994a). Research on human mate choice suggests that some human mating behaviors are not culturally bound; in fact they transcend geographical, racial, and economic boundaries (Buss 1994b). There are some worldwide trends in human mating behavior that seem to correspond to survival problems we have faced in our evolutionary past. The following is an incomplete list: Men prefer to mate with younger women but women prefer to mate with older men; men are more interested in short-term mating than are women; men value physical attractiveness in a mate more than women do (Buss 1994b). Men may prefer younger women because they have a higher reproductive potential than older women, and women may prefer older men because they have more resources for offspring. Women generally prefer men with either good financial resources or high social standing (because a man's ability to provide resources for her offspring is positively related to his social status) (Buss 1994b).

There is also some speculation that the global modal duration of marriage (4 years) is a result of our evolutionary history (Fisher 2000). Because ancestral bipedal female humans had to carry offspring in their arms rather than on their backs, the female needed someone to help protect and provide for her and her offspring during the early years. Because the biological father had a vested interest in the offspring surviving, there was selection for pair bonding between him and the mother. Once the offspring reached a level of independence (around 4 years), the father's role was less critical, and both mother and father were free to find new partners (Fisher 1989, 2000). Not everyone agrees that human mate choice is a result of our evolutionary past. Can you think of alternative hypotheses to describe why women prefer men with good financial resources or why men prefer to mate with younger women? Would it be adaptive for women to prefer men who are younger than themselves? Do you think that men value physical attractiveness more than women do?

Scientists test hypotheses about human mate choice in a variety of ways, which include interviews, questionnaires, observing behavior at bars and health clubs, and reading the personal want ads. In personal ads we describe characteristics that we hope others will find attractive and indicate what characteristics we are looking for in a mate. In today's lab you will look through personal ads and develop hypotheses about mate choice in today's Western society.

MATERIALS

Copies of personal ads from a variety of newspapers and online web sites devoted to personal ads.

HYPOTHESES AND PREDICTIONS

Working in pairs, select one page of want ads from the stack available in the front of the room, or log onto the Internet and select one site that features personal ads. Read over the ads quickly, and begin to write down characteristics that people use to describe themselves. Think of some biological hypotheses about what human females or human males find attractive or about how they describe themselves. For example, women prefer older men because they are more likely to provide resources for offspring. To test this hypothesis, we can examine only ads placed by women and make the prediction that women should advertise for men who are older than themselves; see the sample data sheet (Table 32.1). Alternatively, we can compare men's and women's advertisements and predict that men should advertise for women who are younger and women should advertise for men who are older. Be prepared to explain why you think your biological hypothesis is related to sexual selection theory or to other frameworks for explaining human mate choice (such as cultural or religious practices). Some of your biological hypotheses might be alternative hypotheses to those generated by Buss. Pick one hypothesis to test.

Once you generate your hypothesis from the want ads available, you need to test it with an independent, different set of want ads obtained either from the library or from the Internet. Before looking through the set of want ads that you will use to test your hypothesis, you need to decide on an adequate sample size and devise a way to sample the ads independently. One way to do this is to use a table of random numbers; another is to look only at every second or third ad in the list.

DATA RECORDING AND ANALYSES

Create a worksheet on which to record your data. For example, let's say you are interested in determining whether females prefer males who are older than they are. Your data sheet might look like Table 32.1.

Table 32.1 **Sample data sheet.**

Want ads sampled from: <u>*New York Times*, May 21, 1998</u>
Sample size: <u>75</u>
Method of sampling: <u>We selected every fifth ad.</u>
Biological hypothesis: <u>Females prefer to date older men because they provide more resources for</u>
<u>offspring than younger men.</u>

Category	Older Than They Are	Their Age or Younger	No Age Preference
Number of women who advertised for men...			

Statistical null hypothesis: <u>Women should show no preference for men older than themselves over</u>
<u>men their own age or younger (that is, there should be a 1:1 ratio of preferences in the first</u>
<u>two categories).</u>

Statistical alternative hypothesis: <u>More women should prefer men older than themselves than men</u>
<u>their own age or younger (that is, there should be greater than a 1:1 ratio of preferences in the</u>
<u>first two categories).</u>

Your data sheet may have more or fewer columns than Table 32.1, depending on your hypothesis. Because different students will pose different hypotheses, data sheets will not be the same. For each ad that you read, place a hash mark in the appropriate column of your data sheet.

Analysis of the data requires us to examine two alternative statistical hypotheses in each case. (*Note* that these are different from the biological hypotheses for mate preferences.) The first is traditionally called the statistical null hypothesis, which in this case is that women prefer older and younger men equally. If this null hypothesis is correct, women should be just as likely to state a preference for an older man as to state a preference for a younger man. The alternative hypothesis (that is, the alternative to the null hypothesis) is that more women express a preference for older men than for younger men. If this alternative hypothesis is true, then there should be more hash marks in the "Older than they are" box than in the "Their age and younger" box. Write the null and alternative hypotheses for the data you will test, and check these with your instructor.

Once you have collected all the data, create a bar graph of your results and then use a chi-square test to determine whether your hypothesis was supported. Use the chi-square goodness-of-fit test to test questions about choices made by one sex (as in the sample data sheet). Note that in this chi-square goodness-of-fit test there is no way to predict the proportion of women who state no preference, so although these numbers should be collected, they should not be used in the goodness-of-fit test. To compare women's vs men's choices, you will need to use the chi-square test of independence. Check that your expected values are all greater than 5. If they are less, you will need to collect more data. If you have any questions about what your "expected" values for the chi-square are, be sure you talk to your instructor.

QUESTIONS FOR DISCUSSION

Do you think people who use the want ads are searching for a short-term or a long-term mate? How is a short-term mate different from a long-term mate? Why would this distinction be important for mate choice criteria?

Do you think people are honest in describing themselves in a personal ad? What does a person have to gain (or lose) by being honest?

What are some advantages and disadvantages of using want ads to study human mate choice? What assumptions are you making by looking through the want ads? Do you think that the want ads are an appropriate way to study human mate choice? Can you think of other ways to study human mate choice? If so, what are the advantages and disadvantages of these methods?

Do you think that what people say they are looking for in the want ads corresponds with the things that matter when people fall in love?

Are there examples of human mate choice that do not support evolutionary theory? If so, what are they?

ACKNOWLEDGMENTS

I'd like to thank Jessica Young and Martha Leah Chaiken for critical reviews of previous versions of this lab. I'd also like to thank Coastal Carolina University students for their comments, frank discussions, and ideas for this lab.

LITERATURE CITED

Alcock, J. 1998. *Animal Behavior: An Evolutionary Approach.* 6th ed. Sunderland, MA: Sinauer.

Buss, D. M. 1994a. The strategies of human mating. *American Scientist,* **82**, 238–249.

Buss, D. M. 1994b. *The Evolution of Desire: Strategies of Human Mating.* New York: Basic Books.

Fisher, H. E. 1989. Evolution of human serial pair bonding. *American Journal of Physical Anthropology,* **78**, 331–354.

Fisher, H. E. 2000. Lust, attraction, attachment: biology and evolution of three primary emotion systems for mating, reproduction and parenting. *Journal of Sex Education and Therapy,* **25**, 96–104.

SECTION V

Games

CHAPTER 33

Demonstrating Strategies for Solving the Prisoner's Dilemma

KATHLEEN N. MORGAN
Department of Psychology, Wheaton College, Norton, MA 02766, USA

INTRODUCTION

Animals organize their behavior with respect to conspecifics in a variety of ways. In some species, individuals actively avoid one another, coming together only to mate. In others, individuals gather in groups. For any individual, the advantage of living in a group may be little more than reducing the likelihood of being eaten by a predator (the "dilution effect" hypothesis). However, in some species, group living may provide more benefits than simply improvement of odds in escaping predators. Group members may actively help one another to find and catch food, to harvest food, to build and maintain shelters, and/or to care for young.

Why animals that are unrelated to one another would cooperate has been a puzzle for evolutionists. Behaving **altruistically**—doing something that costs you while benefiting another—simply does not "make sense" in the evolutionary scheme of things. A truly altruistic individual that repeatedly helps unrelated others is jeopardizing its survival and may not survive to pass on any of those "altruistic genes" to the next generation.

Nonetheless, some animals do behave altruistically. For example, female vampire bats roost during the day in communal groups of related and unrelated individuals. An individual bat might not always be successful at night in finding a host for the blood meal it needs, and a vampire bat cannot survive a missed feeding for very long. However, bats that roost together will share food with a roost-mate that has been less successful on a given night by regurgitating blood for her (Wilkinson 1984). In this situation, a bat that fed successfully is sacrificing some of her own gain to benefit another, perhaps unrelated individual.

By definition, behaving altruistically means that the behavior entails a cost of some kind. What is to keep other individuals in a group from "mooching" off the kindness of an altruist? You may have had the experience of working on a group project for a class, in which there was a member of your group who did nothing and yet received the same credit as everyone else in the group. Imagine, for example, that you are assigned a project to work on as a pair. Your own grade, then, is dependent on the joint grade earned for the project that you do with your partner. If you and your partner both work hard, you should do well. However, if your partner does none of the work, you have a dilemma. In order to get a good grade, you would have to do *all* the work yourself, thus giving your partner a "free ride" at your expense. But if you retaliate against your partner's laziness by *also* not doing the work, *both* of you will get a poor grade. (This "damned if you do, damned if you don't" situation is called the "cruel bind" by behavioral ecologists.) How could you maximize the advantages of group living (or in this case, group working) without incurring the cost of being taken advantage of? What might you do if you were assigned to work with that partner again?

The purpose of this investigation is to reveal some possible strategies that humans might use when presented with an opportunity either to cooperate or to behave selfishly, and to explore which strategies appear to be more effective. Your mission is to ask pairs of humans to play a game. The purpose of the game is to move a penny from the center spot of the game board (Figure 33.1) to a goal spot (different for each player). The game is similar to checkers, except that there is only one playing piece, and the number of moves that each player can use is limited. Be sure you read through all of the instructions before collecting your data.

MATERIALS

For half of the class (Team 1): 2 human game players per observer. For the other half of the class (Team 2): 20 human volunteers (10 pairs of game players per observer). Each observer will need a game board (Figure 33.1), a data sheet for each game the volunteers will be playing (Figure 33.2), a penny or similar coin, and something to write with.

PROCEDURE

Your instructor will have you do this lab in one of two ways: either collecting data on games played by people outside of this class or collecting data on games played by class members.

If you are to collect data on people who are not in this class, you will *all* be observers. Your mission as an observer is to find volunteers who are not in this class, explain the game to them, and then collect data on the games your volunteers play.

If data are to be collected on class members, then some of you will be players while others will be observers of the game. In such a case, your instructor will divide the class into two groups. One half of the class will follow procedure 1 (see the following section). The other half of the class will follow procedure 2. *Do not read one another's procedures or directions.* After each half of the class has collected its data, everyone in the class will come together again to share their methods and to compare and discuss their findings.

Procedure 1

For this assignment, you need a penny, a game board (Figure 33.1), and a data sheet (Figure 33.3) for each game played, and a quiet room with no distractions where you can all be alone until you have finished collecting your data. You will find the game board and a data sheet at the end of this exercise. Complete the top part of the data sheet for that game, taking care to record your name, the date, and other relevant information. Sit so that the game board can be placed between the game players, with one player seated before Goalspot A and one player seated before Goalspot B. Place the penny in the Center Spot. Then the observer should read the following instructions to the game players.

> The purpose of this study is to see how people solve the problem of how to play this game so they maximize their winnings and minimize their losses. The game is fairly simple, but winning it can be difficult. I'll ask you to play the game 10 times, taking turns at going first. In each game, each player has only 10 moves to try to move the penny into his or her own "goalspot," for a total of 20 moves per game. The penny can be moved only one circle at a time and only by following the lines connecting two circles. If after 20 moves neither of you has moved the penny into your own goalspot, the game ends in a draw, and neither player wins. I need to count the number of moves you make in this game, so please don't move too quickly! Also, the game must be played in total silence. Please do not talk to each other or otherwise signal one another before or during the game.

The observer should ask the players if they have any questions. Do not continue until everyone understands how to play the game. Toss the penny, and have a player call heads or tails. The player who wins the toss should begin the first game. Figure 33.3 shows how to use a game data sheet to keep track of the moves made during the game. As each player makes a move, the observer should write down the number of that move in the circle that represents it on the data sheet for that game. When one player wins, or when the 20 moves allowed for that game have been made, end that game. Record the outcome of the game in the space provided on the data sheet for Game 1.

Observers should remind the players that the object of the game is "to see how people solve the problem of how to play this game so they maximize their winnings and minimize their losses." Again, remind them not to speak or discuss the game with you or with each other but, rather, simply to play the game again. This time, the person who lost the toss for the first game should go first. Just as for Game 1, as each player makes a move, the observer should write down the number of that move in the circle that represents it on the data sheet for that game. When one player wins, or when the 20 moves allowed for that game have been made, end that game. Record the outcome of the game in the space provided on the data sheet, and then begin the next game by replacing the penny in the Center Spot. The observer should allow the players to take turns starting each new game and should remind them not to communicate with one another until the 10 games are over. If the players are volunteers, the observer should be sure to thank them for helping with this assignment after finishing!

For each of the 10 games, the observer should count the number of moves made and should end the game if it is not won within 20 moves. Record the outcome of each game. At the end of the 10 games, the observer should ask the players to describe any ways in which someone could win the game. The observer should record any answers on the data sheets.

Procedure 2

For this assignment, you need a penny, a game board (Figure 33.1), a data sheet (Figure 33.2) for each game played, and a quiet room with no distractions where you all can be alone until you have finished collecting your data. You will find the game board and a data sheet at the end of this exercise. Complete the top part of the data sheet for that game, taking care to record your name, the date, and other relevant information. Sit so that the game board can be placed between the game players, with one player seated before Goalspot A and one player seated before Goalspot B. Place the penny in the Center Spot. Then the observer should read the following instructions to the game players.

> The purpose of this study is to see how people solve the problem of how to play this game so they maximize their winnings and

minimize their losses. The game is fairly simple, but winning it can be difficult. I'll ask you to play the game only once. In the game, each player has only 10 moves to try to move the penny into his or her own "goalspot," for a total of 20 moves per game. The penny can be moved only one circle at a time and only by following the lines connecting two circles. If after 20 moves neither of you has moved the penny into your own goalspot, the game ends in a draw, and neither player wins. I need to count the number of moves you make in this game, so please don't move too quickly! Also, the game must be played in total silence. Please do not talk to each other or otherwise signal one another before or during the game.

The observer should ask the players if they have any questions. Do not continue until everyone understands how to play the game. Toss the penny, and have a player call heads or tails to determine which player will start the game. Figure 33.3 shows how to use a game data sheet to keep track of the moves made during the game. As each player makes a move, write down the number of that move in the circle that represents it on the data sheet for that game. When one player wins, or when the 20 moves allowed for that game have been made, end that game. Record the outcome of the game in the space provided on the data sheet for Game 1. If the players are volunteers, the observer should thank them for helping with this assignment and then go look for another two volunteers. The observer should repeat this procedure until he or she has tested 10 pairs, each pair playing the game only once. The observer must be sure to treat all pairs identically, including testing each in the same room and reading each the same set of instructions.

For each of the 10 games, the observer should count the number of moves made and should end the game if it is not won within 20 moves. Record the outcome of each game. At the end of each game, the observer should ask the players to describe any way in which someone could win the game. The observer should record any answers on the data sheets.

HYPOTHESES AND PREDICTIONS

Before you find your subjects and begin, take a few moments and make some predictions about what you think will happen.

Team 1

What do you think people will do when confronted with this game for the first time? Do you think they will play in such a way that one person wins while the other loses? Will the players compete or cooperate? How about after they have played a few rounds?

Team 2

What do you think people will do when you confront them with this game? Do you think they will play in such a way that one person wins while the other loses? Will the players compete or cooperate?

Both Teams

What kinds of factors do you hypothesize might influence the decisions that your subjects make? Do you think that your subjects will figure out how to maximize their winnings? Why or why not? Respond to these questions *first*, and then collect all of your data before continuing to read this chapter.

DATA RECORDING AND ANALYSES

DO NOT READ THIS SECTION until after you have completed all data collection.

By now, you have made some thoughtful hypotheses and collected your data. If some class members were game players, these players should help the observers they worked with to analyze the data for the games. To analyze your results, calculate the average number of moves used for all 10 games. Calculate the average number of wins for each player (A or B), and the number of "draws"—games that no one won. This will tell you a little about how well your subjects did at playing the game. The higher the average number of moves, and the higher the average number of draws, the less successful your subjects were at trying to figure out the best strategies for maximizing their winnings. To look for the strategies that your subjects used to win the game, you need to examine your data in a little more detail.

Remember that the purpose of this assignment is to determine some possible strategies that humans might use when presented with an opportunity either to cooperate or to behave selfishly. Under most circumstances and in most animals, natural selection favors essentially selfish behavior and does not favor behavior that benefits another at a cost to one's self. The female vampire bat that offers a blood meal to her starving neighbor does so at a cost to herself. If our generous bat made these offers to unrelated neighbors repeatedly, to the point at which she compromised her own health, it is unlikely that her altruistic genes would be passed on to future generations.

Trivers (1971) suggested that animals would behave altruistically in situations in which it was highly likely that the favor would be returned at some time in the future: sort of an "I'll scratch your back now if you scratch mine tomorrow" scenario. However, such a strategy would be highly susceptible to sabotage by "cheaters"—individuals who gladly accepted aid but failed to return it in the future. Trivers recognized this problem with his idea and considered that **reciprocal altruism** (behaving altruistically and having that behavior reciprocated at some time in the future) would evolve only under conditions in which individuals recognized one another and in which the likelihood of two individuals meeting repeatedly was high.

Still another view of why and under what circumstances animals would cooperate with one another was developed by Axlerod & Hamilton (1981). Their approach was to apply the classic social science model of the "**prisoner's dilemma**" to the problem of reciprocal altruism. The prisoner's dilemma posits that two suspects are captured by the police and held in prison. Neither one may communicate with the other. Each is told that the case against him or her is weak and that, because of this, each will be held in jail for only a little while. However, each is also told that if he or she would be so kind as to inform on the other (that is, to defect), the stool pigeon would be released immediately, while the other would go to jail for a long, long time. Finally, each is told that if both prisoners inform against each other, they will both remain in jail for an intermediate time period.

Figure 33.4 shows the relative "payoff" to a given prisoner—in this case, Prisoner A—in a single instance of the prisoner's dilemma (**PD**). In this situation, prisoners have only one chance to make a decision. Under such circumstances, the **evolutionary stable strategy** (**ESS**, a behavioral strategy that cannot be bettered once it is widely adopted) is for both prisoners to defect (to inform against one another). Imagine that you are one of the prisoners. Remember what the police have said about the consequences of your decision. If you cooperate with your fellow prisoner and do not inform, and he or she is equally cooperative, both of you will be in jail for a little while. This is *R*, the payoff for mutual cooperation. Because the jail term here is not very long, the payoff has an intermediate value. However, if you cooperate and your colleague defects, you will be in jail for a much longer time, while he or she goes free. This situation represents *S*, the "sucker's payoff," with you as the sucker! In this situation, you gain nothing and in fact "lose" the game because you will receive the maximum penalty.

Suppose that you decide to inform on your colleague, however, and defect. If he or she does the same, you will both stay in jail for a while, but not for as long as you would have had you remained loyal while your colleague ratted on you. This is *P*, the "payoff" or punishment for mutual defection. However, if you defect and your colleague does not, you gain the most because under these conditions you will go free while he or she stays in jail. This situation represents *T*, the payoff that results in temptation to defect. In cases in which the prisoners may make only one decision (in other words, in cases in which the "game" of the prisoner's dilemma is played only once), defecting is the best strategy. No matter what the other prisoner does, it always pays in the short run to defect, because T (the reward for defection when the other prisoner cooperates) is greater than R (the reward for mutual cooperation). Thus, evolutionarily speaking, cooperation is a suboptimal strategy when individuals meet only once in a situation in which cooperation is possible.

But what about under those conditions specified by Trivers (1971) as the exceptions? When individuals are able in interact repeatedly, as in a social species, then a different ESS might obtain. Imagine, wrote Axlerod & Hamilton (1981), two individuals who exchange goods as part of a trade system. The two never meet but, rather, leave the goods at a predetermined place in the forest. When each comes to pick up what the other has left, should one simply take what is left and leave nothing in return? Does it profit the individual to reciprocate if there is the possibility of another exchange in the future? And if one individual should opt on occasion to cheat and leave nothing for the other, what should the offended party do when he or she next returns to find an empty larder? Such a situation is essentially an **iterated prisoner's dilemma** (**IPD**)—a repeated playing of the same decision game, over and over again.

Using a series of tournaments in which a computer played a string of IPDs, Axlerod & Hamilton (1981) concluded that the strategy that would result in a cooperator gaining the most while cutting losses was what they called **tit-for-tat**. In this strategy, a player's first move should be to cooperate, but that player on his or her next move should do whatever the other player did. In other words, if after you cooperated the other player responded in kind, your next move should also be cooperative. If, on the other hand, you cooperated but the other player defected and cheated you, then the way to cut your losses is to defect also on your turn.

In this assignment, the class has collected data on two kinds of prisoner's dilemma games: PD and IPD. Team 1, in having the same pair of players play the game repeatedly with one another, essentially have data on IPDs. Team 2, in having different players play each game, have PD data. Each team should analyze its own data and then get together with the other team to compare. Some ideas on how to analyze your data follow.

In solving the problem of how each player can win the game the most times while losing the least, people may have adopted one of several strategies. One strategy involves **cooperating**. In the original prisoner's dilemma, cooperating meant not informing on your fellow prisoner. Effectively, this helps your fellow prisoner (because informing on that person would mean a longer jail term for him or her), while costing you a little (because if you had informed on your comrade, the police would have let you go). It is also a bit of a risk, because if you cooperate but your fellow prisoner defects, you will be in jail for a long, long time. Among animal behaviorists, cooperation similarly means behaving in a way that helps another, sometimes at a cost to oneself. For this game, cooperation is defined as moving the game piece in a direction that directly benefits one's opponent, at an obvious cost to one's own potential success. For example, say you move the penny toward your goalspot, then I cooperate by moving it in the same direction on my turn, and then you move it again into your goalspot. In this case, rather than attempting to win the game for myself, I have cooperated in helping you to win. A game won through pure cooperation can be won in 3 moves (from the Center Spot straight to someone's goalspot). For the purposes of this lab, a game won through pure cooperation is one that is won in 3 moves, in which one player completely cooperates with the other. Calculate how many times (if any) one of your subjects won a game in 3 moves. How many examples of pure cooperation were there among your 10 games?

Another strategy for solving the problem of how each player can win the game the most times while losing the least involves competition. In the single-iteration prisoner's dilemma (PD), the optimal strategy for a given prisoner is to defect (or "compete" for the prospect of an early release) by informing on the other prisoner. In a situation in which both prisoners essentially "compete" for an early release by informing on one another, neither prisoner benefits because each still gets a jail term. However, that jail term is not as long for the defecting prisoner as it would have been if he or she had cooperated with the fellow prisoner while that fellow prisoner defected. Competition, then (in the sense of competing for the possibility of early release), may yield a maximum payoff (freedom) or an intermediate punishment (an intermediate jail term). For the purposes of this assignment, competition might result in a win for a given player (if his or her opponent cooperates), or it could result in a draw with neither player winning. One might think of a draw as a consequence that is intermediate between winning and losing. Pure competition would result in neither player winning. All the moves would be used up, as each player moved the penny back toward her or his goalspot, away from the circle into which the opponent had just moved it. For the purposes of this lab, a game ending in a draw because of pure competition is one in which each player goes head to head with the other, and all moves are spent moving back and forth between the same few circles. Calculate how many times (if any) your subjects failed to win a game and spent all 20 moves in vain. How many examples of pure competition were there among your 10 games?

Still another strategy involves one player giving up or "submitting" after the other directly challenges him or her. In single-iteration (PD) games, this opportunity never arises, because each prisoner has only one chance to make a decision about whether to cooperate or to defect. This is unlike our game of 20 moves, in which each move (and, in fact, each game) is like a new decision for a player: "Do I cooperate, or do I defect?" Except with the first move, each player knows what the other player did just before he or she must make his or her own move or decision. Submitting might be one of those decisions. In the IPD games observed by Team 1, submitting not only could occur multiple times within a game but also could occur in multiple games between the same players. Each game in this case represents an opportunity to decide, on the basis of the outcome of the last game, whether to try to win or just to submit. The games observed by Team 2 are like PD games in that the game is played only once. Even if multiple decisions occur within a game, there is no opportunity to use the outcome of one game to guide future decisions about whether to try to win or just submit.

What would submission look like? Imagine I move the penny toward my goalspot, and you move it back one. On my next move, I might submit to you, and move the penny toward your goalspot yet again, so that it may take as few as 5 moves for you to win the game. For the purposes of this lab, a game showing submission is one in which in at least one instance, a player moves toward another player's goalspot, rather than to the side or toward his or her own goalspot. Calculate how many games (if any) show at least one instance of submission among your 10 games. In how many games did at least one player show signs of "giving up"? *Do not include in this count games that also show "conflict avoidance" (see the next paragraph); rather, count only those games that show at least one instance of submission.*

A final strategy involves trying to avoid conflict by moving the penny in a direction that is neither toward nor away from one's own goalspot but, rather, to the side. In this game, conflict avoidance is sometimes called "staircasing," because it results in the movement of the penny on the game board making a staircase pattern similar to the one that you can see in the demonstration game in Figure 33.3. Again, such a strategy is unique to this game, because PD games (unlike to IPD games) allow players to make only one decision and do not allow players to interact. For the purposes of this lab, a game showing conflict avoidance is one in which in at least one instance, a player moves the penny to the side, rather than toward his or her own goalspot or toward the other player's goalspot. Calculate how many games (if any) show at least one instance of a player trying to avoid conflict by staircasing. How many games of the 10 showed evidence of conflict avoidance? *Do not include in this count games that also show submission (see the previous paragraph); rather, count only those games that show at least one instance of conflict avoidance.*

Sometimes, players will attempt a combination of strategies as they consider how the game might be won. A single game of 20 moves might show evidence of submission *and* conflict avoidance. (Remember that a game in which one player yields completely to the other is an example of pure cooperation for our purposes here; submission is what we are calling an apparent attempt to cooperate after the initial use of some other strategy). Calculate the number of games (if any) in which both conflict avoidance and submission were employed as strategies.

Each observer should summarize his or her data on the individual researcher data summary sheet (Table 33.1). If some class members were game players, the players should help the observers they worked with to analyze the data for their games. Enter, in the place provided on your summary sheet, the number of games that you observed that showed pure cooperation (won in 3 moves) and the number of games that you observed that showed pure competition (ending in a draw with neither party attempting to yield to the other). Similarly, in the places provided on your summary sheet, enter the number of games showing submission, the number of games showing conflict avoidance, and the number of games showing both submission and conflict avoidance. Then, add up the number of games showing submission, the number of games showing conflict avoidance, and the number of games showing both submission and conflict avoidance. Enter this number on the summary sheet under "Mixed strategy." For the purposes of this lab, a game showing a "mixed" strategy is any game that does not show either pure competition or pure cooperation.

Team 1 Members

Get together with the others in your class who followed procedure 1 and pool your data using the team summary data sheet (Table 33.2). What was the average number of wins per player that your team observed? What was the total number of draws observed?

How many of the games that your team observed show pure cooperation? How many show pure competition? How many show a mixed strategy? Did anything change in the strategies employed by subjects as they played one another again and again? To examine this possibility, create some pie charts in which you compare the number of games that showed each of the strategies we discussed (pure cooperation, pure competition, and mixed (submission and/or conflict avoidance) for the first, fifth, and tenth games played. Are there any apparent changes in strategy over time?

Was there any evidence of a "tit-for-tat" strategy being employed? In other words, did the behavior of one player appear to affect the behavior of the other player in a predictable way? Document the frequency of occurrence of tit-for-tat strategies in your team's games.

Team 2 Members

Get together with the others in your class who followed procedure 1 and pool your data using the team summary data sheet (Table 33.2). What was the average number of wins per player that your team observed? What was the total number of draws observed?

How many of the games that your team observed show pure cooperation? How many show pure competition? How many show a mixed strategy?

Create some pie charts in which you display the relative percentages of these strategies in the first, fifth, and tenth games that your team observed.

Both Teams

Compare and contrast the data collected from the two teams (Team 1, with the IPD games, and Team 2, with the PD games). If players of IPD games were beginning to use a tit-for-tat strategy, what would that look like in the pie charts for each team? Can you find evidence of the adoption of such a strategy by Team 1's IPD players as you compare the pie charts for both teams?

If your instructor asks you to do so, you may conduct some statistical analyses of your data. Specifically, you will want to test the null hypothesis that there is no difference between IPD games (run by Team 1) and PD games (run by Team 2) in average number of games won per player. (The alternative hypothesis is that IPD and PD games differ in the average number of games won per player.) If your class has collected data from 30 or more observers per team, use Student's t test for independent groups. Use a second Student's t test to determine the effects of game type on the number of draws observed.

If your class collected data from fewer than 30 observers per team, use the nonparametric Mann–Whitney U test, which does not make as many restrictive assumptions about normality of the underlying data as does the Student's t test. Use a second Mann–Whitney U test to determine the effects of game type on the number of draws observed. Finally, determine whether the two game types differ in the frequencies of different strategies they elicit, using a chi-square test of independence.

Discuss as a class the typical things that players said when you asked them whether they could see any way to win the game. Were those things different for the two kinds of games played (PD vs IPD)?

QUESTIONS FOR DISCUSSION

Get together with the rest of your classmates and exchange your results. Compare the typical outcome of game data collected by Team 1 with the typical outcomes of game data collected by Team 2. Was the average number of wins per player the same for each team? If not, why might that be?

Did subjects appear to play the game differently when they knew they were going to play only once vs when they knew they were going to play the game repeatedly? If so, in what way was the play different?

Among the PD games, was there one strategy that seemed to be most common? What about in the IPD games? Were the most common strategies in the two kinds of games the same or different? How do your results compare with predictions from game theory?

In the IPD games, why might one strategy have "worked" better than another for maximizing each player's winnings while minimizing losses?

From your examination of the pattern of strategies employed by each player during the 10-game bouts, did Player A's behavior appear to influence what Player B chose to do? If so, how?

What do your data tell you about humans and their decisions on when to cooperate with one another? Do you think you might have had different results if you asked humans from a different culture to play this game? If so, why? If not, why not?

Are there any particular conditions under which you think that the strategies a person employed might be different? Why?

ACKNOWLEDGMENTS

The author would like to thank Matthew Miller, the text editors, and an anonymous reviewer for their thoughtful comments on previous drafts of the manuscript.

LITERATURE CITED

Axlerod, R. & Hamilton, W. D. 1981. The evolution of cooperation. *Science*, **211**, 1390–1396.

Trivers, R. L. 1971. The evolution of reciprocal altruism. *Quarterly Review of Biology*, **46**, 35–57.

Wilkinson, G. A. 1984. Reciprocal food sharing in the vampire bat. *Nature*, **308**, 181–184.

SUGGESTED READING

Axlerod, R. & Dion, D. 1988. The further evolution of cooperation. *Science*, **242**, 1385–1390.

Boyd, R. 1988. Is the repeated prisoner's dilemma a good model of reciprocal altruism? *Ethology and Sociobiology*, **9**, 211–222.

Dugatkin, L. A. 1988. Do guppies play TIT FOR TAT during predator inspection visits? *Behavioral Ecology and Sociobiology*, **23**, 395–399.

Dugatkin, L. A. 1998. Game theory and cooperation. In: *Game Theory and Animal Behavior* (ed. L. A. Dugatkin & H. K. Reeve), pp. 38–63. New York: Oxford University Press.

Fishbein, H. D. & Kaminski, N. K. 1985. Children's reciprocal altruism in a competitive game. *British Journal of Developmental Psychology*, **3**, 393–398.

Kagan, S. & Madsen, M. C. 1971. Cooperation and competition of Mexican, Mexican American, and Anglo-American children of two ages under four instructional sets. *Developmental Psychology*, **5**, 32–39.

McDonald, D. B. & Potts, W. K. 1994. Cooperative display and relatedness among males in a lek-mating bird. *Science*, **266**, 1030–1032.

Mesterton-Gibbons, M. & Dugatkin, L. A. 1997. Cooperation and the prisoner's dilemma: towards testable models of mutualism versus reciprocity. *Animal Behaviour,* **54,** 551–557.

Milinski, M., Pfluger, D., Kuelling, D, & Kettler, R. 1990. Do sticklebacks cooperate repeatedly in reciprocal pairs? *Behavioral Ecology and Sociobiology,* **27,** 17–21.

Noee, R. 1990. A veto game played by baboons: a challenge to the use of the prisoner's dilemma as a paradigm for reciprocity and cooperation. *Animal Behaviour,* **39,** 78–90.

Goalspot A

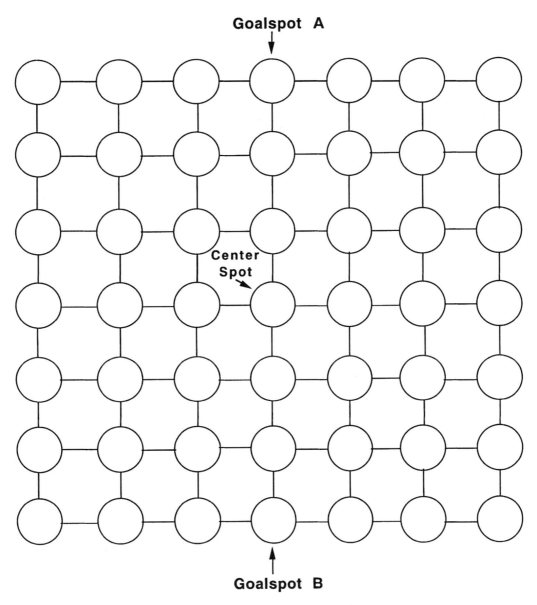

Figure 33.1 **Gameboard for prisoner's dilemma game.**

Observer's name:_____   Game number: ___ of ___

Date:_____ Time:_____ Location:_____

Who went first?   Player A   Player B   Who (if anyone) won?   Player A   Player B   Draw
(Circle one.)   (Circle one.)

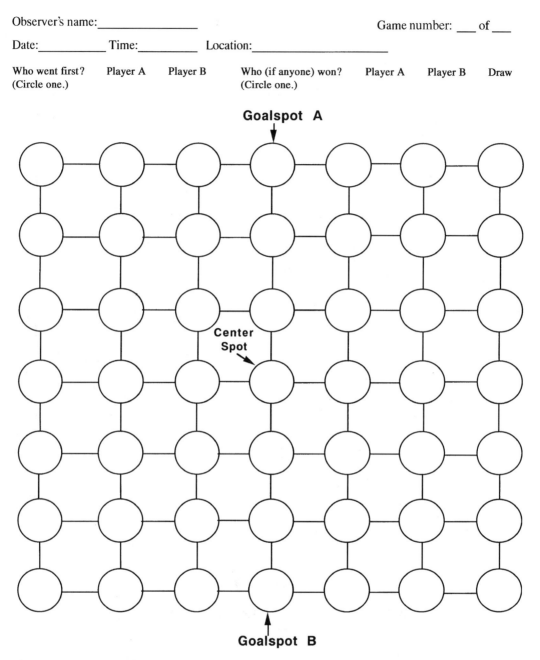

Figure 33.2 **Game data sheet.**

Observer's name: __Kersti__   Game number: _1_ of _1_

Date: _4/11/02_ Time: _23:30_   Location: __My dorm room__

Who went first?   (Player A)   Player B   Who (if anyone) won?   (Player A)   Player B   Draw
(Circle one.)   (Circle one.)

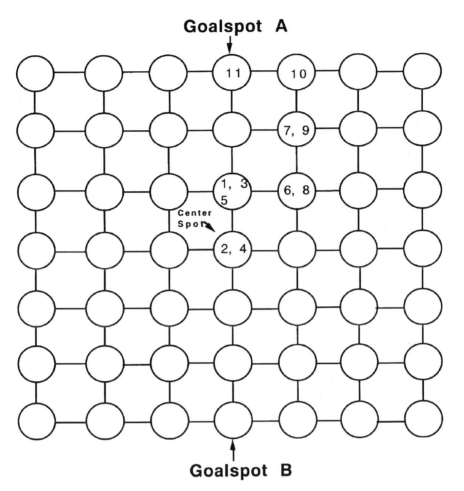

Goalspot A

Goalspot B

Figure 33.3 **Example of a completed game data sheet. Numbers indicate each sequential move in the game. In this game, Player A (move 1) begins by moving the penny directly toward his gamespot. Player B (move 2) retaliates by moving the penny back to the center. A few exchanges of competition take place (moves 3–5), and then Player B tries to avoid directly confronting Player A by moving to the side in move 6 (showing the "staircase" pattern associated with conflict avoidance). Player A responds by moving the gamepiece once more in the direction of his goalspot (move 7). Player B attempts to move it back (move 8), but Player A continues to be competitive (move 9). Player B finally submits to Player A in move 10 so that Player A wins the game in his next move (move 11). This is an example of a game showing mixed strategies; it includes evidence of conflict avoidance and submission.**

Prisoner B Strategy

		C Cooperation	D Defection
	C Cooperation	$R = 3$ Reward for mutual cooperation	$S = 0$ Sucker's payoff
	D Defection	$T = 5$ Temptation to defect	$P = 1$ Punishment for mutual defection

Prisoner A Strategy

Figure 33.4 **Payoffs to prisoner A in the prisoner's dilemma, for each possible strategy employed by prisoner B. Payoffs are represented with arbitrary numerical values for illustrative purposes (Axlerod & Hamilton 1981).**

Table 33.1 Individual researcher data summary sheet.

Observer's name: _____ What team were you on? _____ (Circle one.) Team 1 Team 2

Game Number	Game Outcome			Strategy Shown			
	Player A Won	Player B Won	Draw	Pure Competition	Pure Cooperation	Conflict Avoidance (CA)	Submission (S)	Mixed Strategy (CA + S)
1								
2								
3								
4								
5								
6								
7								
8								
9								
10								
TOTALS:								

Average number of wins per player: _____
[(total number of times Player A won) + (total number of times Player B won)]/2

Table 33.2 **Team summary data sheet.**

Team Name   1   or   2   (Make copies and add more names as needed.)
(Circle one.)

Observer's Name	Total Number of Wins, Player A	Total Number of Wins, Player B	Average Number of Wins per Player	Total Number of Draws	Pure Competition	Pure Cooperation	Other (CA, S, or CA+S)
TOTALS:							
Averages:							

Outcomes in 10 Games — *Total Games (of the 10 observed) with Strategy of:*

CHAPTER 34

Using Empirical Games to Teach Animal Behavior

PHILIP K. STODDARD
Department of Biological Sciences, Florida International University, Miami, FL 33199, USA

INTRODUCTION

Efficient and safe foraging is crucial to an animal's survival and reproduction. The most efficient forager is the one that acquires necessary food resources at the highest rate, while at the same time avoiding becoming another animal's dinner. But food and hazards are rarely distributed uniformly in the environment, so animals need mechanisms for avoiding areas of unacceptably low yield or high risk. A common problem for animals is deciding whether to stay in one resource patch or move to another. This problem has an optimal solution (the **marginal value theorem**) if the animal knows the quality of each patch and the cost of relocating between them (Charnov 1976). The problem is complicated by a variable risk of predation (Sih 1992). If the animal spends a lot of time fleeing or hiding, it may lose so much foraging time that it grows too slowly to reproduce or metamorphose before the season ends. More obviously, if it fails to flee or hide, it may be eaten.

The problem gets more complex—and also more interesting—when we realize that animals do not have perfect knowledge of the resources in their environment. Thus decisions about whether to stay in a given patch or switch to a different patch may be based on simple rules (that is, genetically hard-wired) or on knowledge acquired through experience (that is, learned). More complex animals may start with a fixed strategy and then modify that strategy as they learn about the environment.

If animals make their foraging decisions on the basis of what they know of the environment, two important questions arise: (1) How do they obtain that knowledge? (2) What does it cost them to obtain it? Nobody needs to tell a college student that acquiring knowledge is an expensive and time-consuming proposition. This game is designed to illustrate the trade-offs between the costs of acquiring knowledge and the value of using it to make a living. The game hinges on the problem of making wise foraging decisions without knowing in advance the parameters necessary to make them.

PROCEDURE

In this game you are a forager seeking enough food to reproduce. You have three foraging locations, represented by cups A, B, and C. As you spend time foraging at each location, you may encounter food, nothing, or worse, a predator from which you must flee. Food, nothing, and predators are represented by slips of paper on which is written "food," "noth," or "pred." Each cup has a unique ratio of food slips to noth slips to pred slips. One example of these ratios could be

	A	B	C
food	25	50	10
noth	60	45	80
pred	15	5	10
total	100	100	100

Before launching into the game, first consider the costs and benefits of every behavior involved in foraging. Make a list of these behaviors, with costs and benefits listed to the right of each. You and your fellow students must agree how much each of the following is worth. (I give some sample values for you to try, but don't feel bound to them.)

- Finding, processing, and digesting a food item (+2 points).
- Spending time and getting nothing (0 points, but note that you have lost time).
- Encountering a predator (−2 points).
- Switching between patches (−1 point), although if the cups are far apart, the cost can be dropped because the time of switching is sufficient penalty.

- Resources needed to reproduce (+5 points).
- Resources provided by your parents before you begin life as an independent forager (+2).

Rules of the Game

The rules are simple. Mark your results on the score sheet (Table 34.1).

1. Begin with 2 points (they are marked at the top of the "point total" column on your score sheet).
2. Choose a cup, close your eyes, and draw one slip:
   "food" = 1 point (or whatever valuation you chose)
   "noth" = 0 points,
   "pred" = −2 points
3. Refold the slip and return it to the same cup; then tally your score on your score sheet.
4. If you get above 5 points, you reproduce and get two picks the next time, one for you and one for your progeny. Pass 10 and you get another offspring, and so on.
5. If you fall below zero you die. Draw a line across your score sheet, and start again with a new life. Note that death is a common outcome in nature.
6. You may switch cups at a cost of 1 point.

Although you need only fill out the first four columns of your score sheet, the remaining columns with headings in italics are set up to enable you to keep additional statistics that will help you make intelligent choices if your chosen strategy allows such analysis. The sample score sheet (Table 34.2) gives an example of how those extra columns can be filled out.

HYPOTHESES AND PREDICTIONS

Some animals have genetically programmed rules for making foraging decisions, whereas others base these decisions on knowledge acquired through experimentation or observation. Take a few minutes to settle on your strategy before starting to play. Some examples of basic strategies follow.

Genetically Programmed Rule of Thumb

Set in advance an arbitrary criterion for leaving the patch. Examples: (1) how long you will stay at a patch (e.g., 10 picks), (2) under what circumstances you will leave, perhaps at a score threshold (e.g., 1), (3) some performance criterion (e.g., 3 nothings in a row), or (4) perhaps at some threshold ratios of food:noth:pred results (e.g., worse than 3 foods to every predator or worse than 1 food to every 3 nothings).

Acquired Resource Knowledge

Sample N different patches, calculate the payoff ratios of each, and then focus your efforts on the best of these N.

Local Enhancement

Keep an eye on the other foragers and get a sense of whether they seem successful or not. If they look happier than you, join them.

Gut Intuition

If you choose this strategy, you are obliged to explain after the fact what factors guided your intuition.

Combination

Two strategies could be combined. For example, start with a pro-grammed strategy such as 3 nothings in a row (one of the "Genetically Programmed Rule of Thumb" examples) and amend by observation and knowledge ("Acquired Resource Knowledge" and "Local Enhancement").

Remember to stick to your chosen strategy throughout your "life" as a forager so that you can evaluate its success at the end. When you have finished the game, enter your data in the class tally: (1) your name, (2) the final score for each life, (3) your preferred cup for each life, and (4) your broad strategy type for each life. Be prepared to explain your strategy in detail. While you are waiting for the group to finish, try to estimate the actual ratios of slips in each cup. When the class tally is complete, the instructor will reveal the actual count of foods, nothings, and predators in each patch.

QUESTIONS FOR DISCUSSION

Calculate the relative profitability of each patch from (1) your estimate, obtained while foraging, of the frequency of food and predators in each patch and (2) the actual frequencies of food and predators in each patch given by the instructor. How different is your estimated profitability from the true profitability for each patch? Is your estimated profitability accurate enough to guide you in choosing the best patch?

How would you alter your foraging behavior if the cost of switching were higher? if the frequency of predators were higher or lower? if the frequency of food were higher or lower? if the differences between patches were smaller or larger? if the game ran longer or shorter? if the number of patches were much higher (effectively infinite)?

Does crowding at a patch affect is profitability?

If you could take a simultaneous running tally of profitability at all three patches, you could focus your efforts at the patch with the highest quality estimate at any given time. But you are forced to sample patches sequentially instead of simultaneously. How much time should you spend at each patch, given that switching imparts a cost?

What is the best way to arrive at a really good foraging strategy?

Is there a single best (optimal) strategy for foraging in this game?

LITERATURE CITED

Charnov, E. 1976. Optimal foraging: the marginal value theorem. *Theoretical Population Biology,* **9**, 129–136.

Sih, A. 1992. Forager uncertainty and the balancing of antipredator and feeding needs. *American Naturalist,* **139**, 1052–1069.

ADDITIONAL SUGGESTED READING

Krebs, J. R. & Davies N. B. 1993. *An Introduction to Behavioural Ecology.* 3rd ed. Oxford, England: Blackwell Science.

Table 34.1 Score sheet.

Patch	Pick (f, n, p)	Points	Point Total 2	No. of picks at patch	Cum. no. of foods	Cum. no. of preds	Freq. of food	Freq. of pred	Comments

Table 34.2 **Sample score sheet, filled out with data from a game.**

Patch	Pick (f, n, p)	Points	Point Total 2	No. of picks at patch	Cum. no. of foods	Cum. no. of preds	Freq. of food	Freq. of pred	Comments
A	f	1	3	1	1	0	1.00	0.00	Strategy: 10 picks
	n	0	3	2	1	0	0.50	0.00	at each patch,
	n	0	3	3	1	0	0.33	0.00	then select the
	f	1	4	4	2	0	0.50	0.00	best
	p	−2	2	5	2	1	0.40	0.20	
	n	0	2	6	2	1	0.33	0.17	
	f	1	3	7	3	1	0.43	0.14	
	f	1	4	8	4	1	0.50	0.13	
	f	1	5	9	5	1	0.56	0.11	
	n	0	5	10	5	1	0.50	0.10	
B		−1	4						switch patches
	n	0	4	1	0	0	0.00	0.00	(cost −1)
	n	0	4	2	0	0	0.00	0.00	
	f	1	5	3	1	0	0.33	0.00	reproduce, 2 lives
	n	0	5	4	1	0	0.25	0.00	(2 picks)
	f	1	6	5	2	0	0.40	0.00	
	n	0	6	6	2	0	0.33	0.00	
	f	1	7	7	3	0	0.43	0.00	
	n	0	7	8	3	0	0.38	0.00	
	n	0	7	9	3	0	0.33	0.00	
	f	1	8	10	4	0	0.40	0.00	
C		−1	7						switch
	f	1	8	1	1	0	1.00	0.00	
	p	−2	6	2	1	1	0.50	0.50	
	p	−2	4	3	1	2	0.33	0.67	lost 1 life, back to
	n	0	4	4	1	2	0.25	0.50	1 pick
	f	1	5	5	2	2	0.40	0.40	reproduce, 2 lives
	f	1	6	6	3	2	0.50	0.33	(2 picks)
	f	1	7	7	4	2	0.57	0.29	
	n	0	7	8	4	2	0.50	0.25	
	p	−2	5	9	4	3	0.44	0.33	
	p	−2	3	10	4	4	0.40	0.40	lost 1 life (1 pick)
B		−1	2						Patch B looks
	n	0	2	11	4	0	0.36	0.00	best—switch
	n	0	2	12	4	0	0.33	0.00	resume number
	f	1	3	13	5	0	0.38	0.00	counts from B
	f	1	4	14	6	0	0.43	0.00	
	n	0	4	15	6	0	0.40	0.00	
	f	1	5	16	7	0	0.44	0.00	reproduce, 2 lives
	n	0	5	17	7	0	0.41	0.00	(2 picks)
	n	0	5	18	7	0	0.39	0.00	
									GAME OVER

SECTION VI

Evolution

CHAPTER 35

The Evolution of Behavior: A Phylogenetic Approach

KEN YASUKAWA

Department of Biology, Beloit College, 700 College Street, Beloit, WI 53511, USA

INTRODUCTION

Evolution is the underlying principle of biology, and animal behavior rests on a firm foundation of evolutionary principles (Whitman 1899; Chapin 1917; Wheeler 1919; Friedmann 1929). For some anatomical traits, evolutionary history can be followed in the fossil record, especially when a relatively complete fossil series of species is available. Unfortunately, behavior doesn't fossilize, so until time travel becomes a reality or the behavior of extinct animals can be studied via the methods described in *Jurassic Park*, evolutionary studies of behavior must rely on **comparative methods,** by which ancestral states and the pattern of evolutionary changes in behavior can be inferred (Martins 1996). Comparative studies of behavior gained considerable momentum in the 1940s and 1950s as influential ethologists such as Konrad Lorenz and Niko Tinbergen championed the view that behavior could be used to infer evolutionary relationships (**phylogeny**) among species and that behavior must evolve in a phylogenetic context. For example, in his study of the evolution of courtship displays of "dabbling" ducks (those that feed from the surface), Lorenz (1958) said, "every time a biologist seeks to know why an organism looks and acts as it does, he must resort to the comparative method." Tinbergen (1963) included evolution as one of the now famous four areas of animal behavior and he described the comparative method as follows:

> Through comparison [the naturalist] notices both similarities between species and differences between them. Either of these can be due to one of two sources. *Similarity* can be due to affinity, to common descent; or it can be due to convergent evolution.... The *differences* between species can be due to lack of affinity, or they can be found in closely related species. The student of survival value concentrates on the latter differences, because they must be due to recent adaptive radiation (Tinbergen 1964).

Tinbergen's characterization of comparative biology foreshadowed the basic concepts of modern phylogenetic comparative methods.

Curiously, from this high point in the evolutionary analyses of behavior, phylogeny was increasingly ignored by students of animal behavior, and by the 1970s, most biologists believed that morphological characteristics were much more valuable for comparative analyses than were behavioral characteristics (see Brooks & McLennan 1991). With the rise of the "new" comparative biology, however, behavior has once again been thrust into the evolutionary spotlight.

Comparative studies attempt to identify evolutionary patterns by examining traits of different organisms. The traits can be molecular, morphological, or behavioral characteristics. These days, comparative analyses are based primarily on a careful examination of **shared derived traits**, which are novel traits that are shared by two or more species because they were inherited from a common ancestor in which the trait originated. For example, suppose that calling behavior evolved in a species of frog and that the ancestors of that species did not call. Lack of calling was thus the **ancestral state**, and calling is a **derived trait**. Suppose further that two other species, X and Y, evolved from the calling species. If species X and Y also call, and if they inherited this behavior from their common ancestor, then calling is a shared derived trait in species X and Y and is said to be **homologous** (it shares the evolutionary transformation to calling from the noncalling ancestral state). Of course, not all shared traits are homologous. Shared character states may represent parallel or convergent evolution rather than common ancestry. If calling behavior evolved separately in two species, then calling is a **homoplasy**.

When done carefully, comparative studies can help us to infer **adaptation**, or whether the character of interest has evolved its current function in response to natural selection. For example, calling in frogs would be an adaptation for mate attraction if calling evolved from a noncalling condition in response to differential mate acquisition based on calling. On the other

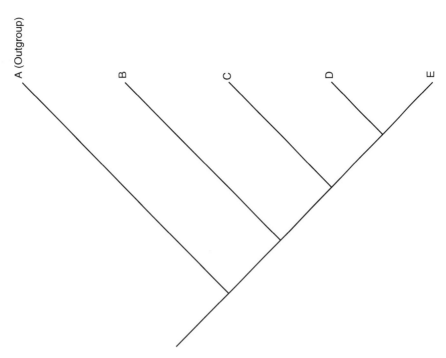

Figure 35.1 **Phylogenetic tree for five hypothetical species. Species A is the outgroup, and species B through E are the ingroup.**

hand, if calling evolved because it improved a male's ability to defeat other males in combat, then it would not be an adaptation for mate attraction even if it plays that role now (see Harvey & Pagel 1991).

In a sense, the "new" comparative methods all center on building and interpreting "trees," or phylogenies, which represent the order of shared ancestry among organisms. In other words, phylogenies are like family trees, except that instead of showing the ancestry of individuals in a family, they are "evolutionary genealogies" (Alcock 1998). The term *tree* describes the physical appearance of these proposed phylogenetic relationships, which are depicted as branching patterns of evolutionary events. An example of such a tree (or phylogeny) is shown in Figure 35.1. This figure shows five hypothetical species (A through E) and their phylogenetic relationships. What can we infer from this tree? Well, for one thing, it seems as though species D and E are each other's closest evolutionary "relatives" (comparative biologists call them **sister species**) because they both diverge from the same **node** (the point at which the **branches** leading to D and E emerge). In the language of comparative biology, D and E are called operational taxonomic units (OTUs), which occur at the **tips** of the branches, and together they form a **clade** (a group of taxa sharing a common ancestor and thus having a closer relationship to each other than to members of any other clade). In tree terminology, a clade comprises all tips on branches that come from the same node. Two other terms that are important in discussing trees are the **root** (base) of the tree and the **internodes** (branches between nodes). In our figure, D and E are a clade, as are C-D-E, and B-C-D-E, and A-B-C-D-E. A clade is also called a **monophyletic group**, which means a group of OTUs descended from a single ancestral taxon and including the ancestral species and all descendant species. Note that there are other ways to construct sets of OTUs from Figure 35.1. The set A-C-E (but not B-D) would be called a **polyphyletic group** (why?), whereas the set A-B-C-D (but not E) is a **paraphyletic group** (explain this name).

Our tree analogy is actually a really good one. Not only do we have branches coming from nodes, but also we know were the root of the tree is. But how can we decide where to root the tree. Conceptually, at least, the tree in Figure 35.1 could be rooted at any of the tips A through E. See whether you can figure out what the tree would look like with the root at E.

The way to root the tree is to use a so-called **outgroup**. An outgroup is composed of taxa that are closely related to, but outside of (that is, not *as* closely related to), your group of interest, which of course is called the **ingroup**. How does the outgroup root the tree? Any trait that exists within the ingroup *and* in species in the outgroup must be an ancestral trait. For example, both mammals and nonmammals have backbones, so vertebrae must have evolved before mammals and nonmammals diverged from each other. Note that we are assuming that vertebrae evolved before mammals, so that the presence of vertebrae in all mammals is a product of their shared common ancestry. We could also hypothesize that each clade of mammals and nonmammalian vertebrates evolved backbones independently, but when you think about it, that explanation seems pretty unlikely (although it's not strictly impossible).

Note that we have made an important transition of our own here. We have moved from talking about OTUs back to discussing traits, or **characters** (a set of possible states or conditions that are thought to evolve one to the other; each alternative condition is called a **character state**) such as backbones. Because character states change in evolution, we can use them to build phylogenetic trees, and our ability to root trees enables us to infer the order (evolutionary sequence) of character changes, or "state transitions." In other words, we have directionality (comparative biologists call it a polarized series), which tells us the direction in which evolution took place (that is, which state was ancestral and which was derived from it).

Although behavioral characters have been used to construct phylogenies, these days trees are constructed primarily from other kinds of characters. Anatomical features have been used as characters for many decades. More recently, molecular sequences (such as those of proteins or DNA or RNA) have been used. The existence of trees produced from such data has been very useful in animal behavior in recent years, because these data provide an independent way of producing a proposed phylogenetic relationship. Using them, we can infer the sequence of events by which behavior evolved. In this exercise we will examine several examples of this kind of analysis.

MATERIALS

You will need access to a Macintosh computer and an application called MacClade (Maddison & Maddison 1992). Alternatively, you can do this exercise with the handouts provided by your instructor, a pencil, and an eraser.

WHOLE-CLASS EXERCISE

Procedure

Form working groups of 3 or 4 students. If you are going to use the computer to do your analysis, your instructor will take some time now to teach you how to use MacClade. Once you are familiar with this program, you can do the following exercise. If you are not using the computer, you can proceed directly to the exercise.

We will start with a simple example. We will use Figure 35.1 again, with the following slight modification: we will assume that species A is the outgroup, which enables us to identify the ancestral state of a given character. Note that all five species are extant (currently living) and that we are trying to infer their evolutionary history from their current character states. The tree depicted in Figure 35.1 is itself a hypothesized evolutionary relationship among species A through E. As students of animal behavior, we are using this hypothesized phylogeny to infer an evolutionary sequence of events. For example, perhaps the hypothesized phylogeny is based on nucleotide sequences of a particular gene in these five species or on an analysis of their leg bones. Use of such a phylogeny enables us to avoid the appearance of circular reasoning. For our purposes, it is better *not* to use the information you want to study (in this case, behavior) to build your phylogenetic tree (although, as you may have guessed, comparative biologists don't agree about this either).

Let's suppose that a number of studies on the breeding behavior of these species have been published. Males of some of the species defend breeding territories, but other species do not. Some of the territorial species use vocalizations to defend their territories, but not all of them have vocal territory defense. And some of the species have vocal mate attraction, but others don't. How can we make sense of the diversity of behavioral states among these five species? What is the evolutionary sequence of events that produced the current distribution of behavioral character states?

The first thing you need to do is to find out what each species's character states are. From the published studies you assemble the following descriptions of each species.

Species A (the Outgroup)

Males gather in specific locations to breed, but they do not defend territories. Males use visual displays and physical combat to establish dominance hierarchies. High-ranking males acquire females, but low-ranking males do not.

Species B

Males defend specific areas against intrusion by conspecific males. Territory defense involves visual displays and physical combat, but not vocalization. Females are attracted by the visual displays of males, but males do not vocalize during mate attraction.

Species C

Males defend territories with a combination of visual and vocal displays. Physical combat between males is rare. Males do not vocalize during mate attraction.

Species D

Males defend territories with vocal displays but not visual displays. Males also have a vocalization to which females are attracted.

Species E

Males gather in small, traditional breeding locations and are not aggressive with one another. In these breeding gatherings, males call and females seem to be attracted by this vocal behavior.

You will use the foregoing information and MacClade (or your pencil and paper) to infer the evolutionary history of breeding behavior in this hypothetical group of animals.

Hypotheses and Predictions

In our whole-class exercise, our general research question is "What was the evolutionary sequence of events (that is, the order of character states) for male breeding behavior?" To answer this general question, you will have to infer the ancestral state of each behavioral character and when (on which internodes) each derived state evolved.

Table 35.1 **States of three behavioral characters for four species in the ingroup and one outgroup species.**

	Species				
Character	*A*	*B*	*C*	*D*	*E*
Male territoriality					
Vocal territory defense					
Vocal mate attraction					

Data Recording and Analyses

The first and at this point the most important step in your analysis is to choose the behavioral characters of interest and their possible states. The simplest characters are those with only two possible states (so-called **binary characters**). Discuss within your group the characters and their states. Once your group has agreed on characters and states, compare them with those of another group. It is not likely that the two groups will agree. See whether you can find a common system of characters and states.

For the purposes of this whole-class exercise, the author has chosen three binary characters, but your class could choose to use a different set. Use the information on each of the five species to fill in the appropriate spaces in Table 35.1, which lists the author's three binary characters. In Table 35.1 the possible states for each character are "no" and "yes" (or 0 and 1). Use the information from Table 35.1 and the hypothesized phylogeny of Figure 35.1 to "map" the behavioral characteristics onto the branches and internodes of the phylogeny. Your instructor will suggest ways to do this mapping using MacClade or pencil and paper.

Once your group is finished with your analysis, the class will discuss this exercise. It is possible (likely?) that your group's mapping of behavioral character transitions will differ from that of another group. Be prepared to defend your proposed evolutionary sequence and to question the sequences of other groups. By what criterion can the class decide which proposed sequence of behavioral state transitions is "best"? It may interest you to know that disagreements in phylogenetics (this kind of analysis) can and do become extremely heated, both in print and in person.

SMALL-GROUP EXERCISE

Hypotheses and Predictions

Now that you have had some practice, your group will attempt a second analysis with a somewhat more complex example. Our general research question, however, remains the same: What was the evolutionary sequence of events for the behavior of interest? You will once again have to map behavioral character state transitions onto a phylogenetic tree to infer the ancestral states and when each derived state evolved.

Procedure

Your instructor will either provide several data sets for analysis or tell you how to find appropriate phylogenies and behavioral descriptions. Choose one set and have at it! Follow the steps from the whole-class exercise to map the behavioral states onto the proposed phylogeny and then to infer the evolutionary history of behavior in the ingroup.

Data Recording and Analyses

Devise a table to record the states of the behavioral characters for each species. Use MacClade or pencil-and-paper methods to generate one or more proposed evolutionary sequences for the behavioral characters of interest.

QUESTIONS FOR DISCUSSION

According to some comparative biologists, you should *never use the information you want to study to build your phylogenetic tree.* Why is it important to follow this rule? What happens when you do use the information you want to study to build your phylogenetic tree?

Parsimony (simple explanations are better than complicated ones) is the most commonly used criterion in building phylogenetic trees. In practice, the parsimony criterion means trying to minimize the number of character state changes needed to explain the distribution of character states on the tree. Why is this criterion used? Do you think evolution necessarily operates according to this principle? Explain.

Why is an outgroup important in tree construction? What are the consequences to tree building when you don't use an outgroup?

Originally, ethologists saw behavior as another set of characters that could be used to construct phylogenies. These days, however, phylogenies are usually constructed from molecular or anatomical information. Could you use behavior to construct a phylogeny? Why or why not? Which source of information is better? Why?

Sometimes "the comparative method" is taken to mean any comparison of two or more species. But if two species share a behavioral characteristic, we don't necessarily know why. What are some possible explanations of why two species have the same behavior? What implications do these different explanations have for evolutionary analyses of behavior?

ACKNOWLEDGMENTS

I thank Sam Donovan for sharing his phylogeny exercise with me and for providing feedback on an earlier draft of this exercise. I also thank the BioQUEST Curriculum Consortium for their help and cooperation. Finally, I thank the students at Beloit College who "field-tested" this exercise.

LITERATURE CITED

Alcock, J. 1998. *Animal Behavior: an Evolutionary Approach.* 6th ed. Sunderland, MA: Sinauer.

Brooks, D. R. & McLennan, D. A. 1991. *Phylogeny, Ecology, and Behavior: a Research Program in Comparative Biology.* Chicago: University of Chicago Press.

Chapin. J. P. 1917. The classification of the weaver birds. *Bulletin of the American Museum of Natural History,* **37**, 243–280.

Friedmann, H. 1929. *The Cowbirds.* Springfield, IL: Charles C Thomas.

Harvey, P. H. & Pagel, M. D. 1991. *The Comparative Method in Evolutionary Biology.* Oxford: Oxford University Press.

Lorenz, K. 1958. The evolution of behavior. *Scientific American,* **199**, 67–78.

Maddison, W. P. & Maddison, D. R. 1992. *MacClade: Analysis of Phylogeny and Character Evolution.* Sunderland, MA: Sinauer.

Martins, E. P. (Ed.) 1996. *Phylogenies and the Comparative Method in Animal Behavior.* Oxford: Oxford University Press.

Tinbergen, N. 1963. *Social Behaviour in Animals, with Special Reference to Vertebrates.* London: Methuen.

Tinbergen, N. 1964. On aims and methods of ethology. *Zeitschrift für Tierpsychologie,* **20**, 410–433.

Wheeler, W. M. 1919. The parasitic *Aculeata,* a study in evolution. *Proceedings of the American Philosophical Society,* **58**, 1–40.

Whitman, C. O. 1899. Animal behavior. In: *Biological Lectures, Woods Hole* (ed. C. O. Whitman), pp. 285–338. Boston: Ginn.

APPENDIX A

Guidelines for the Treatment of Animals in Behavioural Research and Teaching[1]

THE ANIMAL BEHAVIOR SOCIETY AND
THE ASSOCIATION FOR THE STUDY OF ANIMAL BEHAVIOUR

Behavioural studies are of great importance in increasing our understanding and appreciation of animals. In addition to providing knowledge about the diversity and complexity of behaviour in nature, such studies also provide information crucial to improvements in the welfare of animals maintained in laboratories, zoos and agricultural settings. The use of animals in behavioural research and teaching does, however, raise important ethical issues. While many behavioural studies are noninvasive and involve only observations of animals in their natural habitat, some research questions cannot be answered adequately without manipulation of animals. Studies of captive animals necessarily involve keeping animals in confinement, and manipulative procedures and surgery may be necessary to achieve the aims of the research. Studies of free-living animals in their natural habitats can cause disruption, particularly if feeding, capture, marking or experimental procedures are involved.

While the furthering of scientific knowledge is a proper aim and may itself advance an awareness of human responsibility towards animal life, the investigator must always weigh the potential gain in knowledge against any

[1] Reprinted with permission from Academic Press Ltd., 32 Jamestown Road, London NW1 7BY, UK.

adverse consequences for the animals and populations under study. This is equally true for the evaluation of animal use in animal behaviour teaching activities. In fact, animal behaviour courses provide an excellent opportunity to introduce students to the ethical obligations a researcher accepts when animals are studied.

In order to help the members make what are sometimes difficult ethical judgments about the procedures involved in the study of animals, the Association for the Study of Animal Behaviour and the Animal Behavior Society have formed Ethical and Animal Care Committees, respectively. These committees jointly produced the following guidelines for the use of all those who are engaged in behavioural research and teaching activities involving vertebrate and invertebrate animals. These guidelines are general in scope, since the diversity of species and study techniques used in behavioural research precludes the inclusion of specific details about appropriate animal care and treatment. The guidelines will be used by the Editors of *Animal Behaviour* in assessing the acceptability of submitted manuscripts. Submitted manuscripts may be rejected by an Editor, after consultation with the Ethical or Animal Care Committees, if the content violates either the letter or the spirit of the guidelines. These guidelines supplement the legal requirements in the country and/or state or province in which the work is carried out. They should not be considered an imposition upon the scientific freedom of individual researchers, but rather as helping to provide an ethical framework which each investigator may use in making decisions related to animal welfare.

1. LEGISLATION

Investigators are accountable for the care and well-being of animals used in their research and teaching activities, and must therefore abide by the spirit as well as the letter of relevant legislation. For those who reside in Great Britain, a summary of the laws designed to ensure the welfare of animals is given by Crofts (1989); detailed guidance on the Operation of the Animals (Scientific Procedure) Act, 1986 is provided by the Home Office (1990). In the U.S.A., federal, state and local legislation and guidelines may apply. In particular, the care and use of many vertebrate laboratory animals are regulated under the Animal Welfare Act and its amendments and regulations (Code of Federal Regulations, Title 9) and/or the policies of the Public Health Service (PHS 1986; NRC 1996). Guidelines for farm animals used in research and teaching may also be applicable (Guide Development Committee 1988). In Canada, guidance can be obtained from the Guide to the Care and Use of Experimental Animals (Canadian Council on Animal Care 1992).

In Britain, lists of threatened species and laws aiming to protect them can be obtained from the International Union for the Conservation of Nature, Species Conservation Monitoring Unit (219C Huntingdon Road,

Cambridge CB3 0DL, U.K.). In the U.S.A., information pertaining to the Endangered Species Act of 1973 may be found in the Code of Federal Regulations (Title 50, 1973). Lists of endangered species can be obtained from the Office for Endangered Species, U.S. Department of Interior, Fish and Wildlife Service (Room 430, 4401 N. Fairfax Drive, Arlington, VA 22203), or from the Committee on the Status of Endangered Wildlife in Canada, Canadian Wildlife Service (Environment Canada, Ottawa, Ontario K1A 0E7).

Investigators working in other countries must familiarize themselves with legislation both on animal welfare and on threatened and endangered species and conform with the spirit and letter of the laws. When submitting manuscripts to *Animal Behaviour*, all authors must confirm in their cover letter that they have adhered to the legal requirements of the country in which the study was conducted.

2. CHOICE OF SPECIES AND NONANIMAL ALTERNATIVES

Investigators should choose a species for study that is well suited for investigation of the questions posed. Choosing an appropriate subject usually requires knowledge of a species's natural history and phylogenetic level. Knowledge of an individual animal's previous experience, such as whether or not it has spent a lifetime in captivity, is also important. When research or teaching involves procedures or housing conditions that may cause pain, discomfort or stress to the animal, and when alternative species can be used, the researcher should employ the species which, in the opinion of the researcher and other qualified colleagues, is least likely to suffer (OTA 1986). Live animal subjects are generally essential in behavioural research, but nonanimal alternatives such as video records from previous work or computer simulations can sometimes be used (Smyth 1978). Material of this kind also exists for teaching purposes and can be used instead of live animals to expand the range of behavioural subjects available to students.

3. NUMBER OF INDIVIDUALS

The researcher should use the smallest number of animals necessary and sufficient to accomplish the research goals, especially in studies which involve manipulations that are potentially detrimental to the animal or the population. The number of animals used in an experiment can often be dramatically reduced by pilot studies, good experimental design and the use of statistical tests that enable several factors to be examined simultaneously. Hunt (1980), Still (1982) and McConway (1992) discuss ways of reducing the number of animals used in experiments through alternative designs. Useful reference works are Cox (1958) and Cochran & Cox (1966).

4. PROCEDURES

Investigators are encouraged to discuss with colleagues both the scientific value of their research proposals as well as possible ethical considerations. There are several models for evaluating animal research which can be of use when making ethical decisions (Bateson 1986; Orlans 1987; Shapiro & Field 1988; Donnelley & Nolan 1990; Porter 1992). If procedures used in research or teaching involve exposure to painful, stressful or noxious stimuli, the investigator must consider whether the knowledge that may be gained is justified. Bateson (1991) discusses the assessment of pain and suffering. Additional information can be obtained from the U.S. National Academy of Sciences Publication, "Recognition and Alleviation of Pain and Distress in Laboratory Animals" (NRC 1992), and from the American Veterinary Medical Association panel report on animal pain and distress (AVMA 1987). Researchers are urged to consider the use of alternative procedures before employing techniques that are likely to cause physical or psychological discomfort to the animal. Pain or suffering should be minimized both in duration and magnitude to the greatest extent possible under the requirements of the experimental design. Attention should be given to proper pre- and postoperative care in order to minimize preparatory stress and residual effects. Unless specifically contraindicated by the experimental design, procedures that are likely to cause pain or discomfort should be performed only on animals that have been adequately anesthetized. Analgesics should be used after such procedures to minimize pain and distress whenever possible (Flecknell 1985; Benson et al. 1990).

The following more specific points may be of use.

(a) Fieldwork

Investigators studying free-living animals must take precautions to minimize interference with individuals as well as the populations and ecosystems of which they are a part. Capture, marking, radiotagging, collection of physiological data such as blood or tissue samples or field experiments may not only have immediate effects on the animal, but may also have consequences such as a reduced probability of survival and reproduction. Investigators should consider the effects of such interference, and use less disruptive techniques such as individual recognition by the use of natural features rather than marking (Scott 1978) where possible. Cuthill (1991) discusses the ethical issues associated with field experiments and recommends pilot investigations to assess potential environmental disruption and follow-up studies to detect and minimize persistent effects. Investigators should weigh the potential gain in knowledge from field studies against the adverse consequences of disruption for the animals used as subjects and also for other animals and plants in the ecosystem. When an experimental protocol requires that animals be removed from the population either temporarily or on a long-term basis, investigators should ensure that suffering or discomfort are minimized not only for the removed animals but for others dependent on them (e.g., dependent offspring). Removed individuals and their dependents must be housed and cared for appropriately. Sources of further information on field techniques are the books edited by Stonehouse (1978) and Amlaner & Macdonald (1980).

(b) Aggression, Predation and Intraspecific Killing

The fact that the agent causing harm may be another nonhuman animal does not free the experimenter from the normal obligations to experimental animals. Huntingford (1984) and Elwood (1991) discuss the ethical issues involved and suggest ways to minimize suffering. Wherever possible, field studies of natural encounters should be used in preference to staged encounters. Where staged encounters are necessary, the use of models should be considered, the number of subjects should be kept to the minimum needed to accomplish the experimental goals, and the experiments made as short as possible. Suffering can also be reduced by continuous observation with intervention to stop aggression at predefined levels, and by providing protective barriers and escape routes for the subjects.

(c) Aversive Stimulation and Deprivation as Motivational Procedures

Aversive stimulation or deprivation can cause pain or distress to animals. To minimize suffering, the investigator should ascertain that there is no alternative way of motivating the animal, and that the levels of deprivation or aversive stimulation used are no greater than necessary to achieve the goals of the experiment. Alternatives to deprivation include the use of highly preferred foods and other rewards which may motivate even satiated animals. Use of minimal levels requires a knowledge of the technical literature in the relevant area: quantitative studies of aversive stimulation are reviewed by Church (1971) and Rushen (1986) and the behaviour of satiated animals is considered by Morgan (1974). Further comments on reducing distress due to motivational procedures are to be found in Lea (1979) and Moran (1975).

(d) Social Deprivation, Isolation and Crowding

Experimental designs that require keeping animals in over-crowded conditions, or that involve social deprivation or isolation, may be extremely stressful to the animals involved. Because the degree of stress experienced by the animal can vary with species, age, sex, reproductive condition, developmental history and social status, the natural social behaviour of the animals concerned and their previous social experience must be considered in order to minimize such stress.

(e) Deleterious Conditions

Studies aimed at inducing deleterious conditions in animals are sometimes performed in order to gain scientific knowledge of value to human or animal problems. Such conditions include inducing diseases, increasing parasite loads, and exposing animals to pesticides or homeostatic stressors. Where feasible, studies inducing a deleterious condition in animals should address the possible treatment or alleviation of the condition induced. Animals exposed to deleterious conditions that might result in suffering or death should be monitored frequently and, whenever possible, considering the aims of the research, treated or humanely killed as soon as they show signs of distress. If the goals of the research allow it, the investigator should also consider experimental designs in which the deleterious condition is removed (e.g., removing rather than adding parasites as the experimental treatment) or in which naturally occurring instances of deleterious conditions are observed.

5. ENDANGERED SPECIES

All research on endangered or locally rare species must comply with relevant legislation and be coordinated with official agencies responsible for the conservation effort for the particular species under study. Legislation and sources of help in identifying endangered species have been outlined in Section 1. Members of threatened species should not be placed at risk except as part of a serious attempt at conservation. Observation alone can result in serious disturbance, including higher predation rates on nests or young, or their abandonment, and should only be undertaken after careful consideration of techniques and of alternative species. Investigators should also consider further adverse consequences of their work, such as opening up remote areas for subsequent access or teaching techniques of anesthetization and capture which might be misused (e.g., by poachers).

6. PROCUREMENT OF ANIMALS

When it is necessary to procure animals either by purchase or donation from outside sources, only reputable suppliers should be used. For workers in the U.K., advice about purchasing animals may be obtained from the Laboratory Animal Breeder's Association, Charles River (U.K.) Ltd, Manson Research Centre, Manson Road, Margate, Kent CT9 4LP. In the U.S.A., information on licensed animal dealers can be obtained from the local office of the U.S. Department of Agriculture (USDA). Other sources of information on laboratory animal suppliers in North America are the American Association for Laboratory Animal Science (70 Timber Creek Drive, Suite 5, Cordova, TN 38018) and the Canadian Association for Laboratory

Animal Science (M524 Biological Sciences Building, University of Alberta, Edmonton, Alberta T6G 3E9). If animals are procured by capture in the wild, this must be done in as painless and humane a manner as possible and must comply with any relevant legislation. Individuals of endangered species or populations should not be taken from the wild unless they are part of an active conservation program. So far as is possible, the investigator should ensure that those responsible for handling purchased, donated or wild-caught animals en route to the research facilities provide adequate food, water, ventilation and space, and do not impose undue stress.

7. HOUSING AND ANIMAL CARE

The researcher's responsibilities extend also to the conditions under which the animals are kept when not being studied. Caging conditions and husbandry practices must meet, at the very least, minimal recommended requirements of the country in which the research is carried out. Guidance can be obtained from the Universities Federation for Animal Welfare (U.F.A.W.) handbook (Poole 1987), the National Research Council guide (NRC 1996), the U.S.D.A. Animal Welfare Act Regulations (Code of Federal Regulations), the Guide for the Care and Use of Agricultural Animals in Agricultural Research and Teaching (Guide Development Committee 1988), and the Canadian Council on Animal Care's Guide to the Care and Use of Experimental Animals (1992).

Although these publications provide general guidance, the housing and care regimes established for the commonly used laboratory animals are not necessarily suitable for wild animals or for individuals of wild species born in captivity. Special attention may be required to enhance the comfort and safety of these animals. Normal maintenance of captive animals should incorporate, as much as possible, aspects of the natural living conditions deemed important to welfare and survival. Consideration should be given to providing features such as natural materials, refuges, perches and dust and water baths. Companions should be provided for social animals where possible, providing that this does not lead to suffering or injury. Frequency of cage cleaning should represent a compromise between the level of cleanliness necessary to prevent diseases and the amount of stress imposed by frequent handling and exposure to unfamiliar surroundings, odors, and bedding. Researchers in the United States should also ensure that the requirements outlined under the 1985 Amendment to the Animal Welfare Act to provide exercise for laboratory-housed dogs and to ensure the psychological well-being of captive nonhuman primates are met.

The nature of human–animal interactions during routine care and experimentation should be considered by investigators. Depending upon species, rearing history and the nature of the interaction, animals may perceive humans as conspecifics, predators or symbionts (Estep & Hetts 1992).

Special training of animal care personnel can help in implementing procedures that foster habituation of animals to caretakers and researchers and minimize stress. Stress can also be reduced by training animals to cooperate with handlers and experimenters during routine husbandry and experimental procedures (Biological Council 1992).

8. FINAL DISPOSITION OF ANIMALS

When research projects or teaching exercises using captive animals are completed, it may sometimes be appropriate to distribute animals to colleagues for further study or breeding, if permitted by local legislation. However, if animals are distributed care must be taken to ensure that the same animals are not used repeatedly in stressful or painful experiments, and that they continue to receive a high standard of care. Animals should never be subjected to major surgery more than once unless it is an unavoidable element of a single experiment. Except as prohibited by national, federal, state, provincial, or local laws, researchers may release field-trapped animals if this is practical and feasible, especially if it is critical to conservation efforts. However, the researcher should assess whether releases into the wild might be injurious or detrimental both to the released animal and to existing populations in the area. Animals should be released at the site where they were trapped (unless conservation efforts dictate otherwise), and only when their ability to survive in nature has not been impaired and when they do not constitute a health or ecological hazard to existing populations. If animals must be killed subsequent to a study this must be done as humanely and painlessly as possible; death of the animals should be confirmed before their bodies are discarded. A veterinarian should be consulted for advice on methods of euthanasia that are appropriate for the particular species being used. Additional information on euthanasia methods can be found in the report of the AVMA panel on euthanasia (AVMA 1993).

9. OBTAINING FURTHER INFORMATION

There are a number of organizations that provide publications and detailed information about the care and use of animals. These include The Canadian Council on Animal Care (1105-151 Slater Street, Ottawa, Ontario, K1P 5H3 Canada), the Scientists Center for Animal Welfare (4805 St. Elmo Avenue, Bethesda, MD 20814, U.S.A.), and the Universities Federation for Animal Welfare (The Old School, Brewhouse Lane, Wheathampstead, Hertfordshire AL4 8AN, U.K.). The Animal Welfare Information Center at the National Agricultural Library (Room 205, Beltsville, MD 20705, U.S.A.) publishes a series of bibliographies on special topics, and can also provide individualized database searches for investigators on potential alternatives, including techniques for replacement with nonanimal models or alternative species, methods for reducing the total number of animals necessary to address the research question, and experimental refinements which can reduce pain and stress.

For those with access to it, the Internet provides a wealth of information on animal care and welfare issues. Many of these are government web pages, particularly those of NIH, USDA and the U.K. Home Office. Good starting places are: http://www.nih.gov/and http://www.aphis.usda.gov/ while U.K. legal requirements and Home Office Codes of Practice for the housing and care of animals and for humane killing can be obtained from http://www.open.gov.ukfhome_off/abcu.htm. Additional information in the APHIS site can be found at/reac/reachome.html. Further on in the NIH site is/grants/oppr/library, from which one can gain access to the 1996 Institute of Laboratory Animal Resources (ILAR) Guide for Care and Use of Laboratory Animals (published by the National Academy Press), as well as get information on the IACUC Guidebook published by ARENA (Applied Research Ethics National Association). The Animal Welfare Information Center (AWIC) at the National Agriculatural Library (NAL) can be reached via the USDA home page, and they also have available a Compendium of Animal Resources (CARE) CD ROM. For more information, contact Michael Kreger at the NAL: email: mkreger@nal. usda.gov or write to AWIC, National Agricultural Library, 5th floor, 10301 Baltimore Avenue, Beltsville, MD 20705. AAALAC International (Association for Assessment and Accreditation of Laboratory Animal Care) also has a home page: http://www.aaalac.org and a toll-free phone number: 1-800-926-0066.

The Scientists' Center for Animal Welfare (SCAW) is at 7833 Walker Drive, Suite 410, Greenbelt, MD 20770. Their email is: scaw@erols.com. Their web site is: www.scaw.com.

Additional sources of information are NetVet at http://netvet.wustl.edu/, the National Academy of Sciences at http://www.nas.edu/homepage/ pus/pubs.html, the National Academy Press at http://www.nap.edu/ readingroom/and the Universities Federation for Animal Welfare (UFAW) at http://www.users.dircon.co.uk/~ufaw3/index.htm/.

REFERENCES

Amlaner, C. L. J. & Macdonald, D. G. 1980. *A Handbook on Biotelemetry and Radio Tracking.* Oxford: Pergamon.

AVMA (American Veterinary Medical Association). 1987. Colloquium on recognition and alleviation of animal pain and distress. *Journal of the American Veterinary Medical Association*, **191**, 1184–1296.

AVMA (American Veterinary Medical Association). 1993. Report of the Panel on Euthanasia. *Journal of the American Veterinary Medical Association*, **202**, 222–249.

Bateson, P. 1986. When to experiment on animals. *New Scientist*, **1496**, 30–32.

Bateson, P. 1991. Assessment of pain in animals. *Animal Behaviour*, **42**, 827–839.

Benson, C. J., Thurman, J. C. & Davis, L. E. 1990. Laboratory animal analgesia. In: *The Experimental Animal in Biomedical Research*, Vol. 1, *A Survey of Scientific and Ethical Issues for Investigators* (ed. B. E. Rollin & M. L. Kessel), pp. 319–329. Boca Raton, FL: CRC Press.

Biological Council. 1992. *Guidelines on the Handling and Training of Laboratory Animals*. Potters Bar, Herts: U.F.A.W. (Universities Federation for Animal Welfare).

Canadian Council on Animal Care. 1992. *Guide to the Care and Use of Experimental Animals*, Vols I and 2. Ottawa, Ontario: Canadian Council on Animal Care.

Church, R. M. 1971. Aversive behaviour. In: *Woodworth and Schleosberg's Experimental Psychology*. 3rd ed. (ed. J. W. Kling & L. A. Riggs), pp. 703–741. London: Methuen.

Cochran, W. C. & Cox, G. M. 1966. *Experimental Designs*. 2nd ed. New York: Wiley.

Code of Federal Regulations. *Title 9 (Animal and Animal Products), Subchapter A (Animal Welfare), Parts 1–3*. Available from: Regulatory Enforcement and Animal Care, APHIS, U.S.D.A., Federal Building, 6505 Belcrest Road, Hyattsville, MD 20782.

Code of Federal Regulations. *Title SO (Wildlife and Fishenes), Chapter 1* (Bureau of Sport Fisheries and Wildlife Service, Fish and Wildlife Service, Department of Interior). Washington, DC: U.S. Government Printing Office.

Cox, D. R. 1958. *Planning of Experiments*. New York: Wiley.

Crofts, W. 1989. *A Summary of the Statute Law Relating to the Welfare of Animals in England and Wales*. Potters Bar, U.F.A.W.

Cuthill, I. 1991. Field experiments in animal behaviour: methods and ethics. *Animal Behaviour*, **42**, 1007–1014.

Donnelley, S. & Nolan, K. (Eds.). 1990. *Animals, Science and Ethics*. New York: The Hastings Center.

Elwood, R. W. 1991. Ethical implications of studies on infanticide and maternal aggression in rodents. *Animal Behaviour*, **42**, 841–849.

Estep, D. Q. & Hetts, S. 1992. Interactions, relationships, and bonds: the conceptual basis for scientist–animal relations. In: *The Inevitable Bond: Examining Scientist–Animal Interactions* (ed. H. Davis & D. Balfour), pp. 6–26. Cambridge: Cambridge University Press.

Flecknell, P. A. 1985. The management of post-operative pain and distress in experimental animals. *Animal Technology*, **36**, 97–103.

Guide Development Committee. 1988. *Guide for the Care and Use of Agricultural Animals in Agricultural Research and Teaching*. Washington, DC: Consortium for Developing a Guide for the Care and Use of Agricultural Animals in Agricultural Research and Teaching. Available from: Association Headquarters, 309 West Clark Street, Champaign, IL 61820.

Home Office. 1990. *Guidance on the Operation of the Animals (Scientific Procedures) Act, 1986*. London: H.M.S.O.

Hunt, P. 1980. Experimental choice. In: *The Reduction and Prevention of Suffering in Animal Experiments*, pp. 63–75. Horsham, Sussex: Royal Society for the Prevention of Cruelty to Animals.

Huntingford, F. A. 1984. Some ethical issues raised by studies of predation and aggression. *Animal Behaviour*, **32**, 210–215.

Lea, S. E. F. 1979. Alternatives to the use of painful stimuli in physiological psychology and the study of behaviour. *Alternatives to Laboratory Animals Abstracts*, **7**, 20–21.

McConway, K. 1992. The number of subjects in animal behavior experiments: is Still right? In: *Ethics in Research on Animal Behaviour* (ed. M. Stamp Dawkins & L.M. Gosling), pp. 35–38. London: Academic Press.

Moran, C. 1975. Severe food deprivation: some thoughts regarding its exclusive use. *Psychological Bulletin*, **82**, 543–557.

Morgan, M. J. 1974. Resistance to satiation. *Animal Behaviour*, **22**, 449–466.

NRC (National Research Council). 1992. *Recognition and Alleviation of Pain and Distress in Laboratory Animals*. A report of the Committee on Pain and Distress in Laboratory Animals. Institute of Laboratory Animal Resources, Commission on Life Science, National Research Council, Washington, DC: National Academy Press.

NRC (National Research Council). 1996. *Guide for the Care and Use of Laboratory Animals*. A Report of the Institute of Laboratory Animal Resource Committee on the Care and Use of Laboratory Animals. NIH publication no. 85-23. Washington, DC: U.S. Department of Health and Human Services.

Orlans, F. B. 1987. Research protocol review for animal welfare. *Investigations in Radiology*, **22**, 253–258.

OTA (Office of Technology Assessment), U.S. Congress. 1986. Alternatives to Animal Use in Research, Testing and Education. Washington, DC: U.S. Government Printing Office, OTA-BA-273.

PHS (Public Health Service). 1986. *Public Health Service Policy of Humane Care and Use of Laboratory Animals*. Washington, DC: U.S. Department of Health and Human Services. Available from: Office for Protection from Research Risks, Building 31, Room 4809, NIH, Bethesda, MD 20892.

Poole, T. B. (Ed.). 1987. The *UFAW Handbook on the Care and Management of Laboratory Animals*. 6th ed. Harlow, England: Longman Scientific and Technical.

Porter, D. G. 1992. Ethical scores for animal experiments. *Nature, Lond.*, **356**, 101–102.

Rushen, J. 1986. The validity of behavioral measures of aversion: a review. *Applied Animal Behaviour Science*, **16**, 309–323.

Scott, D. K. 1978. Identification of individual Bewick's swans by bill patterns. In: *Animal Marking: Recognition Marking of Animals in Research* (ed. B. Stonehouse), pp. 160–168. London: Macmillan.

Shapiro, K. J. & Field, P. B. 1988. A new invasiveness scale: its role in reducing animal distress. *Humane and Innovative Alternatives to Animal Experiments*, **2**, 43–46.

Smyth, O. H. 1978. *Alternatives to Animal Experiments*. London: Scolar Press, Research Defence Society.

Still, A. W. 1982. On the number of subjects used in animal behaviour experiments. *Animal Behaviour*, **30**, 873–880.

Stonehouse, B. (Ed.). 1978. *Animal Marking: Recognition Marking of Animals in Research*. London: Macmillan.

APPENDIX B

Ethical Use of Human Subjects[1]

<small>AMERICAN PSYCHOLOGICAL ASSOCIATION</small>

6. TEACHING, TRAINING SUPERVISION, RESEARCH, AND PUBLISHING

6.06 Planning Research

(a) Psychologists design, conduct, and report research in accordance with recognized standards of scientific competence and ethical research.

(b) Psychologists plan their research so as to minimize the possibility that results will be misleading.

(c) In planning research, psychologists consider its ethical acceptability under the Ethics Code. If an ethical issue is unclear, psychologists seek to resolve the issue through consultation with institutional review boards, animal care and use committees, peer consultations, or other proper mechanisms.

(d) Psychologists take reasonable steps to implement appropriate protections for the rights and welfare of human participants, other persons affected by the research, and the welfare of animal subjects.

[1] Copyright © 1992 by the American Psychological Association. Reprinted with permission.

6.07 Responsibility

(a) Psychologists conduct research competently and with due concern for the dignity and welfare of the participants.

(b) Psychologists are responsible for the ethical conduct of research conducted by them or by others under their supervision or control.

(c) Researchers and assistants are permitted to perform only those tasks for which they are appropriately trained and prepared.

(d) As part of the process of development and implementation of research projects, psychologists consult those with expertise concerning any special population under investigation or most likely to be affected.

6.08 Compliance with Law and Standards

Psychologists plan and conduct research in a manner consistent with federal and state laws and regulations, as well as professional standards governing the conduct of research, and particularly those standards governing research with human participants and animal subjects.

6.09 Institutional Approval

Psychologists obtain from host institutions or organizations appropriate approval prior to conducting research, and they provide accurate information about their research proposals. They conduct the research in accordance with the approved research protocol.

6.10 Research Responsibilities

Prior to conducting research (except research involving only anonymous surveys, naturalistic observations, or similar research), psychologists enter into an agreement with participants that clarifies the nature of the research and the responsibilities of each party.

6.11 Informed Consent to Research

(a) Psychologists use language that is reasonably understandable to research participants in obtaining their appropriate informed consent (except as provided in Standard 6.12, Dispensing with Informed Consent). Such informed consent is appropriately documented.

(b) Using language that is reasonably understandable to participants, psychologists inform participants of the nature of the research; they inform participants that they are free to participate or to decline to participate or to withdraw from the research; they explain the foreseeable consequences of declining or withdrawing; they inform participants of significant factors that may be expected to influence their willingness to participate (such as risks, discomfort, adverse effects, or limitations on confidentiality, except as provided in Standard 6.15, Deception in Research); and they explain other aspects about which the prospective participants inquire.

(c) When psychologists conduct research with individuals such as students or subordinates, psychologists take special care to protect the prospective participants from adverse consequences of declining or withdrawing from participation.

(d) When research participation is a course requirement or opportunity for extra credit, the prospective participant is given the choice of equitable alternative activities.

(e) For persons who are legally incapable of giving informed consent, psychologists nevertheless (1) provide an appropriate explanation, (2) obtain the participant's assent, and (3) obtain appropriate permission from a legally authorized person, if such substitute consent is permitted by law.

6.12 Dispensing with Informed Consent

Before determining that planned research (such as research involving only anonymous questionnaires, naturalistic observations, or certain kinds of archival research) does not require the informed consent of research participants, psychologists consider applicable regulations and institutional review board requirements, and they consult with colleagues as appropriate.

6.15 Deception in Research

(a) Psychologists do not conduct a study involving deception unless they have determined that the use of deceptive techniques is justified by the study's prospective scientific, educational, or applied value and that equally effective alternative procedures that do not use deception are not feasible.

(b) Psychologists never deceive research participants about significant aspects that would affect their willingness to participate, such as physical risks, discomfort, or unpleasant emotional experiences.

(c) Any other deception that is an integral feature of the design and conduct of an experiment must be explained to participants as early as is feasible, preferably at the conclusion of their participation, but no later than at the conclusion of the research. (See also Standard 6.18, Providing Participants With Information About the Study.)

6.18 Providing Participants with Information about the Study

(a) Psychologists provide a prompt opportunity for participants to obtain appropriate information about the nature, results, and conclusions of the research, and psychologists attempt to correct any misconceptions that participants may have.

(b) If scientific or humane values justify delaying or withholding this information, psychologists take reasonable measures to reduce the risk of harm.

APPENDIX C

Introduction to Statistics

BONNIE J. PLOGER[1] AND KEN YASUKAWA[2]
[1]*Department of Biology, Hamline University, 1536 Hewitt Avenue, St. Paul, MN, PA 55104, USA*
[2]*Department of Biology, Beloit College, 700 College St, Beloit, WI 53511, USA*

INTRODUCTION

In rare cases behavioral research provides results that are clearcut and easy to interpret, but in most cases things aren't so easy. Behavior is complex and highly variable even in the simplest of situations, so people who study the behavior of animals need tools to help them with their hypothesis testing. Statistical analysis is one of the most important and most frequently used items in the behavioral toolbox. For this reason, it is important for students of animal behavior to have some knowledge and understanding of statistics.

Simply put, a **statistic** is a number that summarizes information about a group of numbers. For example, if you measured the body lengths of all the spring peeper frogs (*Hyla crucifer*) in Clear Lake, you could summarize their lengths by listing the length of each frog, or, alternatively, you could give their average length. The average (or mean) is an example of a **descriptive statistic**, a single number that summarizes information about a group of numbers and is used to describe that group. Descriptive statistics are used to summarize information about (1) the sample of data that you

ISBN 0-12-558330-3
Copyright 2003, Elsevier Science (USA). All rights reserved.

observed and (2) the total population from which the data were drawn. Consider the following example.

Suppose there are 100 adult male spring peepers in a small lake named Clear Lake. You are interested in knowing the mean length of male frogs in this entire adult population. You *could* measure the total adult male population, but this would be impractical and probably impossible. Instead, you would have to restrict your observations to a subset (that is, a **sample**) of the total population. When you measure a sample of 18 males, the mean and other descriptive statistics that you calculate from this sample constitute **sample statistics**. Sample statistics enable you to make precise descriptions of your sample.

But the population of all adult males in the lake cannot be described precisely from the statistics that you derived from your sample. Remember that although you measured only 18 animals, you really wanted to know about the lengths of the entire population in the lake. The population of all 100 adult males in the lake is your **statistical population**, the collection of all elements about which you seek information. Had you measured the lengths of all the adult males in the lake, you could have calculated the **parametric mean**—the mean for the entire population. The parametric mean is an example of a **population parameter**, which is simply a descriptive statistic that is derived from the entire population of interest (the statistical population).

For example, the mean length from your sample of 18 males is an estimate of the mean for the 100 males in the Clear Lake population. The mean and other parameters of the total population may differ from the descriptive statistics of a sample of the population because by chance you may have sampled lots of really large males (or lots of really small ones). Although what you really want to know are the population parameters, what you usually have are sample statistics, which are only estimates of the parameters. But the sample statistics become better and better estimates of the population parameters as the sample size is increased. For example, you would get a much better estimate of the population parameters if you measured and observed 50 males instead of 18. In this case, if you took your data correctly, then your sample statistic should be a good representation of the population as a whole.

The preceding example demonstrates the two purposes for which statistics are computed from data: (1) to describe the data obtained in a sample, and (2) to make inferences about the characteristics of a population on the basis of a sample of observations drawn from that population. The calculation of the mean from a sample is an example of statistical description. The use of the sample mean as an estimate of the population mean (a parameter) is an example of statistical inference. Descriptive and inferential statistical analyses are discussed in further detail in the sections that follow. But the first step in data analysis is to summarize your data.

SUMMARIZING DATA

You can tell little from one observation (one datum) of some phenomenon. Therefore, you always want to summarize data from many different individuals or observations. You must use the right kind of summary for the data that you have collected.

Ways To Express Data
Frequency

The **absolute frequency** is the total number of times you observed some characteristic or phenomenon (e.g., number of hops or number of frogs); the **frequency** is the number in a unit of time (e.g., hops/minute; also called a **rate**) or space (e.g., frogs/puddle; also called a **density**).

Percentage

The absolute frequency divided by the total number of observations times 100 is the **percentage** at which some observation occurred. Percentages can range from 0 to 100%.

Probability

The **probability** is the absolute frequency divided by the total number of observations (it is also called the **relative frequency**). Possible probabilities range from 0 to 1. If every time your professor goes near your animal, it hides, you could say that the chances are 100% that it will happen again or that it happened 10 out of 10 times or that the probability of its happening again is 1. If 75 times out of 100, your animal turned black when you turned on the light, you could say that the chance of such an occurrence happening again is 75% or that the probability is 0.75.

Presenting Data
Tables

A table presents the detailed numerical findings of a study, but it never presents raw data (unless it is in an appendix). Every table must have (1) a descriptive title at its *top*, (2) headings to all rows and columns (including units), and (3) the summarized data (such as descriptive statistics) that make up the body of the table. Tables are numbered sequentially in the order in which you refer to them in the text. You want to present a summary that includes some measure of central tendency and the amount of variation (see the explanation that follows). A helpful "rule of thumb" is that a table is something that you could make with an ordinary typewriter.

Figures

The clearest and easiest way to get the reader to understand your important findings is to present your data in a graphical form (where you may present either raw or summarized data). In scientific writing, graphs, photographs, maps, drawings, and any other illustrations are called figures and are numbered sequentially in the order in which you refer to them in the text. Several kinds of graphs, such as bar graphs, histograms, scatterplots, and box-and-whisker plots, are commonly used in animal behavior. Others kinds, such as pie charts, are not typically used.

When you make a graph, you should follow these rules: (1) Plot the independent variable on the x-axis and the dependent variable on the y-axis. (2) Always plot your data in such a way as to maximize the chances that the reader will see the point you are trying to make. (3) Use the proper scale for the data—for example, a log scale for growth data. (4) Always include clearly labeled axes with the units you used. (5) For every figure, be sure you have an explicit and clear title placed *below* the figure.

You may be tempted to present the same data in both tables and figures, but you should avoid this temptation. Data should be reported only once (in a table or in a figure, but not in both), so you must think carefully about the *most effective* way to present them. Unless you have limited data to present, a good figure is usually more effective than a table. One guideline to consider when you design a table or a figure is to present it in such a way that it can *stand alone*. In other words, someone should be able to understand the information given in your table or figure by reading it and nothing else. This means that axes and legends must be clearly labeled and that titles may need to be fairly detailed, including names of species and sample sizes, where these are not obvious from the data presented.

DESCRIPTIVE STATISTICS

As the name implies, descriptive statistics are used simply to describe a group of numbers. Suppose that you are doing a study of courtship patterns in male spring peepers in Clear Lake. You successfully trapped, measured, and individually marked 18 of the 100 males in the lake. Say you also followed each marked male for one night and determined the number of times that each animal called during the night. These hypothetical data are presented in Table C.1.

Before calculating any statistics, it is often wise to plot the data. To plot the frequency of numbers of calls, simply plot the number of frogs that called 1, 2, 3, etc. times per night, as shown in Figure C.1 (i.e., plot the number of frogs that fall into each category of calling). This figure is called a **bar graph**. Note that the bars do not touch each other because the

Table C.1 **Hypothetical data on male spring peeper (*Hyla crucifer*) calling frequency.**

Frog I.D.	Number of Calls
1	9
2	10
3	6
4	7
5	5
6	11
7	8
8	10
9	9
10	6
11	7
12	8
13	8
14	9
15	9
16	7
17	10
18	9

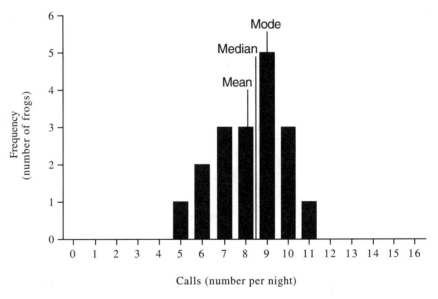

Figure C.1 **Hypothetical data for the number of calls per night by male spring peepers (*Hyla crucifer*) in Clear Lake.**

distribution is **discrete** (i.e., a frog can call 1 time or 2 times a night, but it cannot call 1.5 times in a night). If we had plotted **continuous** data (e.g., length or speed), then we would have used a **histogram** in which the bars touch each other, with each bar representing the frequency of occurrence within a range of possibilities (e.g., the number of frogs between 3 and 4 centimeters in length).

Figure C.1 is a **frequency distribution**, a graph of the distribution of data points that represents how often each value occurred in the sample. To describe your data with descriptive statistics alone, you need to find a way to capture the shape of the frequency distribution. A complete description of the frequency distribution must include a description of both the center of the curve (distribution) and its width. That is to say, when you characterize your data, you must use two descriptive statistics: one that measures the **central tendency** (location) and one that measures the **variability** (dispersion) in your data.

Measures of Central Tendency

Measures of central tendency are descriptive statistics that represent the common values in the distribution. Each statistic of central tendency is a single number that represents the value of the variable where the majority of the data lie (the center of the distribution).

Mean

The **arithmetic mean**, or average, $\bar{X}$ (pronounced "ex bar"), is calculated by taking the sum of the values obtained, ΣX, and dividing by the total number of values, n. For example, the mean number of frog calls is 8.22 (from data in Table C.1; see Figure C.1). Calculate the mean as follows:

$$n = 18$$
$$\sum X = 9 + 10 + 6 + 7 + 5 + 11 + 8 + 10 + 9 + 6 + 7 + 8 + 8$$
$$+ 9 + 9 + 7 + 10 + 9 = 148$$
$$\sum X = 148$$
$$\bar{X} = \sum X/n = 148/18 = 8.22$$

Median

The **median** is the value that divides a frequency distribution into two equal halves such that the same number of items fall on each side of the median value. For example, in the series 1, 2, 3, 4, 5, the value "3" divides the data such that there are the same number of points, 2, on either side. If there are an even number of values, you must take the mean of the middle two values. Thus, in the series 3, 4, 5, 6, 7, 8, the median is 5.5.

Similarly, the median number of frog calls is 8.5 (from the series in Table C.1 of 5, 6, 6, 7, 7, 7, 8, 8, 8, 9, 9, 9, 9, 9, 10, 10, 10, 11; see Figure C.1).

Mode

The **mode** is the most common value in a series. In a frequency distribution, the mode is the value of the variable at which the distribution peaks. For example, the modal number of frog calls is 9 (see Table C.1 and Figure C.1).

In a **symmetric distribution** (one in which the left and right sides are mirror images of each other), the mean, median, and mode are identical. In such cases we typically use the mean as the measure of central tendency. In contrast, the three measures of central tendency differ when a distribution is **asymmetric**. Suppose you had a distribution that was **skewed** to the right (in other words, one with a big "hump" on the left-hand side and a long "tail" on the right-hand side). What do you think would be the left-to-right order of the three measures of central tencency? It would be mode < median < mean, but can you figure out why? Here is a hint: Why is the mean to the right of the median in this case? (That is, why is it closer to the side with the long tail?) How can you remember this order? It's alphabetical from the side with the long tail, which is also the side toward which the distribution is said to be skewed.

Measures of Variability

Measures of variability or dispersion are descriptive statistics that represent the spread of values in the distribution on either side of the center.

Range

The **range** is the difference between the largest and smallest values. It represents the maximum spread in the data. For example, for the series 21, 15, 13, 24, 18, 19, the range is $24 - 13 = 11$. Similarly, the number of frog calls (Table C.1) spanned from 5 to 11, so the range is $11 - 5 = 6$. Note that it would be incorrect in these two examples to say that the range was 13–24 or 5–11 (that is, the range is the *difference* between the minimum and maximum values, not the minimum and maximum values themselves).

Variance

The **variance**, s^2, measures the amount of variability in your sample. Variance differs from the range in that the variance takes into account the distribution of all data points, whereas the range simply describes the distance between the lowest and highest extremes. For example, imagine that you have the following two sets of data:

Dataset A: 5, 8, 8, 8, 8, 8, 8, 8, 8, 8, 9, 9, 9, 9, 9, 9, 9, 11
Dataset B: 5, 5, 6, 6, 6, 7, 7, 7, 8, 8, 8, 9, 9, 10, 10, 10, 11, 11

In both sets, the lowest value is 5 and the highest value is 11 (range = 6). But in Dataset A, the values are all clustered around 8 and 9, except for the two odd points, 5 and 11, whereas the data are spread widely through the entire range in Dataset B. Clearly, Dataset B with its many dissimilar values is far more variable than Dataset A, with its cluster of data between 8 and 9. Variance is a way of comparing the degree of variability among different sets of data.

To calculate variance, take the **deviation** (or difference) of each value, X, from the mean, $\bar{X}$ (that is, calculate $\bar{X} - X$ for each value). Then calculate the **squared deviation** by squaring each deviation [$(\bar{X} - X)^2$ for each value], to eliminate any negative signs. Add all the squared deviations together [that is, compute $\Sigma(\bar{X} - X)^2$] to calculate the **sum of squares**, and divide by the number of values minus one $(n - 1)$, which is called the **degrees of freedom**. In other words,

$$s^2 = \frac{\Sigma(\bar{X} - X)^2}{n - 1}.$$

An easier but equivalent formula is

$$s^2 = \frac{\Sigma X^2 - \frac{(\Sigma X)^2}{n}}{n - 1}.$$

For example, the variance in the number of frog calls can be calculated from Table C.1 as follows:

$$\Sigma X^2 = 9^2 + 10^2 + 6^2 + 7^2 + 5^2 + 11^2 + 8^2 + 10^2 + 9^2 + 6^2 + 7^2$$
$$+ 8^2 + 8^2 + 9^2 + 9^2 + 7^2 + 10^2 + 9^2$$
$$= 81 + 100 + 36 + 49 + 25 + 121 + 64 + 100 + 81 + 36 + 49$$
$$+ 64 + 64 + 81 + 81 + 49 + 100 + 81$$

$$\Sigma X^2 = 1{,}262$$

$$\left(\Sigma X\right)^2 = (148)^2 = 21{,}904$$

$$s^2 = \frac{1{,}262 - \frac{21{,}904}{18}}{18 - 1} = 2.65$$

Similarly, Dataset A in the preceding example has $s^2 = 1.31$, whereas for Dataset B, $s^2 = 3.82$.

Standard Deviation

The variance is based on the sum of the squared deviations and so has units of measure that are squared. To convert this measure of dispersion to one that uses the original unit of measure, we could take the square root of the variance. That is how to calculate the **standard deviation**

(s, where $s = \sqrt{s^2}$). Another advantage of the standard deviation emerges if your data conform to a **normal distribution**, familiar to most as symmetric, bell-shaped curves around the mean. In a normal distribution, about 95% of the values fall within 2 standard deviations on either side of the mean.

Percentile

Percentiles are points that divide the distribution of the data into hundredths. The 95th percentile is the point on a distribution below which 95% of the data fall.

INFERENTIAL STATISTICS

Experimental Design

Sampling

According to Webster, *to infer* means "to derive as a conclusion from facts or premises." For example, when you see a person driving a police car, you may infer that the person is a police officer. In this example, you made conclusions about a person's employment and lifestyle on the basis of one observation of the vehicle being driven. In statistical inference, you draw conclusions about a large number of events on the basis of your observations of a subset of them, your **sample**.

Suppose that you sampled two sets of 18 male spring peepers in Clear Lake rather than just one as described earlier. Let's say that one sample had a mean of 8.2 calls and the second sample had a mean of 8.8. Both samples came from the same population of adult male frogs in Clear Lake, so why do the two samples differ? There are several possible reasons.

Biased Samples

One reason for the difference might be that different selection criteria were used to choose study animals for the two samples. An example of such **biased sampling** would be if, for your first sample, you selected small males that call rarely, whereas for your second sample, you selected large males that call frequently. In this example, rather than being estimates of the total adult male population, your first sample would be an estimate of the population of small adult males, and your second sample an estimate of the population of large adult males, in Clear Lake. Had you selected small males for both samples, you would still not have a good estimate for the total male population, but only an estimate for small males. Why would someone choose to gather biased samples? It usually occurs by accident. For example, perhaps the first sample was gathered near shore and the second from deeper water. It just so happened that small, quiet males were hanging out in shallow water and big noisy males in deeper water. Unintended bias in sampling is a serious problem. Proper sampling methods are designed to avoid it, among other things.

Random Samples

Another reason for the differences between your two samples might be that *by chance alone*, your first sample happened to consist of animals that called infrequently, whereas your second sample happened to consist of animals that called frequently. This would be an unlikely outcome if chance differences were the only differences between your samples. If you looked at more samples, and the samples were random, differing only by chance, then the calling rates in most samples would be fairly similar, although a few would be quite different. When you do a study, what you want is to have **random samples**. In your study of frog calls, to make sure that the differences between the means of the two samples are the result of chance alone, you must choose the sample males randomly by applying objective, preset criteria for selecting the animals (rather than choosing the easiest ones to catch or the loudest callers).

To do any type of statistical inference, it is *essential* that the sample be random so that it will be truly representative of the population. To ensure randomness in the sampling procedure, make all decisions about the experiment before the experiment begins. Before beginning the experiment, you must decide (1) what individual animals you will use for the experimental and control groups, and (2) what data you will take and how you will take them. You must be very careful *not* to say, "Oh, this looks like a nice aggressive animal, so I'll test this one" or "I didn't quite see what happened that time, so I'll use it as a control."

How do you randomize the collection of data? You have to apply preset, objective, and clearly specified criteria to obtain a random sample. This may involve assigning numbers to your animals and treatment groups and using a table of random numbers to assign the animals to the different treatment groups.

Data Independence

Independent data are those in which the presence of one value or data point has not influenced the presence of another value in your sample. Within any sample, the data must always be independent. For an example of nonindependent data, imagine that you watched a frog in an aquarium and recorded at 15-second intervals where the animal is located. Your data were not independent because the position of the animal in the previous 15 seconds drastically influenced where it was in the following 15 seconds. Similarly, if you always present the animal with a red stimulus after a blue one, then your data will not be independent because the blue stimulus may influence the response to the red one. Your data are useful only when they are independent.

The most common way in which data independence is a problem is through the **pooling fallacy**, which is sometimes called **pseudoreplication**. We often treat repeated observations of the same individual as though they were independent. This is incorrect. To make repeated observations on one individual is not a substitute for making observations on many individuals. All that these repeated observations of one individual do is increase your knowledge about that one individual. In other words, one individual is one data point, and by observing the same individual over and over, you simply increase the reliability of your estimate of that one data point. For example, suppose you want to know if ponds next to highways have fewer frogs than ponds far from highways. You could pick two ponds, one by a highway and one far from any highways, and count the number of frogs in the two ponds once a month for 6 months. This is an example of pseudoreplication. You would generate lots of data, but at the end of the study, all you would know would be whether these two particular ponds differed. You would not know whether the differences had anything to do with being near or far from highways. To avoid pseudoreplication, a better experimental design would be to pick a bunch of different ponds, half near highways and half far from highways. You would want to pick ponds that were as similar as possible (in size and vegetation type, for example) and differed mainly in their proximity to highways. You could then count the frogs in each of these ponds and compare the counts in ponds near highways to those in ponds far from highways. This experimental design would enable you to draw conclusions about the relationship between proximity to highways and number of frogs in ponds.

Independent and Dependent Variables

The goal of many scientific investigations is to determine whether changes in some condition(s) (the **independent variable**) result in different effects (the **dependent variable**). The dependent variable is the score that you measure for each of your study subjects. For example, you might want to know whether frog calling rate (dependent variable) depends on distance from highways (independent variable). You might start by simply measuring the calling rates of frogs at ponds adjacent to highways vs those 4 kilometers from highways. Here, your independent variable would include two **comparison ("treatment") groups**, adjacent to highways and 4 kilometers from highways. Such an **observational comparison** would not involve any experimental manipulation. Next, you might conduct a **controlled experiment**, in which you manipulate the location of the frogs. For example, you might randomly assign equal numbers of frogs to the following three treatment groups: aquaria adjacent to a highway, aquaria 4 kilometers from a highway, and "control" aquaria at the same distance (such as 2 kilometers) from a highway as was the pond from which all the frogs originally were captured.

Observational Comparisons

To make an observational comparison, a researcher simply compares two or more groups that already exist naturally and measures some dependent variable to see whether it differs among these comparison groups. In other words, these are observations that do not involve any experimental manipulation of the independent variables. Examples include comparing the calling rates of frogs in ponds near and far from highways, or calling rates by frogs with different skin colors (certain colors may provide better camouflage so that cryptic frogs may call more because of lower risk of being killed by predators). Observational comparisons are often made at the beginning of a study to detect possible relationships. With such observations you can detect potentially interesting relationships for further study, but they do not enable you to make conclusions about what *caused* an observed relationship. To identify causes requires controlled experiments.

Consider the following example of an uncontrolled observation. Let's say that from your initial investigations of the effects of highways on frogs, you hypothesized that there were fewer frogs near highways because oil from the highways was affecting the hatching success of the frogs' eggs. You then revisited your bunch of ponds that were near and far from highways, this time measuring the amount of oil on the water and the hatching success of frog eggs. If you found that fewer eggs hatched in areas with high amounts of oil than in areas with low amounts of oil, you might be tempted to conclude that the oil *caused* low hatching success. But such a conclusion would be valid only if you had controlled for all other factors. In the case just described, you did not control for all other factors. Thus, although it might be that oil caused low hatching, the low hatching in oily ponds could just as easily have been caused by some other factor (a **confounding factor**) that you did not measure. For example, it might be that the ponds that had more oil in them also happened to have higher levels of lead in them. Perhaps the lead was what was killing the eggs. Or perhaps the ponds that had more oil in them happened to be in more open areas such that the water received more sunlight. High light levels and/or higher water temperatures, rather than oil, might have been responsible for lower hatching success in these ponds. Your observation that fewer eggs hatched in ponds with more oil is still useful in that it indicates that there might be a relationship between oil and hatching success. Such uncontrolled observations are a good first step in figuring out what is going on. But in order to find out what actually *caused* the reduction in hatching success in oily ponds, you must do a controlled experiment.

Controlled Experiments

To do a controlled experiment, the researcher manipulates a factor of interest and holds *all* other factors constant (the **rule of one variable** or one difference among treatment groups). Often, these independent treatment groups include a **control group** that experiences normal conditions, plus one (or more) **experimental group**(s) that receive(s) some manipulation.

To keep all other factors as constant as possible, the researcher should conduct a **laboratory experiment**, which results in high **internal validity** (i.e., if there is an effect, you know exactly what *caused* it). But laboratory conditions can be so artificial that their results may not be generalized to the real world (a lab experiment has low **external validity**). Clearly, then, a researcher cannot achieve both high internal validity and high external validity—these 2 characteristics of experimental design "trade-off" against one another. As a compromise, a researcher may choose to do a **field experiment** by observing organisms in their natural habitat while manipulating some factor of interest (such an experiment would have moderate internal and external validity). The following are examples of a field experiment and a laboratory experiment that you might choose to do to find out whether oil really caused the reduction in hatching success of frog eggs that you observed in the uncontrolled observation described above.

As a field experiment, you could pick a set of ponds that have either no oil or very similar, low levels of oil. The ponds should also be very similar in other potentially important respects, such as size, water temperature and chemistry, and vegetation along shore. Best would be ponds that also had similar hatching success of frog eggs. You could then *randomly* assign half of the ponds to be the experimental group, which would receive a certain dose of oil, and assign the other ponds to the control group. Control ponds could receive pond water instead of oil, to control for any effects of disturbance by experimenters. If you then found that more eggs hatched in the ponds with no oil (control group) than in the ponds that received oil (experimental group), you could conclude that oil *caused* the reduction in hatching success. Concluding that oil caused the difference is reasonable because other differences between the two sets of ponds were controlled by random assignment of treatments. If the two sets of ponds were otherwise identical, then the conclusion that oil *caused* the difference is valid. Keep in mind that if you actually planned to conduct such an experiment that involves adding a pollutant like oil, you would need to use the lowest appropriate dose, for example, similar to that released by roadsides. By doing so, you will not only make your results applicable to your question about the effects of roadside oil, but also minimize harm to the pond organisms. Exposing organisms to deleterious conditions like these may be necessary to advance knowledge needed to reduce harm from pollution, but must also minimize suffering in experimental animals. Of course you would not conduct a field study like this in an ecologically sensitive area that contained locally rare organisms!

The difficulty with field experiments is that you can never find situations in the field where all other factors are truly identical. There are sure to be some variations among your ponds, some of which might have influenced hatching success. If, despite random assignment, the experimental group happened to have more ponds with higher temperatures than the control ponds, then your conclusion that oil caused lower hatching would be questionable; high temperatures might have been the cause. But if you pick large enough samples—that is, if you do the experiment with a large enough number of different ponds—and if you assign ponds randomly to treatment groups, the chances are that differences among the ponds will appear equally in both the control group and the experimental group. For example, if you picked enough ponds and if your assignment of treatments was unbiased, then there would be about the same number of warm and cold ponds in both the experimental group and the control group. In this situation, average temperatures would be the same in both groups, so temperature differences would not be a possible explanation for lower hatching success in the experimental group. With large enough samples, you could reasonably conclude that oil caused the low hatching success. But what if getting large samples is impractical or you cannot make your control and experimental groups similar to each other in all factors other than the one you are manipulating? In this situation, you need a laboratory experiment.

As a laboratory experiment, you could set up in your lab a bunch of aquaria that all had identical conditions. You could then put an equal number of fertilized, healthy frog eggs into each aquarium. You could then randomly assign half of the aquaria to be the experimental group, giving each a prescribed dose of oil. The other aquaria would be assigned to be the control group and would receive no oil (but you could add an equal volume of aquarium water instead). Because you were doing this in your laboratory, you would have (at least theoretically) complete control over all of the variables, so you could set up the experiment in such a way as to be sure that the only difference between the control and experimental aquaria was the presence or absence of oil. If you found that hatching success was higher in the control aquaria than in the experimental aquaria to which oil was added, you could conclude that the oil *caused* the observed reduction in hatching success. However, it is risky to assume that you have controlled all the factors except the one of interest. Even in the laboratory, you might have overlooked a factor that might affect the outcome of your experiment. For example, if you put all your experimental tanks along the window and all your control tanks along the wall, you would have uncontrolled differences in light levels and temperature (confounding factors) between the two groups. The higher light or temperature levels in the tanks along the windows might have caused lower hatching success in the experimental tanks. In a room without windows, tanks might still experience differences in air currents or noise levels if some tanks were placed near air ducts or doors and others were far from such passageways. To avoid these problems, you must be sure that an equal number of tanks in the experimental and control groups are near ducts or doors and that an equal number are far from ducts and doors. In short, the mere fact that you are working in a laboratory does not mean you automatically have controlled conditions. Be aware of easily overlooked differences in variables such as light levels, air currents, noise levels (some walls may be near machine or construction noises), background complexity or color (which might affect animal behavior), and temperature differences along inner and outer walls. The easiest way to avoid these confounding factors is to assign aquaria randomly to control and experimental groups.

The Logic of Hypothesis Testing

Statistical inference is used most frequently to test **hypotheses**. When studying animal behavior, we seek explanations for the patterns that we observe. We learn about the patterns by forming hypotheses and testing specific **predictions** that are based on the hypotheses. For example, suppose we are interested in finding out why male frogs call. We may hypothesize that one **function** of calling by male frogs is to attract females. From this general, **biological hypothesis**, we can deduce specific, testable predictions. Some of these predictions may involve observations of undisturbed animals, whereas others may involve field or laboratory experiments. For example, one prediction of this hypothesis might be that among undisturbed animals in the field, more females will be found near males that called than near males that did not call. Another prediction might be that fewer females will be found near experimentally muted males than near sham-operated, unmuted males (the control treatment). (Similar field experiments involving temporary muting are relatively easy to conduct with songbirds but may be impossible to do with frogs. Keep in mind that the frog study in this handout involves hypothetical examples, not real data.)

To discuss the validity of our biological hypothesis (here, that male calling attracts females), we must first determine which (if any) of the predictions are correct. To test each prediction, we must collect data and determine whether the data fit the prediction. If all the predictions are correct, we may conclude that the data support the hypothesis. (Note that we can never say that the data *prove* the hypothesis—we encourage you to declare a moratorium on use of the word *prove*.) If the data do not support one or more of the predictions, then we must conclude that the data do not support the hypothesis, and thus we reject it. To understand the phenomenon, we will have to form a new hypothesis or modify the old one and then start all over again by designing new experiments and collecting new data to test the predictions of the new or revised hypothesis.

How do we decide whether our data meet our prediction? How can we tell whether the experimental differs from the control? For example, say that in testing the hypothesis that male calling attracts female frogs, you operated on 20 frogs. You muted 10 so that they could not call for a few weeks. You did a sham operation on the other 10, your control, so that they experienced surgery but were still able to call following the procedure. After you released the animals, you found that there were an average of 1.3 ± 0.24 (mean ± 1 standard deviation) female frogs within 5 meters of muted males, whereas there were an average of 3.2 ± 1.4 females within 5 meters of unmuted males.

From these results, you might be tempted to say that there were more females around males that could call than around those that could not. But remember that two groups can differ by chance alone. How can you tell whether the apparent differences between two groups are due to real differences rather than to chance alone? How can we tell whether a mean of 1.3 females is really different from a mean of 3.2 females? This is where statistics are used in testing hypotheses. Statistical inference tests determine how large the observed differences must be before we can be reasonably sure that they represent real differences in the populations from which only a few events were sampled. We can never be *certain* that two groups differ, but we can use inferential statistics to find out how likely (how probable) it is that the differences represent real differences between the groups rather than differences based on chance alone.

Null and Alternative Hypotheses

All statistical tests involve discriminating between pairs of alternative hypotheses (note that these **statistical hypotheses** are distinct from our original, biological hypothesis, which we are attempting to test). The **null hypothesis** is that there are no differences among groups or no effects—that is, any apparent differences are the result of chance alone. The **alternative hypothesis** (the alternative to the null) is that there *are* differences or effects. The alternative hypothesis includes all possible alternatives to the null. Because only one of the two hypotheses can be true, we call these hypotheses **mutually exclusive**. When testing these hypotheses, we accept the simpler, null hypothesis unless there is good reason to reject it. Therefore, when we are doing a statistical test, we usually say that we are testing the null hypothesis. The goal of such testing is to figure out how likely it is that our study would produce our results when the null hypothesis is true.

How are the null and alternative hypotheses related to our original biological hypothesis and its predictions? The null and alternative hypotheses are simply ways of restating one prediction of a biological hypothesis. There are separate null and alternative hypotheses for each test of each prediction. For example, in testing the biological hypothesis that male calling attracts female frogs, several experiments might be conducted. For each experiment, we can make one or more prediction(s) about what the outcome would be if the biological hypothesis were correct. To do a statistical test of each prediction, we first must restate each prediction in the form of a null and alternative hypothesis. We then conduct the statistical test, which is actually a test of whether we should accept or reject the null hypothesis.

The following examples are null (H_0) and alternative (H_a) hypotheses for the prediction that more females will be near unmuted than will be near muted male frogs. There are two forms for these hypotheses: **one-tailed** and **two-tailed**.

Two-Tailed Hypotheses

H_0: There is no difference between the numbers of females near unmuted and muted male frogs.

H_a: There is a difference between the numbers of females near unmuted and muted male frogs.

A two-tailed hypothesis does *not* specify the direction of the difference; thus a difference toward either tail of the distribution means H_0 is rejected. Rejection of the null hypothesis simply means that two groups differ. Two-tailed hypotheses are the most appropriate to use when you have reason to expect groups to differ but have no reason to expect the difference to be in a particular direction.

One-Tailed Hypotheses

H_0: There are not more females near unmuted than near muted male frogs.

H_a: There are more females near unmuted than near muted male frogs.

When you have good reason to expect groups to differ in a particular direction, a one-tailed hypothesis is appropriate. In our example, rejection of the 1-tailed null hypothesis means that there are more females near unmuted than near muted male frogs. Failure to reject the null means either that (1) there were similar numbers of females near unmuted and near muted males or that (2) there were more females near muted than near unmuted males. We cannot distinguish between these two possibilities if we failed to reject the one-tailed null hypothesis. Had we done a two-tailed test, we would have detected a difference if there had been more females near muted than near unmuted males. The selection of a one-tailed test must be based on a good reason for expecting differences in a particular direction, such as past studies of the same or related species showing differences in one direction. (A one-tailed test may also be appropriate if you are testing a prediction of a theoretical model that exists in the literature and the predicted differences must be in one direction for the model to be valid.)

Suppose that you tested the foregoing one-tailed hypothesis and discovered that there appeared to be more females near muted than near unmuted males. Could you switch to a two-tailed hypothesis "after the fact"? No, absolutely not. The null and alternative hypothesis and whether they involve one or two tails must be stated *before* doing the statistical test. In other words, just as in a conversation, you must ask the question before you can answer it. The set of statistical hypotheses to be tested must be chosen before, not after, the analysis and final decision.

Significance Level

Remember that when we have not actually measured the entire population, we do not know which is true, H_0 or H_a. We may decide to accept H_0 when it is true (a correct decision) or when it is false (an incorrect decision). Alternatively, we may decide to reject H_0 when it is true (an incorrect decision) or when it is false (a correct decision). These possibilities are listed in Table C.2.

Table C.2 illustrates the two types of mistakes that we might make when we decide whether to accept or reject H_0. Accepting a null hypothesis when it is actually false is a **type II error**; rejecting a null hypothesis when it is really true is a **type I error**. We try to minimize both types of errors. Type II errors are minimized by increasing our sample size. When we fail to reject the null hypothesis but have only a small sample, we must consider the possibility that a larger sample would have caused H_0 to be rejected (statisticians call it a lack of **statistical power**).

Statistical inference tests are designed to calculate the probability (**P**) that chance alone produced your results if the null hypothesis is true. A P-value equal to 0.05 means that the likelihood of a type I error is 5%. In other words, if you took 100 samples, in 5 of the 100 you might, by chance alone, incorrectly reject H_0 even though H_0 is actually correct.

Table C.2 **Results of decisions to accept or reject the null (H_0) and alternative (H_a) hypotheses.**

	Actual Condition in Nature	
Your Decision	H_0 *is Really True*	H_a *is Really True*
Do not reject H_0 (i.e., accept H_0)	Correct decision	Type II error
Reject H_0 (i.e., accept H_a)	Type I error (α level)	Correct decision

To decide whether a particular result supports H_0, the calculated P-value (type I error) is compared with a predetermined maximal level called the **significance level** or **alpha (α) level**. Typically in animal behavior, $\alpha = 0.05$; we will use $\alpha = 0.05$ in this manual.

The significance level is what you use to assess how confident you can be that you are making a correct decision when you reject H_0. With an α of 0.05, you can have 95% confidence that your decision is correct. Similarly, with $\alpha = 0.01$, you can have 99% confidence that your decision is the correct one. Because 99% confidence seems better than 95%, and 99.9% is better still, why use 95%? It's because of another trade-off, this time between type I and type II error (β). The higher our confidence that we will not incorrectly *reject* the null hypothesis (that we will not make a type I error), the more likely it is that we will incorrectly *accept* the null hypothesis (make a type II error). In other words, you can't have it both ways. An α of 0.05 is a good compromise between the two kinds of error.

Test Statistic

After stating our hypotheses and selecting an α level, we must select and carry out the appropriate statistical test (see the next section). By plugging the values of our sample into a formula for the test statistic (some common ones are t, F, and χ^2), we end up with one number that summarizes the sampled data. To make our decision, we need the P-value associated with this number (such as the value of t, F, or χ^2). If we knew how, we could use integral calculus to figure out the P-value. Luckily for those of us who are somewhat calculus-impaired, we can also look up the P-value in a published table. There is one slight difficulty, however. For any test statistic that we may wish to look up in a table of P-values, there are an infinite number of test statistic values and P-values. Unfortunately, there aren't any publishers who are willing to print a table of infinite length. Instead of *all* of the values, only a few, representative ones are tabulated. These tabulated values are called **critical values**.

For a particular test statistic, one critical value is associated with a particular P-value and sample size. Thus, to compare the numbers of females near muted and near unmuted male frogs, you would look up the critical value for the test statistic when P = 0.05 and sample size (n) = 10. Why look for the critical value corresponding to P = 0.05? Because your level of significance (α) was 0.05, and you are tying to decide whether your results are significant or not significant at that level.

Some test statistics make use of degrees of freedom (abbreviated "d.f.") instead of sample size (n). Degrees of freedom vary with sample size. Critical values for these test statistics are uniquely associated with a particular P-value and d.f. To calculate degrees of freedom, see your instructor or a statistics book.

To say that there is only one critical value for each P-level is not quite correct. Although there is only one value per P-level listed in a statistical table, for most test statistics, critical values actually come in pairs that have the same absolute value but are opposite in sign. Thus, even though only the absolute value is listed in a table of statistics, there are really two critical values, one positive and one negative, for each combination of P-level and sample size.

The absolute value of the critical value is the largest value of the test statistic that you should expect to observe if H_0 is true. Observed values larger than the critical value mean that H_a is probably true.

Decision Rule

We are finally ready to make a decision about whether to accept or reject H_0. To do this, we use the following **decision rule:**

> If the observed test statistic $\geq$ the critical test statistic, then you should reject H_0, accept H_a, and conclude that your results are significant at the $\alpha = 0.05$ level.

> If the observed test statistic $<$ the critical test statistic, then you have failed to reject H_0 and should either (1) conclude that the results were no different from random, or (2) suspend judgment because your sample sizes were too small to reject H_0.

> *Exceptions*: In some statistics books (e.g., Seigel 1956 but not Sokal & Rohlf 1981), the instructions for the Wilcoxon matched–pairs test and the Mann–Whitney test (see below), call for rejecting H_0 when the observed test statistic is *less than* or equal to the critical value.

> When calculating correlations, in addition to determining the statistical significance, you must look at the value of the correlation coefficient (*r*), which describes the strength of the association. You can conclude that you have a high correlation (a strong association) if the correlation is statistically significant *and* $r > 0.7$. You should be aware that if you have a sufficiently large sample size, you might get statistical significance even if $r < 0.2$, which you should interpret as a negligable relationship (Martin & Bateson 1993).

These days most students have access to computer programs that do statistical analyses. Some examples of such statistical software are SPSS, SAS, SYSTAT, Statview, and JMP. The common spreadsheet program EXCEL also does statistical analyses, although recent reports claim that the results are sometimes incorrect. It's a good idea to try an example with a known answer (such as a worked-out example from a textbook) to test a particular calculation. The great advantage of most computerized statistical tests is that they automatically compare the observed value of the test statistic to the critical value. In other words, you don't have to use a table to figure out the P-value.

The computer program reports as the P-value the probability that the observed value of the test statistic will lead you to reject H_0 when H_0 is actually true (type I error). You want to keep your chances of making this type of mistake low, so you want $P \leq 0.05$, which allows you to be at least 95% certain that your decision to reject H_0 is correct (i.e., $\alpha = 0.05$). Thus, when using a computer, we use the following decision rule:

If the computer produces from your data an observed $P \leq 0.05$, then you should reject H_0, accept H_a, and conclude that your results are significant at the $\alpha = 0.05$ level.

If the computer produces from your data an observed $P > 0.05$, then you have failed to reject H_0 and should either (1) conclude that the results were no different from random, or (2) suspend judgment because your sample sizes were too small to reject H_0.

Choosing the Appropriate Statistical Test

Although they all do basically the same thing, which is help you decide whether to accept or reject H_0, there are many different kinds of test statistics. Each is created by a mathematical formula that produces one number from the set of values in your sample. Each test (with rare exceptions) uses the logic outlined in the preceding sections. The steps that you follow, from forming your general hypothesis through the final decision whether to accept or reject H_0, will be the same for all the tests that you will normally encounter. All the tests provide a way of deciding whether differences in samples result from real differences or from chance alone, on the basis of how likely it is that the value of the test statistic that you observed from your sample(s) could have been produced by chance.

With so many tests to choose from, how can you decide which test is the most appropriate for your data? The choice depends on the type of question you are asking and on the way your data will be measured. These topics are discussed in the sections that follow.

Type of Question

Statistical questions can be divided into four basic groups: (1) questions about one sample, (2) questions about two or more related samples, (3) questions about two or more unrelated ("independent") samples, and (4) questions about correlation and regression.

Questions about a single sample concern whether a particular sample could have come from some specified population. One-sample statistical tests answer questions such as the following: Is it likely that the sample was drawn from a population with a particular distribution (e.g., normal, Poisson, binomial). Is there a significant difference between the observed frequencies and the frequencies that we would expect on the basis of some principle, such as expectations from transmission genetics or from events occurring at random? For example, you might ask whether the sex ratio of frogs in Clear Lake is 50:50, as expected by chance. To find the answer, you could capture a single sample of 30 adult frogs, count the males and females in the sample, and use a statistical test to decide whether the difference in the numbers of males and females in your sample was significantly different from 50:50.

In contrast, we might ask questions about differences between two or more comparison groups: Is there a difference between the effects on the treatment and control groups? Is one treatment better than the other(s)? Does the effect differ among different types of observational groups? To answer these questions, we must sample from different groups, so we refer to each group as a different *sample*. The samples may be related to each other (dependent) or unrelated to each other (independent).

Questions about two or more related samples arise from experimental designs in which the same individuals are measured more than once. When the same individual is exposed to two treatments at different times, we say that the two samples are **related** or **matched**. For example, you would have two related samples if you compared the calling rates of 20 male frogs before and after they were confined in buckets. The two samples would be (1) before confinement and (2) after confinement. The samples are related because the same individuals are used in both samples. If the same individuals are measured more than twice, the design is usually called a **repeated-measures design**. Occasionally, samples are matched by other criteria, such as body length, past experience, or age.

When different, unrelated individuals are used in each comparison group, we say that the two samples are **independent**. Such designs lead us to ask questions about two or more unrelated ("independent") samples. For example, all the unmated male frogs and all the unmated female frogs in Lake Minnetonka belong to two independent groups because different, unrelated individuals are in each group. This example specifies unmated frogs because, for some questions, members of mated pairs would have to be considered related, not independent.

Clearly, *samples* (the treatment or comparison groups that are the independent variables) can be related to or independent of one another. But remember that *within* any one sample, each *data* point that you measure must always be independent of every other data point (see the foregoing "Experimental Design" section for a discussion of independent *data*).

Questions about **correlation** and **regression** concern whether one type of score that you measured varies with another type of score. These questions ask whether there is some sort of relationship between the two types of scores, which must both vary continuously rather than being measured in discrete categories. Although they seem quite similar, the questions asked by correlation and regression analyses are actually very different, and you should be very careful in choosing one or the other.

We may ask whether two types of scores vary together in a systematic way. In other words, are two variables **correlated** (associated)? For example, you might compare the body sizes of male frogs to their calling rates. If you found a significant **positive correlation** between body size and rate of calling, you could conclude that big frogs call more than little frogs. Or, if you found a significant **negative correlation**, you could conclude that big frogs call less than little ones. It is very important to note here, however, that *correlation does not imply causation*. In other words, you could *not* say that large size *causes* high calling rates; there may be a third factor that affects both size and calling. If calling takes a lot of energy, then frogs that are eating well might be able to call more often, *and* would attain larger size, than those catching fewer, poorer-quality prey. In this example, the third, causal factor would be caloric intake. For another example, a positive correlation between countries in which food is spicy and countries where there are frequent earthquakes would not mean that eating spicy food causes (external) earthquakes! You may laugh, but there are examples almost as naïve in the biological literature, some current. So *be careful* about what you infer from correlation.

In contrast to correlation, regression analysis tests for a functional relationship between an independent variable, x, and a dependent variable, y (i.e., $y = f(x)$). Note that in both correlation and regression analysis, we are measuring a pair of continuous variables to determine whether they vary together in some way. But in correlation analysis, we cannot say which variable is dependent and which is independent. By contrast, in regression analysis we explicitly test whether one type of score, the dependent variable, depends on the other, independent variable.

In regression analysis, functional relationships can take on many shapes. In this manual, we introduce the simplest form of this type of analysis by asking whether the relationship is linear. You may remember the following equation from high school: $y = mx + b$, where m is the slope (rise/run) and b is the y-intercept. For some reason unknown to us, however, statisticians write the same equation as: $y = bx + a$, with b for the slope and a for the y-intercept. For example, we might shine light of 10 different intensities on 10 different tadpoles (larval frogs), measure the amount of growth of each tadpole, and ask whether tadpole growth is a function of light intensity (it doesn't make sense to say it the other way around). Here, light intensity is the independent variable, plotted on the x-axis, and tadpole growth is the dependent variable, plotted on the y-axis. If we were to find a significant regression, we would say that tadpole growth is a function of light intensity. The relationship could be positive (as indicated by a positive slope), meaning that tadpoles grow more rapidly with brighter light, or negative (negative slope), meaning that tadpoles grow more slowly with brighter light. Note again that saying that y is a function of x almost seems like saying x causes y, but we are *not* necessarily justified in inferring a causal relationship. We would need to do a controlled experiment to test for causation.

Levels of Measurement

All kinds of (independent and randomly obtained) data are useful and interesting, but different kinds of data require different techniques for presentation and analysis.

Nominal data result when the dependent variable is a measure that simply classifies objects or characteristics into separate categories, but the categories cannot be ranked in any particular order. For example, the sex of a spring peeper and the kind of marsh plant on which these frogs can be found are nominal variables. The categories male and female cannot be ranked, and cattails, reeds, and sedges could be placed in any order with equal validity. Nominal measurement variables are somewhat uncommon in animal behavior, but we often use nominal categories as a **classification variable** (that is, to identify comparison groups).

In the two-sample example that follows, imagine that you wanted to compare two groups to find out whether frogs belonging to one group generally occurred on different types of marsh plants than frogs belonging to the other group. To answer this question, you recorded the capture location (plant type) of eight frogs belonging to two different groups, where group 1 and group 2 could be males and females (or drug A vs drug B, or large frogs vs small frogs). You recorded the following data:

Group 1	Group 2
cattails	cattails
reeds	cattails
sedges	cattails
sedges	cattails

Here the independent (classification) variable is group (1 vs 2) and the dependent (measurement) variable is plant type (cattails, reeds, sedges). Once again, because the scores for the dependent variables cannot be ordered or ranked, these are nominal data.

Ordinal data are collected when the dependent variable is a measure that classifies objects or characteristics into mutually exclusive categories, *and* these categories can be put in a ranked order. You can think of *ordinal* as meaning "ordered" (ranked). For a one-sample example, imagine that you wanted to find out how attractive spring peeper calls are to females. You captured a bunch of females and observed their responses to calls played from a tape recorder. Some females showed no reaction, and a few tried to climb onto the tape recorder. Some others turned toward the sound but did not approach it, whereas others hopped toward the source of the calls. It would be reasonable to rank these responses in the following order: (1) no reaction, (2) orient toward call, (3) approach call, (4) mount speaker. Although we gave these categories numbers (1–4), they are really ordinal values (first, second, third, fourth) rather than integers. Because the categories can be ranked, these are called ranked, or ordinal, data.

In the two-sample example that follows, imagine that you wanted to compare 2 types of spring peeper calls to find out whether one is more attractive to females than the other. To answer this question, you observe the responses of five females to call type 1 and of five others to call type 2. You recorded the following data:

Call Type 1	Call Type 2
1	3
1	4
2	4
3	3
1	2

In this case the independent (classification) variable is call type (1 vs 2) and the dependent (measurement) variable is female response (1 = no reaction, 2 = orient toward call, 3 = approach call, 4 = mount speaker). Once again, because the scores for the dependent variable can be ordered, these are ordinal data.

Although there are more precise mathematical definitions of **interval data** and **ratio data**, you will have these types of data when the dependent variable consists of numbers that can be ranked (such as ordinal data) and when, in addition, the distance between each number and the next is of known size. With interval and ratio data, you can associate each object with a unique number along some continuous measurement scale. Measurements such as length, height, weight, volume, temperature, and time are continuous variables because theoretically, if you could measure accurately enough, an infinite number of measurements would be possible between any two measurements. Rates, such as the number of events per unit time or per bout of behavior, can also be treated as interval/ratio data.

For example, imagine that you wanted to find out whether certain sizes of male spring peepers were more common than other sizes. For a one-sample example, let's say you measured the body lengths of nine frogs as follows: 2.42, 2.43, 2.50, 2.51, 2.52, 2.55, 2.57, 2.60, and 2.65 centimeters.

Had you measured less accurately, or if you combined these lengths into categories, your data table might look like this:

Frog Body Length (cm)	Number of Frogs of that Length
2.4	2
2.5	3
2.6	4

Note that when you measure less accurately or when you combine more accurate measurements into categories, it may not look as though there is a unique number associated with each object (here, each frog). But in theory you could always associate a unique number with each object.

In the two-sample example that follows, imagine that you wanted to compare two groups to find out whether frogs belonging to one group had a different body length than frogs belonging to the other group. To answer this question, you recorded the body lengths of 10 frogs belonging to two different groups, where group 1 and group 2 could be males and females (or drug A vs drug B, or heavy frogs vs light frogs). You recorded the following data:

Group 1	Group 2
1.80	1.98
1.90	2.08
2.40	2.19
2.38	2.28
2.30	2.42

Here, the independent variable is group, and the dependent variable is body length. Once again, because the scores for the dependent variables can be ordered along a continuous scale, with known distances between each number and the next, these are interval/ratio data.

CHANGING NOMINAL DATA INTO ORDINAL DATA

When you have data that are counts of how many times each subject reacted to mutually exclusive treatments, you may be tempted just to count how many individuals reacted to each treatment. You would then have **enumeration data** (counts) that you would analyze with a chi-square type of test (see below). But by just counting how many individuals reacted, you will lose considerable information. Martin & Bateson (1993 pp. 72–73) suggest two more powerful ways to handle these types of data—methods that you can use whenever you can associate specific reactions with specific individuals. The following example illustrates these different ways.

Example

You put peanuts without shells and M&Ms in a pile and use the raw data table that follows to record each peanut and M&M taken by each individually identifiable squirrel. You want to know whether squirrels show a preference for one of these food types.

Squirrel Name	Peanuts	M&Ms
Unmarked A	0	1
Two-nicks	1	0
Two-nicks	1	0
Two-nicks	0	1
Broken-paw	1	0
Scar-face	1	0
Scar-face	1	0
Broken-paw	1	0
Two-nicks	1	0
Two-nicks	0	1
Broken-paw	1	0
Unmarked B	0	1
Unmarked B	0	1
Unmarked B	1	0
Unmarked C	1	0
Unmarked C	1	0
TOTALS	11	5

Possible Analyses

You could analyze these data with a chi-square goodness-of-fit test by pooling all the squirrels, but with this approach, information about the behavior of individuals is lost and you face problems of pseudoreplication. Because you know the number of nuts taken by each individual squirrel, you can use the following methods to explore the data more thoroughly.

Absolute Differences

In this method, for each subject you subtract the response to treatment 2 (M&Ms) from the response to treatment 1 (peanuts). To do this for the raw data given earlier, you would need to sum the scores for each individual squirrel (see below) and then calculate d_i, the difference of the total M&Ms − total peanuts taken by each squirrel.

Squirrel Name	Peanuts	M&Ms	d_i	Unsigned Rank	Signed Rank
Unmarked A	0	1	−1	2	−2
Two-nicks	3	2	1	2	+2
Broken-paw	3	0	3	6	+6
Scar-face	2	0	2	4.5	+4.5
Unmarked B	1	2	−1	2	−2
Unmarked C	2	0	2	4.5	+4.5

Analyze these data with the Wilcoxon matched-pairs signed-ranks test by ranking the differences (d_i) and following the instructions given for this test.

For example, for these data: sum + = 17 (that is, the sum of all ranks with positive signs is 17), sum − = 4 (the sum of all ranks with negative signs is 4), and T = smaller of these sums of like-signed ranks, so T = 4. For $n = 6, P < 0.05, T(\text{critical}) = 0$. Because $T(\text{observed}) > T(\text{critical})$, conclude that the differences are not significant (NS). (In the statistical tables from Siegel 1956 used for this example, Wilcoxon has the atypical decision rule: reject H_0 when $T(\text{observed}) < T(\text{critical})$.

Response Ratios

In this method, first sum the scores for each individual squirrel, as you did above. Then let the response ratio = response to treatment 1/(response to treatment 1 + treatment 2) as follows:

Squirrel Name	Peanuts	M&Ms	Response Ratio
Unmarked A	0	1	0
Two-nicks	3	2	0.6
Broken-paw	3	0	1
Scar-face	2	0	1
Unmarked B	1	2	0.33
Unmarked C	2	0	1

To analyze the data when you have just two treatment groups, as here, you would use the Wilcoxon matched-pairs signed-ranks test to compare the observed response ratio to the chance level of response of 0.5. Be careful, especially if you do your analysis on computer, that you analyze the *response ratios* compared to the chance response values, as given below. Do *not* analyze the actual scores of number of peanuts and M&Ms taken.

Squirrel Name	Response Ratio	Chance Response	d_i	Unsigned Rank	Signed Rank
Unmarked A	0	0.5	−0.5	4.5	−4.5
Two-nicks	0.6	0.5	+0.1	1	+1
Broken-paw	1	0.5	+0.5	4.5	+4.5
Scar-face	1	0.5	+0.5	4.5	+4.5
Unmarked B	0.33	0.5	−0.2	2	−2
Unmarked C	1	0.5	+0.5	4.5	+4.5

In this case, sum + = 14.5, sum − = 6.5, T = smaller of these sums = 6.5. For $n = 6$, $P < 0.05$, $T(\text{critical}) = 0$. Because $T(\text{observed}) > T(\text{critical})$, conclude that the differences are NS. (Remember, in the tables used for this example, Wilcoxon has the odd decision rule: reject H_0 when $T(\text{observed}) < T(\text{critical})$.

Because results obtained using absolute differences may differ from those obtained using response ratios, you should use both methods to analyze your data (Martin & Bateson 1993). If results are contradictory, be sure to discuss reasons that could explain the contradiction. Note also that analysis of absolute differences is more sensitive to individuals who respond strongly to a treatment. Analysis of ratios is more sensitive to variation within individuals (Martin & Bateson 1993). These sensitivity differences might help explain contradictory results of these tests.

SUMMARY OF STATISTICAL PROCEDURES

Before Data Collection

1. State the biological question or hypothesis.
2. Make specific predictions of what will happen if this hypothesis is correct.
3. Design experimental or observational comparisons to test each prediction. You must collect separate data and do a separate statistical test for each prediction. Remember that for each test, the *data must be independent and sampling must be random.* The steps that follow assume you are testing a single prediction.
4. Set up the H_0 (null) and H_a (alternative) hypotheses for your prediction.
5. Decide whether the test will be one-tailed or two-tailed. (Make H_0 and H_a consistent with your decision about the number of tails. That is, if you decide to do a two–tailed test, make sure that both H_0 and H_a are phrased as nondirectional, two-tailed hypotheses.)
6. Determine whether your question involves (a) one sample, (b) two related samples, (c) two unrelated samples, (d) k related samples, (e) k unrelated samples, (f) an association (correlation), or (g) a regression.
7. Determine what level of measurement you will use. That is, decide whether your data are nominal, ordinal, or interval/ratio.
8. Decide what statistical test you will use. Use Table C.3 to pick possible tests, and then determine which is most appropriate. Where more than two tests are listed for the same type of data, the more powerful (and therefore preferable) test is underlined. Additional nonparametric tests are discussed in Siegel & Castellan (1988) and in Conover (1980). If your data are interval/ratio, you must decide whether to use a parametric or a nonparametric test. You should use a parametric test if the data meet the assumptions of such tests; otherwise, use a nonparametric test. Additional parametric tests are presented in Sokal & Rohlf (1981).
9. Specify a significance (α) level. Published studies of animal behavior generally use $\alpha = 0.05$.

Table C.3 **Choosing a statistical test.**

Type of Data	Statistical Test
One sample	
Nominal	χ^2 goodness-of-fit test, binomial test
Ordinal or interval/ratio	Kolmogorov–Smirnov one-sample test
Two related samples	
Nominal	McNemar test for significance of changes
Ordinal or interval/ratio	Sign test, <u>Wilcoxon matched-pairs test</u>
Interval/ratio only*	Student's t test for matched samples
Two unrelated samples	
Nominal	χ^2 test of Independence (of two samples)
	Fisher exact test
Ordinal or interval/ratio	<u>Mann–Whitney U Test</u>
	(to detect differences in central tendency such as means, modes)
	Kolmogorov–Smirnov two-sample test
	(to detect *any* differences, including difference in variability)
Interval/ratio only*	Student's t test for independent samples
k related samples	
Nominal	Cochran Q test
Ordinal or interval/ratio*	Friedman two-way analysis of variance
Interval/ratio only*	Two-way analysis of variance without replication
k unrelated samples	
Nominal	χ^2 test of independence (of k samples)
Ordinal or interval/ratio	Kruskal–Wallis test
Interval/ratio only*	One-way analysis of variance (ANOVA)
Association (correlation)	
Nominal	χ^2 test of independence (of k samples)
Ordinal	Spearman rank correlation
Interval/ratio*	Pearson correlation coefficient
Functional relationship	
Interval/ratio only*	Least-squares regression

*Tests marked "Interval/ratio data only" are parametric tests, which should be used only if sample sizes are large ($n > 30$) and meet the assumptions of parametric tests, especially the assumption that the data are normally distributed. All other tests in this table are nonparametric tests, which are more appropriate than parametric tests when sample sizes are small and avoid the restrictive assumptions of parametric tests. Note that χ^2 tests for nominal data can be used for all other types of data, but they are not as powerful and so are less likely to detect significant differences when such differences are real.

During Data Collection

Be sure *data are independent* and *sampling is random*.

After Data Collection

1. Summarize the data by computing descriptive statistics (e.g., mode, mean, standard deviation) and graphing. The appropriate descriptive statistics and graphical methods for various levels of measurement of data are as follows:

Type of Data	Measure of Central Tendency	Measure of Dispersion	Type of Graph
Two or more samples of data that are			
Nominal	Mode	None	Bar graph (frequency or %)
Ordinal	Median	Percentiles Range	Bar graph (frequency or %)
Interval or ratio	Mean	Standard deviation	Frequency distribution
		Variance (percentile, range)	Probability distribution Box-and-whisker plot
Measures of association			Scatter plot (of data points) or point graph (medians or means)
Measures of functional relationship			Scatterplot (of data points) with or without regression line

2. Do the statistical test (compute the value of the test statistic).
3. Decide whether to accept or reject the null hypothesis (H_0). To make this decision, use the decision rule that is associated with the statistical test that you used.

   In most but not all tests, the decision rule is

   If the observed test statistic $\geq$ the critical test statistic, then reject H_0, accept H_a, and conclude that your *results are significant*.

   If the observed test statistic $<$ the critical test statistic, then either (1) accept H_0 (reject H_a) and conclude that your *results are not significant* or (2) suspend judgment if you believe that the *sample sizes were too small*.

   Two exceptions are the Wilcoxon matched-pairs test and the Mann–Whitney U test, for which some statistics books call for rejecting H_0 when the observed test statistic is *less* than or equal to the critical value.

When calculating correlations, you can conclude that you have a high correlation (a strong association) if the correlation is statistically significant *and* the correlation coefficient (r) is greater than 0.7. Beware: with large sample size, you might get statistical significance even if $r < 0.2$, which you should interpret as a negligible relationship (Martin & Bateson 1993).

For *computer-based analyses*, the decision rule is

If the computer gives a P-value ≤ 0.05, then reject H_0, accept H_a, and conclude that your *results are significant*.

If the computer gives a P-value > 0.05, then either (1) accept H_0 (reject H_a) and conclude that the *results are not significant*, or (2) suspend judgment if you believe that the *sample sizes were too small*.

4. Report the results of your statistical test. Use an appropriate standard format, which includes a verbal description of the observed pattern (in the past tense) and a parenthetical statement that includes the value of the test statistic that you calculated from your data, the sample size or degrees of freedom, and the P-value associated with your test statistic. For example, you might write, "Tadpole growth rates increased significantly with increasing light intensity (Spearman $r = 0.79$, $n = 68$, P < 0.0001)" or "Female frogs were significantly larger than male frogs (Mann–Whitney $U = 130.5$, $n = 49$, P $= 0.001$)."

LITERATURE CITED

Conover, W. J. 1980. *Practical Nonparametric Statistics*. 2nd ed. New York: Wiley.

Martin, P. & Bateson, P. 1993. *Measuring Behaviour: An Introductory Guide*. 2nd ed. Cambridge: Cambridge University Press.

Siegel, S. 1956. *Nonparametric Statistics for the Behavioral Sciences*. New York: McGraw-Hill.

Siegel, S. & Castellan, N. J. 1988. *Nonparametric Statistics for the Behavioral Sciences*. 2nd ed. New York: McGraw-Hill.

Sokal, R. R. & Rohlf, F. J. 1981. *Biometry*. 2nd ed. New York: Freeman.

GLOSSARY

absolute frequency: The total number of times that a phenomenon occurred in a sample.

action pattern: The minimum identifiable unit of behavior in an organism.

ad libitum **sampling:** A method of sampling behavior in which the observer opportunistically records the behavioral acts of any individuals in a group.

adaptation: The process of adjustment of an organism to environmental conditions. The process of evolutionary modification that results, in a species, in improved survival and reproductive efficiency. Any morphological or behavioral **character** that enhances survival and **reproductive success** of an organism. Any character that has evolved by **natural selection** for its current **function**.

agonistic: Pertaining to behavior having to do with fighting, including aggression (attack, threat behavior, defense) and fleeing.

air space: The region between the shell membrane and the hard surface of an egg.

alarm call: Vocalization that is given in response to the appearance of a predator and that alerts other individuals to the presence of the threat.

allelochemics: Chemicals that are transmitted and detected between species.

alpha (α) level: The probability of a **type I error** (see also **significance level**).

alternative hypothesis: The **statistical hypothesis** against which the **null hypothesis** is tested.

altricial: Young animals that are relatively undeveloped and highly dependent on parental care at birth or hatching (compare **precocial**).

altruism: Performance of a behavior that reduces the altruist's **fitness** while increasing the fitness of another.

ancestral state: The primitive **character state** present or assumed to be present in an ancestor.

anthropomorphism: The attribution of human motivation, characteristics, or behavior to nonhuman animals.

antiaphrodisiac: A **pheromone** that males apply to females during copulation and that tends to stop other males from mating with those females.

anuran tadpoles: Larval stage of frogs and toads.

arithmetic mean: See **mean**.

asocial aggregation: Aggregation resulting from individuals being independently attracted to a location with a preferred environmental condition.

asphyxia: Unconsciousness or death caused by lack of oxygen.

association: See **familiarity**.

assortative mating: A mating system in which males and females with like characteristics mate with each other.

asymmetric distribution: A probability density function or **frequency distribution** that is not identical on opposite sides of some central value.

bar graph: A graphical representation of a **frequency distribution** of **nominal** or **ordinal data** classified into a number of discrete categories. Equal-width rectangular bars are constructed over each category with height equal to the observed frequency of the category. Adjacent bars are not in contact with each other.

behavior sampling: A method of sampling behavior in which the observer elects to record only specific types of behavior performed by individuals in a group.

behavioral thermoregulation: Behavioral methods of **thermoregulation**, such as modification of body posture to alter the amount of exposed surface area, huddling, and movement into preferred microenvironments to take advantage of heat exchange by convection, conduction, and radiation.

benthic: Pertaining to the sea bed, river bed, or lake floor.

biased sample: An inaccurate sample that was drawn using a flawed sampling design.

binary character: A **character** with two alternative states.

bioassay: An analysis that uses the behavior of living organisms to measure response to a variable.

biological hypothesis: A proposed explanation or description of a biological phenomenon of interest that makes testable predictions; an assumption that suggests an explanation of observed facts, proposed in order to test its consequences.

biparental care: A system of **parental care** in which both parents provide care.

bout: A series of occurrences of the same activity preceded and followed by periods during which other types of behavior occur.

box-and-whisker plot: A graphical method of displaying the important characteristics of a set of observations, such as the **mean** (or **median**) [the line inside the box], the **standard deviation** (or interquartile range) [the box], and the 95% confidence limits (or range) [the whiskers extending above and below the box]. Also called a box plot.

branch: A part of a phylogenetic tree that connects a **node** to a terminal taxon.

central tendency: A measure of the center of the distribution of a variable (such as **mean, median**, and **mode**).

certainty of paternity: A male's perception of his relatedness to particular offspring. Also has been used to mean the average probability that a particular offspring was sired by a given male, and the absence of variation in the average proportion of offspring sired per mating.

character: Any observable part or attribute of an organism.

character state: One of two or more possible expressions of a **character**.

choice: The act of choosing; the existence of alternatives from which to choose.

clade: A group of taxa encompassing an ancestral species and all of its descendants. A **monophyletic** group.

classification variable: A characteristic or attribute used to place each measurement into its appropriate group.

coevolution: Evolution of two or more interacting species in which each influences the other's evolution such that a change in one species acts as a selective force on the other, and vice versa.

comfort movements: Action patterns such as bathing, preening, stretching, scratching, and shaking movements that are involved in care of the body (also known as **self-maintenance behavior**).

communication: Signal transmission in which the behavior of another animal is influenced by means of signs or displays.

comparative method: The attempt to identify evolutionary trends by examining the diversity of expression of particular characteristics across a range of taxa.

comparison group: A group that yields a set of measurements against which another set is compared (sometimes used to mean **control group**). Comparison groups are conditions of the **independent variable**.

competition: In game theory, a strategy in which a player acts in its own best interest and at the expense of the other player. In ecology, a detrimental interaction that occurs between two or more individuals of the same or different species when they use or defend a common resource that is in short supply.

conflict avoidance: A strategy in which a player avoids conflict by choosing a response that benefits neither player.

confounding factor: An uncontrolled or unaccounted for factor that may produce effects that are erroneously interpreted as resulting from the factor of interest.

conspecific: Belonging to the same species (compare **heterospecific**).

conspecific detection hypothesis: This hypothesis posits that group members may be scanning the environment in order to detect other individuals of their own species that are not currently part of the group, as opposed to individuals of other species who might be predators or competitors.

continuous distribution: A distribution in which any conceivable value could occur within any observed range (e.g., height, weight, running speed).

continuous recording: A recording method in which subjects are observed and data are recorded continuously.

control condition: In experimental studies, a stimulus or situation that lacks the experimental manipulation but contains all other elements of the experimental condition.

control group: In experimental studies, a collection of individuals to which the experimental procedure of interest is not applied but is like the experimental group in all other respects.

controlled experiment: A method of investigation in which at least one group (the **experimental group**) receives a manipulation and is then compared with a group that was not manipulated (the **control group**). The groups to be compared should be otherwise identical.

cooperating: A strategy in which two players act in their mutual best interests.

copulation: The sexual coupling of male and female in which spermatozoa from the male are deposited into the body of the female.

correlation: An interdependence between pairs of variables.

critical period: A time with relatively fixed boundaries, usually early in development, during which some event, such as an experience or the presence of a hormone, has a long-lasting effect.

critical value: The value with which a statistic calculated from sample data is compared in order to decide whether a **null hypothesis** should be rejected. The value is related to the sample size and to the particular significance level (such as $\alpha = 0.05$) chosen.

decision: The act of reaching a conclusion or of choosing among alternatives.

decision rule: A formal statement by which results are to be interpreted.

degrees of freedom (df): The number of independent units of information in a **sample** relevant to the estimation of a **parameter** or the calculation of a **statistic**.

density: A measurement of the number of objects in an area of specified size.

dependent variable: The variable of primary importance in investigations; the variable that is measured to study treatment or other effects.

derived state: The **character state** that was not present in the ancestral stock.

descriptive statistic: A numerical summary of some aspect of a **statistical population** or a **sample**.

deviation: The difference between the **mean** of the **sample** and an observed value in that sample.

dilution hypothesis: An individual's chance of being caught by a predator during a given predator attack decreases as the group size increases.

dimorphic: Having two distinct morphological types within a population or species.

discrete distribution: A distribution having only certain values, with no intermediate values in between. Examples include integers (such as number of offspring or number of spines).

display: A behavior pattern that evolved to serve a signal function in communication.

dive cycle: A foraging trip consisting of a dive and a pause on the surface made by a diving bird.

dive duration: The length of the underwater portion of a **dive cycle**.

dominance relationship: A social relationship based on outcomes of **agonistic** encounters.

dominant: A superior position in a rank order or social hierarchy.

ectotherm: An organism that depends primarily on external sources of heat for **thermoregulation**.

egg tooth: A hard calcification on the tip of the beak of a hatching bird.

electric organ: Main, Hunter's, or Sachs' organ of an electric fish that produces electric discharges.

electric organ discharge (EOD): An electric discharge produced by specialized organs of electric fish.

electrocommunication: A conspecific communication system that uses electric signals.

electrolocation: An orientation system that uses sensory input from **electroreceptors**.

electroreceptor: A sensory organ that detects electric discharges in the environment (such as from other electric fish).

elytra: The thickened, horny forewings of beetles, which cover and protect the membranous hindwings when the insect is at rest.

endotherm: An organism that uses internally generated metabolic heat to **thermoregulate**.

energy maximization: Foraging strategy in which net energy gain from foraging is maximized.

enumeration data: Measurements in the form of counts of items or events.

ethogram: A behavioral inventory; a catalog of actions; a survey, as complete and precise as possible, of all the behavior patterns characteristic of a species.

ethology: The zoological, evolutionary approach to animal behavior that includes examination, generally under natural conditions, of the proximate causes and development of behavior, as well as its ultimate evolution and adaptive value.

events: Clearly distinguishable, discrete acts of fairly short duration that can be recorded instantaneously and can be quantified by counting each occurrence.

evolution: A cumulative change in characteristics of species or populations from generation to generation; descent with modification.

evolutionary stable strategy (ESS): A behavioral strategy that, once it is widely adopted, cannot be bettered by an alternative strategy.

experimental group: The set of subjects or observations that received an experimental manipulation in a **controlled experiment**.

external validity: The applicability of results to situations or circumstances beyond those of the study at hand.

familiarity: A mechanism of **kin recognition** in which individuals treat as kin any **conspecifics** with whom they were raised; also known as **association**.

feral: Animals of domesticated varieties that are existing in a wild or untamed state.

field experiment: A controlled experiment performed outside of the laboratory, in an animal's natural habitat (that is, in the field).

fledgling: A young bird that has recently acquired its flight feathers (also used to refer to a young bird that has left its nest but is still receiving care from a parent).

flower constancy: Trait exhibited by foragers that bypass rewarding flowers to restrict visits to a single plant species.

focal-animal sampling: A sampling method in which an observer records the behavior of one individual in a group and its interactions with other group members, exclusively, for a certain time. At the end of that time, another member of the group may be chosen as the focal animal.

food competition hypothesis: The observed decrease in **vigilance** with increasing group size is a result of the need to spend more time foraging in the face of competition with other individuals.

frequency: The number of occurrences of a phenomenon per unit of time (**rate**) or space (**density**).

frequency distribution: The division of a **sample** of observations into a number of classes, together with the number of observations in each class. Acts as a useful summary of the main features of the data such as location, shape, and spread.

function: The benefit or fitness advantage that results from the possession of a specific characteristic or the performance of a particular behavior.

functional analysis: An attempt to identify the benefit or fitness advantage conferred by a specific characteristic or behavior.

graded signal: Information sent from a sender to a receiver that runs through a continuous range of values or degrees of expression.

group-size effect: Decrease in scanning behavior of individuals as group size increases.

Gymnotiformes: An order of weakly electric, freshwater teleost fish from South and Central America.

H_a: See **alternative hypothesis**.

H_0: See **null hypothesis**.

heterospecific: Pertaining to a different species (compare **conspecific**).

histogram: A graphical representation of the **frequency distribution** of a **continuous variable** in which class frequencies are represented by the areas of rectangles centered on the class interval. Adjacent rectangles are in contact with one another.

homeostasis: The maintenance of a relatively steady state or equilibrium by intrinsic regulatory mechanisms.

homeotherm: An animal characterized by a high and constant core body temperature regardless of environmental temperature.

homologous: Said of two traits found in different taxa that are the same as in the common ancestor, or that are genealogically linked by passing through the transformation from an **ancestral** condition to a **derived** condition (compare **homoplasy**).

homoplasy: A trait shared among taxa that resembles, but is not the same as, that found in the common ancestor and is not genealogically linked by passing through the transformation from an **ancestral** condition to a **derived** condition; not **homologous**. Resemblance of traits shared among taxa that originated independently in different lineages.

hormones: Chemicals that operate within a single individual, delivering a message from one part of the organism to another.

Hunter's organ: An electric organ of electric fish that produces a powerful electric shock.

hypothesis: See **biological hypothesis** and **statistical hypothesis**.

hysteresis effect: The effect observed when a biological or physical system produces different output values for a given input value, depending on whether the input variable increases or decreases during the experiment.

imprinting: A phenomenon in which animals form social attachment to specific objects during a **sensitive period** early in life. These objects thereafter become important in eliciting particular behaviors.

independent data: Data are said to be independent if knowing the value of one datum tells us nothing about the value of another datum.

independent samples: Comparison groups that include different, unrelated individuals. For example, different individuals receive experimental and control treatments. (compare **matched samples**).

independent variable: A variable that is thought to predict or explain variation in another (dependent) variable.

individual constancy: Exists when individual foragers exhibit **flower constancy**.

inferential statistics: The use of statistical methods to draw conclusions about a population on the basis of measurements or observations made on one or more **samples** of individuals.

ingroup: Any group (more than two) of closely related organisms of interest to an investigator.

innate behavior: Behavior that is complete when first performed and hence is not **learned**.

instantaneous recording: Discontinuous time sampling in which the observer records behavior instantaneously at specific times that are spaced evenly throughout the observation period. For example, the observer might record behavior every 5 minutes, recording only the behavior pattern observed at the instant that 5 minutes has elapsed.

instantaneous sampling: Sampling in which the action pattern of each individual is recorded at the instant that individual is observed during a sample.

instrumental learning: See **operant conditioning**.

internal validity: The ability of a particular result to be explained by a particular factor.

internode: A part of a phylogenetic tree that connects **nodes**.

interobserver reliability: The degree of agreement between two observers when they are measuring the same thing at the same time.

interspecific: Involving members of different species.

interval data: Measurements from a **continuous variable** with an arbitrary zero point (for example, temperature in degrees Celsius (centigrade) or degrees Fahrenheit).

intraspecific: Involving members of the same species.

intraspecific competition: The demand by two or more individuals of the same species for an essential common resource that is in limited supply, or a detrimental interaction that occurs between two or more individuals of the same species when they use or defend a common resource that is in short supply.

iterated prisoner's dilemma (IPD): A repeated playing of the **prisoner's dilemma (PD)** game.

key stimulus: See **sign stimulus**.

kin discrimination: Treating kin differently from unrelated **conspecifics**.

kin recognition: The ability to distinguish between relatives and nonrelated **conspecifics**.

kin selection: Selection among individuals differing in traits, such as parental care and helping behavior, that affect the survival and reproduction of relatives that possess the same genes by common descent.

kinematic graph: A "box-and-arrow" diagram that illustrates the relative frequency of transitions between behaviors.

kleptoparasite: A species that steals the prey or food stores of another species.

laboratory experiment: A controlled experiment performed in the laboratory.

learned behavior: Behavior that is modified as a result of experience and hence is not **innate**.

limiting resource: Any resource or environmental factor that exists in suboptimal levels and thereby prevents an organism from reproducing at its optimal level. Of all the resources required for an organism, the resource that is scarcest relative to an organism's need.

main organ: An electric organ of electric fish that produces a powerful electric shock.

many-eyes hypothesis: The amount of individual antipredator scanning can decrease as group size increases without decreasing the probability of detecting a predator if all group members scan.

marginal value theorem: A predator with precise knowledge of the qualities and locations of the best available patches in the environment can do better by leaving its current patch if the cost of moving to a better patch is less than benefit conferred by the increased foraging efficiency possible there.

matched samples: Comparison groups each of which comprises the same individuals, or paired individuals who are individually matched on a number of variables such as age, sex, status, and size. For example, the same individuals receive experimental and control treatments on separate occasions (compare **independent samples**).

mate choice: Preferences by members of one sex for mates with specific characteristics such as large size, elaborate display structures, conspicuous vocal behavior, or specific chemical signals.

mean: The average; a measure of location or **central tendency** calculated as the sum of measurements divided by the number of those measurements. It is most useful for data with a **symmetric distribution**.

median: A measure of location or **central tendency** that is suitable for **asymmetric distributions**. The value in a set of ranked observations that divides the data into two parts of equal size. When there is an odd number of observations, the median is the middle value. When there is an even number of observations, the median is calculated as the average of the two central values.

message analysis: In the study of animal communication, the attempt to identify what a signal expresses about the sender (the message) and what the signal, in context, conveys to a recipient (the meaning).

mode: A **descriptive statistic** occasionally used as a measure of location or **central tendency**; it is the most frequently occurring value in a set of observations.

monogamous: Pertaining to the condition in which a male and a female form a prolonged and more or less exclusive breeding relationship.

monogamy: A mating system in which a male and a female form a prolonged and more or less exclusive breeding relationship.

monophyletic group: A group of taxa encompassing an ancestral species and all of its descendants. A **clade**.

mormyriformes: An order of weakly electric, freshwater teleost fish from Africa.

multiple-brooded: Having two or more broods in a single breeding season.

mutually exclusive: A situation in which only one of two or more possibilities can be correct or can apply.

natural selection: The nonrandom and differential reproduction of different genotypes acting to preserve favorable variants and to eliminate less favorable variants.

negative correlation: A correlation in which one measurement increases as another decreases.

negative reinforcement: The cessation of an aversive stimulus in order to induce the desired behavioral response.

negative punishment: The witholding of a desirable stimulus when an incorrect behavioral response is given.

node: A part of a phylogenetic tree that represents a speciation event.

nominal data: A measurement that is classified by a quality or an attribute rather than by a number (examples include subject sex, genetic phenotype, and taxonomic category).

nondescendant kin: Relatives other than offspring.

nonverbal signals: Information passing from a sender to a receiver via channels other than vocalization.

normal distribution: A specific, bell-shaped probability density function of a random variable that is assumed by many statistical (parametric) methods.

null hypothesis: The **statistical hypothesis** that postulates no difference, no effect, or no relationship, which is to be tested against an **alternative hypothesis**.

observational comparison: A comparison based on natural or existing differences between comparison groups rather than on differences between manipulations by an experimenter.

observation period: An interval of time during which animals are observed.

one-tailed hypothesis: A significance test for which the alternative hypothesis is directional (e.g., one population mean is greater than another). The choice whether to use a one-tailed or a two-tailed hypothesis must be made before any **test statistic** is calculated.

operant conditioning: Experimental procedures in which an action that is rewarded increases in frequency through trial-and-error learning; also known as instrumental learning.

operational definition: A definition that is sufficiently specific and measurable that it can be used in an observational or experimental study (for example, a description of how a researcher measured particular behavior patterns).

operationalize: To make operational (for example, to identify a testable prediction of a hypothesis).

optimal foraging theory: an attempt to explain how animals make choices among the available options when presented with choices among foods having different energetic profitabilities. The theory assumes that the behavior exhibited by animals has persisted because it enhances the animal's fitness and has thus been favored by natural selection, and that maximizing the net energy gain or other relevant "currency" enhances fitness.

ordinal data: A measurement that allows a **sample** of individuals to be ranked with respect to some characteristic but wherein differences at different points of the scale are not necessarily equivalent.

outgroup: Any taxonomic group that is compared to a different but closely related group that is undergoing phylogenetic analysis (the **ingroup**); the outgroup cannot contain any members that are part of the ingroup.

overlap of successive broods: Adults begin a second brood while the young of the first brood are still dependent.

pacemaker nucleus: A neuronal oscillator in the brainstem that controls the discharge rate of the electric organ of an electric fish.

parametric mean: The **mean** for the statistical population of interest.

paraphyletic group: An artificial taxon in which one or more descendants of an ancestor are excluded from the group.

parental care: Protecting, nourishing, and nurturing of young, including supplying food; attending to the state of the skin, feathers, or hair; removing feces and other waste; camouflaging, such as covering eggs; transporting the young to places of safety; and guarding, defending, and leading the young.

parental investment (PI): Any behavior toward offspring that increases the probability of their survival to the detriment of possible investment in other offspring.

parsimony: The principle of invoking the minimum number of evolutionary changes to infer phylogenetic relationships.

passerine: Perching bird; a member of the largest order of birds, the Passeriformes.

pause duration: The length of the on-the-surface portion of a **dive cycle**.

percentage: A portion or share in relation to a total of 100, calculated as the fraction of occurrence multiplied by 100.

percentile: Rank on a set of divisions that produce exactly 100 equal parts in a series of continuous values. A value in the 80th percentile is higher than 80% of the other values.

phenotype matching: A mechanism of **kin recognition** in which individuals treat as kin any **conspecifics** whose phenotypes are similar to those of the individuals with whom they were raised.

pheromone: A chemical released by an individual that influences conspecifics.

philopatric: Exhibiting a tendency to remain in or to return to the same locality.

phylogeny: A proposed model of evolutionary relationships among species or groups of species.

physiological ecology: The study of how the normal processes and metabolic functions of living organisms affect their interactions with their biotic and abiotic environments.

physiological thermoregulation: The use of internal metabolic heat to maintaining a high and constant core body temperature.

pipped: An early stage of hatching in which a chick makes a small hole on the blunt side of the egg.

piscivorous: Feeding on fishes.

planktivore: An eater of plankton, which are minute aquatic organisms that passively float or swim weakly.

poikilotherm: An animal characterized by a variable body temperature that fluctuates with daily and seasonal variations in environmental temperature and by a very limited ability to generate metabolic heat for thermoregulatory purposes.

polyandry: A mating system in which a female pairs or **copulates** with more than one male.

polyphyletic group: An artificial taxon in which taxa that are separated from each other by more than two ancestors are placed together without the inclusion of all the descendants of that common ancestor.

pooling fallacy: The mistaken attempt to treat nonindependent observations as independent, resulting in an inflated sample size. See **pseudoreplication**.

population parameter: A numerical characteristic of a statistical population.

positive correlation: A correlation in which two measurements increase together.

positive reinforcement: Strengthening of a response by making the presentation of some reward (such as food, water, or stimulation of a pleasure center in the brain) contingent on its performance. As a result, the response is more likely to occur.

postcopulatory mate guarding: A male remains near a female or in physical contact with her after copulation, while attempting to prevent her from mating with another male.

precocial: Young animals that exhibit a high level of development and independent activity at birth or hatching (compare **altricial**).

predation: The consumption of one animal (the prey) by another (the predator). Prey are said to be **depredated**.

prediction: A testable statement deduced from an **hypothesis**.

prisoner's dilemma (PD): A two-player game that compares the payoffs of cooperation and defection as a function of the strategy of the other player.

probability (P): The quantitative expression of the chance ($0 \leq P \leq 1$) that an event will occur.

pronotum: The dorsal section of the prothorax of an insect.

protocol: A detailed, step-by-step description of an experimental procedure.

proximate mechanism: The immediate cause of a behavior; an explanation of how a behavior is caused or controlled.

pseudoreplication: The use of **inferential statistics** to test for differences among **treatment groups** using data from experiments where either treatments are not replicated (though samples may be) or replicates are not statistically independent. See **pooling fallacy**.

pulse-type fish: Weakly electric fish that produce electric discharges in pulses, which are followed by relatively long intervals of silence.

random sample: Either a set of independent random variables, or a sample of individuals selected in such a way that each individual in the population is equally likely to be included in the sample.

range: A simple measure of dispersion that is calculated as the difference between the largest and smallest observations in a data set.

rate: A measure of the frequency of some phenomenon of interest that is expressed as the number of events occurring per unit time, in a specified sampling period.

ratio data: Measurements from a **continuous variable** that has a fixed rather than an arbitrary zero point (examples include height, weight, and temperature in Kelvin but not in degrees Celsius (centigrade) or Fahrenheit).

reciprocal altruism: Exhibited when an animal behaves altruistically and that behavior is reciprocated by another at some time in the future.

regression: A method of analyzing the relationship between independent and dependent variables in which a dependent variable is expressed as a function of one or more independent variables.

reinforcement: See **positive reinforcement**.

related samples: See **matched samples**.

relative frequency: The frequency of occurrence expressed as a fraction of the total ($0 \leq f_{rel} \leq 1$).

releaser: A stimulus that serves to initiate a fixed **action pattern** or **innate** behavior; a **sign stimulus** used in communication.

repeated matings: A female **copulates** more than once with the same male.

repeated-measures design: An experimental design in which measurements are taken from the same subjects on successive occasions.

reproductive strategy: A set of rules for behavior used by an organism to achieve **reproductive success**.

reproductive success: The success of an organism in reproduction, measured as the number of surviving offspring produced by an individual.

resource-holding potential (RHP): The ability to win encounters over important resources in which winning is determined by characteristics such as size, strength, or fighting ability.

ritualization: A process by which nondisplay behavior evolves into **display** behavior via the enhancement of its signal function.

root: The **internode** at the base of a phylogenetic tree.

rule of one variable: A goal of experimental design in which only one factor can explain differences between control and experimental groups.

Sachs' organ: An electric organ of electric fish that generates weak electric discharges.

sample: A selected subset of a **statistical population** chosen by some process, usually with the objective of investigating particular properties of the population.

sample statistics: Descriptive statistics of a **sample**.

scan sampling: A method of sampling behavior in which the observer quickly censuses ("scans") a group of animals and records the action pattern of each individual at the instant it was observed.

self-maintenance behavior: See **comfort movements**.

selfish-herd hypothesis: An individual's chance of being caught by a predator during a given predator attack decreases because the individual uses the other members of the group as a living shield.

sender–receiver matching: Mutual adaptation of the physiological properties of the sensory system of a receiver and the motor system of a sender to ensure optimal perception of the signal transmitted in the process of communication.

sensitive period: A stage in life, usually early in development, in which an animal is especially susceptible to certain learning experiences.

sex attractant: A **pheromone** released by females that attracts males and releases copulatory behavior from them, or a **pheromone** released by males that stimulates locomotion in females and causes them to aggregate in the vicinity of the male.

sex pheromone: A **pheromone** used in communication between individuals of opposite sexes in a mating context.

sex role reversal: Reversal of the roles usually played by males and females (e.g., males care for young and females compete aggressively for access to males).

sexual dimorphism: Males and females of a species being distinguishable to a human observer on the basis of external morphology.

sexual monomorphism: Males and females of a species not being distinguishable to a human observer on the basis of external morphology.

shaping: Use of operant conditioning techniques to get an animal to perform movement patterns that were not originally in its repertoire. Shaping consists of reinforcing any movement that more resembles the desired pattern, so that the animal comes progressively closer to the desired movement patterns and eventually performs them completely.

shared derived trait: A **character state** that is restricted to some members of the study group but that was not present in the ancestral stock.

sign stimulus: An external stimulus that elicits specific, stereotyped behavior patterns that are generally called fixed **action patterns**. Also known as a **key stimulus**.

significance level: The level of **probability** at which it is agreed that the **null hypothesis** will be rejected. Conventionally set at 0.05 (see also α **level**).

sister species: The species that is most closely related genealogically to the species of interest.

skew: The degree of symmetry in a probability density function. Asymmetric distributions are said to be skewed in the direction of the longer tail.

smile: A facial expression characterized by an upward curving of the corners of the mouth.

social aggregation: Aggregation resulting from attraction of individuals to **conspecifics**.

sperm competition: Sperm from two or more males competing within the reproductive tract of the female for fertilization of the eggs.

spontaneous (individual) constancy: Individual foragers choose a flower on the basis of a suite of cues on the first visit to the patch, and they continue to visit the same flower type on repeated visits, regardless of the alternatives available and of what their hivemates are doing.

squared deviation: The square of the deviation of a measurement from the mean of its **sample**.

standard deviation: The most commonly used measure of dispersion, or the spread of observations; calculated as the square root of the **variance**.

state: Actions that last for a relatively long time and can best be quantified by timing the duration of the action.

statistic: A numerical characteristic of a sample (e.g., **mean** and **variance**).

statistical hypothesis: One of two mutually exclusive statements (the null hypothesis and the alternative hypothesis) used in statistical inference testing.

statistical population: Any finite or infinite collection of units of interest.

statistical power: The **probability** of correctly rejecting the **null hypothesis** when it is false. Among other things, power gives a method of discriminating between competing tests of the same **hypothesis**, the test with the highest power being preferred.

statistically significant: Unlikely to have occurred by chance alone; significance is usually accepted at the 0.05 α **level**.

strategy: A set of rules used by an organism to meet a particular set of conditions; in game theory, a plan that completely specifies the choices a competitor would make in all circumstances that could arise.

strongly electric: Voltages of 20–600 volts generated by electric fish.

submitting: A strategy in which one player gives up in response to a direct challenge by the other player.

sum of squares: The value resulting from adding all of the **squared deviations** from a sample.

symmetric distribution: A probability density function or **frequency distribution** that is identical on opposite sides of some central value.

test statistic: a number calculated from the sampled data by applying the formula for an inferential statistical test (examples include t, F, and χ^2).

thermoregulation: The establishment and maintenance of a core body temperature that differs from the environmental temperature.

thermotaxic: Approaching a heat source.

tip: A terminal taxon in a phylogenetic tree.

tit-for-tat: A strategy in which the player's first move is to cooperate, but in all subsequent moves the player does whatever was done by the other player.

treatment group: In an experimental design, the group that receives the experimental manipulation. Treatment groups are conditions of the **independent variable**.

trichromatic vision: Color vision with three primary color receptors.

two-tailed hypothesis: A significance test for which the **alternative hypothesis** is not directional (for example, the hypothesis that one population mean is not the same as another). The decision whether to use a one-tailed or a two-tailed hypothesis must be made before any test statistic is calculated.

type I error: An error that results when the **null hypothesis** is falsely rejected.

type II error: An error that results when the **null hypothesis** is falsely accepted.

ultrasonic vocalizations: Vocalizations above the frequencies that can be heard by humans.

ultimate mechanism: The historical explanation of a behavior; an explanation of why a behavior evolved, including past and current functions of the behavior.

variability: The ability to vary. Measures of variability in statistical samples include **range**, **variance**, **standard deviation**, and standard error.

variance: A commonly used measure of the dispersion, or spread, of observations. The unbiased sample variance is calculated as the **sum of squares** divided by $n - 1$.

vigilance: Scanning and alert behavior by which animals monitor the environment, especially to detect predators.

vigilant behavior: The frequency and/or duration of scans by which animals monitor the environment, especially to detect predators.

wave-type fish: Weakly electric fish that produce electric discharges that resemble continuous waves when displayed on an oscilloscope screen.

weakly electric: Voltages generated by electric fish that are too low in amplitude to be detected unless they are amplified.

References

Alcock, J. 2001. *Animal Behavior: an Evolutionary Approach.* 7th ed. Sunderland, MA: Sinauer.

Drickamer, L. C., Vessey, S. H. & Jakob, E. M. 2002. *Animal Behavior.* 5th ed. Boston, MA: McGraw-Hill.

Everitt, B. S. 1998. *The Cambridge Dictionary of Statistics.* Cambridge: Cambridge University Press.

Immelmann, K. & Beer, C. 1989. *A Dictionary of Ethology.* Cambridge, MA: Harvard University Press.

Krebs, J. R. & Davies, N. B. 1997. *Behavioural Ecology: an Evolutionary Approach.* 4th ed. London: Blackwell Science.

Lincoln, R. & Boxshall, G. 1995. *The Cambridge Illustrated Dictionary of Natural History.* Cambridge: Cambridge University Press.

Lincoln, R., Boxshall, G. & Clark, P. 1998. *A Dictionary of Ecology, Evolution and Systematics.* 2nd ed. Cambridge: Cambridge University Press.

Martin, E. & Hine, R. S. (Eds.). 2000. *Oxford Dictionary of Biology.* 4th ed. Oxford: Oxford University Press.

Martin, P. & Bateson, P. 1993. *Measuring Behaviour: An Introductory Guide.* 2nd ed. Cambridge: Cambridge University Press.

Sokal, R. R. & Rohlf, F. J. 1995. *Biometry.* 3rd ed. New York: Freeman.

Zar, J. H. 1999. *Biostatistical Analysis.* 4th ed. Upper Saddle River, NJ: Prentice-Hall.

INDEX

463